Sustainable Textiles: Production, Processing, Manufacturing & Chemistry

Series Editor

Subramanian Senthilkannan Muthu, Head of Sustainability, SgT and API, Kowloon, Hong Kong

This series aims to address all issues related to sustainability through the lifecycles of textiles from manufacturing to consumer behavior through sustainable disposal. Potential topics include but are not limited to: Environmental Footprints of Textile manufacturing; Environmental Life Cycle Assessment of Textile production; Environmental impact models of Textiles and Clothing Supply Chain; Clothing Supply Chain Sustainability; Carbon, energy and water footprints of textile products and in the clothing manufacturing chain; Functional life and reusability of textile products; Biodegradable textile products and the assessment of biodegradability; Waste management in textile industry; Pollution abatement in textile sector; Recycled textile materials and the evaluation of recycling; Consumer behavior in Sustainable Textiles; Eco-design in Clothing & Apparels; Sustainable polymers & fibers in Textiles; Sustainable waste water treatments in Textile manufacturing; Sustainable Textile Chemicals in Textile manufacturing. Innovative fibres, processes, methods and technologies for Sustainable textiles; Development of sustainable, eco-friendly textile products and processes; Environmental standards for textile industry; Modelling of environmental impacts of textile products; Green Chemistry, clean technology and their applications to textiles and clothing sector; Eco-production of Apparels, Energy and Water Efficient textiles. Sustainable Smart textiles & polymers, Sustainable Nano fibers and Textiles; Sustainable Innovations in Textile Chemistry & Manufacturing; Circular Economy, Advances in Sustainable Textiles Manufacturing; Sustainable Luxury & Craftsmanship; Zero Waste Textiles.

More information about this series at https://link.springer.com/bookseries/16490

Subramanian Senthilkannan Muthu
Editor

Sustainable Approaches in Textiles and Fashion

Fibres, Raw Materials and Product Development

 Springer

Editor
Subramanian Senthilkannan Muthu
SgT Group and API
Hong Kong, Kowloon, Hong Kong

ISSN 2662-7108   ISSN 2662-7116 (electronic)
Sustainable Textiles: Production, Processing, Manufacturing & Chemistry
ISBN 978-981-19-0880-4   ISBN 978-981-19-0878-1 (eBook)
https://doi.org/10.1007/978-981-19-0878-1

This Springer imprint is published by the registered company Springer Nature Singapore Pte Ltd.
The registered company address is: 152 Beach Road, #21-01/04 Gateway East, Singapore 189721,
Singapore

Contents

About the Editor

Dr. Subramanian Senthilkannan Muthu currently works for SgT Group as Head of Sustainability and is based out of Hong Kong. He earned his Ph.D. from the Hong Kong Polytechnic University, and he is a renowned expert in the areas of Environmental Sustainability in Textiles & Clothing Supply Chain, Product Life Cycle Assessment (LCA) and Product Carbon Footprint Assessment (PCF) in various industrial sectors. He has five years of industrial experience in textile manufacturing, research and development and textile testing and over a decade of experience in life cycle assessment (LCA), carbon and ecological footprints and assessment of various consumer products. He has published more than 100 research publications and written numerous book chapters and authored/edited over 100 books in the areas of carbon footprint, recycling, environmental assessment and environmental sustainability.

Importance of *Asclepias Syriaca* (Milkweed) Fibers in Sustainable Fashion and Textile Industry and Its Potential End-Uses

Ece Kalayci, Ozan Avinc, and Kemal B. Turkoglu

Abstract As a result of the increasing world population, rapidly changing consumption habits and fashion trends, the textile and fashion industry has become one of the industries with the highest annual global environmental burden. It is of great significance to overcome the possible negative effects of textile production on the environment and to contribute to the sustainable fashion and textile industry by ensuring the responsible production of textile products. For this reason, the selection and use of sustainable, renewable, and biodegradable textile materials for each product produced in the textile industry can be seen as the first step toward sustainable production. Milkweed (*Asclepias syriaca*) fibers constitute an important raw material potential in every field of textile as daily textile products, composite textiles, and technical and functional textiles in terms of the properties they exhibit at this point. In this chapter, it was aimed to examine in detail the topics such as the place, importance, and application areas of milkweed fibers in sustainable fashion and textile production.

Keywords *Asclepias syriaca* · Milkweed · Sustainability · Sustainable fashion · Sustainable textile

1 Introduction

As we approach the end of the first quarter of the twenty-first century, we are witnessing a global climate crisis. Months of forest fires, major hurricanes, storms, deadly floods, volcanic eruptions, drying up rivers, lakes, polluted seas, and more, we receive news of a new disaster caused by the climate crisis every day. Our planet is now sounding the alarm and it has now been moved to dimensions that cannot be ignored. For the future of our planet and humanity, it has become mandatory for every individual breathing today to strive for the solution of this problem. In this

E. Kalayci · O. Avinc (✉) · K. B. Turkoglu
Textile Engineering Department, Engineering Faculty, Pamukkale University, Denizli 20160, Turkey
e-mail: oavinc@pau.edu.tr

© The Author(s), under exclusive license to Springer Nature Singapore Pte Ltd. 2022   1
S. S. Muthu (ed.), *Sustainable Approaches in Textiles and Fashion*, Sustainable Textiles: Production, Processing, Manufacturing & Chemistry,
https://doi.org/10.1007/978-981-19-0878-1_1

context, as a result of years of research, the concept of sustainability has entered our lives. The applicability of this concept in every field of production and consumption from agriculture to fashion is investigated under various sub-titles.

It is possible to summarize the concept of sustainability as meeting the requirements of today without compromising the requirements of future generations. This is a system created not only ecologically, but also by considering three main systems consisting of ecological, economic, and social conditions. The textile and fashion industry is accepted as one of the most basic needs of humanity, and unfortunately, raw materials and production techniques in the textile industry are very difficult to meet the requirements of the values examined under the concept of sustainability. Due to the expanding world population, rapidly changing consumption habits and fashion trends, the textile and fashion industry is one of the sectors with the highest annual global environmental burden.

As a result of this heavy consumption, enormous amounts of clean water and energy resources are consumed, and the resulting chemical wastes and greenhouse gases harm the environment by causing potential climate change and global warming [1, 2]. For this reason, it is of great importance to overcome all these possible negative impacts of textile production on the environment and to contribute to the sustainable fashion and textile industry by ensuring responsible production of textile products.

There are many methods proposed so far that can create solutions for environmentally friendly design and production in sustainable textile and fashion production:

- Selection and use of sustainable, renewable, and biodegradable textile materials [3–32],
- Reusability [1, 2, 33, 34],
- Recyclability [1, 2, 33–37],
- Less water consumption [38–40],
- Use less chemicals [38–46]
- Less energy consumption [38–40, 45, 46]
- Using natural or natural-based auxiliaries instead of synthetic chemicals [9, 10, 38, 43–54],
- Using environmentally friendly textile production and finishing techniques [7, 11, 39, 40, 42, 45, 46, 49, 50, 54–62].

For example, there are critical problems in the life cycles of cotton and polyester, which are the most widely used textile fibers [34, 63, 64]. During the cultivation of cotton, pesticides, insecticides, etc., which cause significant damage to the soil, many different chemicals and enormous amounts of clean water are needed. In the production of polyester fibers, unsustainable and non-renewable petroleum resources are raw materials [65–67]. In addition, the non-recycling of synthetic fibers and the accumulation of waste create a major pollution problem. For this reason, it is of great importance to search for renewable, biodegradable, environmentally friendly, low-cost natural fiber sources that can be an alternative to synthetic fibers [68]. In this context, it is seen that there are many studies in the literature on various lignocellulosic fibers extracted from agricultural wastes or existing natural sources [69–74].

Milkweed (*Asclepias syriaca*) is an herbaceous, perennial herb commonly grown in the southern regions of the Americas, Asia, and Southern Africa [75–78]. Milkweed fibers are very light fibers with a thin cell wall and hollow structure. These fibers are also buoyant on the water and have a hydrophobic character [79]. In this way, materials such as mattresses, pillows, and life jackets can be used as filling materials [80, 81]. Milkweed and kapok fibers have been the most widely used fiber types for many years thanks to their similar structural and physical properties, while these fibers have been replaced by synthetic fiber fillings with the spread of synthetic fibers [79, 82–84]. However, in recent years, as an outcome of the increasing interest in sustainable production, natural fiber products and nature-friendly, anti-allergic, organic products, and milkweed and kapok fibers have started to attract attention again in terms of their properties and usage areas [84]. In this chapter, it was aimed to summarize and examine in detail the research carried out on topics such as milkweed fibers, the place and importance of milkweed fibers in sustainable fashion and sustainable textile industry.

2 History of the *Asclepias Syriaca* (Milkweed)

Asclepias, milkweed, recorded as an American genus of herbaceous perennial dicotyledonous plants from the family *Apocynaceae,* containing more than 140 known species [75, 76]. Its Latin name is thought to come from the Greek god of medicine and healing, Asclepius [73, 75]. Asclepias are known as persistent hardy weeds containing cardiac glycosides, which are generally poisonous when consumed by livestock [85, 86]. It is stated that in the past years, milkweed plant, especially the root bark, has been utilized in the treatment of numerous diseases such as leprosy, fever, menorrhagia, malaria, and snake bites [87, 88]. The fragrant flowers of milkweed plants, the latex, and hard roots in its structure have attracted attention not only by researchers but also by farmers engaged in agriculture. Milkweed seeds were first sent from Canada to France by the Frenchman Louis Hebert, where they were studied by the botanist and physician Philippe Cornut [89, 90]. In 1746, Gleditsch from Germany was the first to use milkweed fibers for filling material and later for fabrics and shiny velvets. Limitations in the cultivation of milkweed, such as poor seed germination, resistance to dyeing of milkweed fibers, and the fiber being weak and brittle for spinning, are summarized by Lichtenstein [91].

A big deal of study has been conducted in Germany and the USA on the cultivation, harvesting process and applications of milkweed fiber. During the First World War, Schuroff from Germany stated that milkweed can't be adapted as a domestic fiber source. Schuroff, however, revealed that milkweed has the potential to replace kapok fiber in upholstery [91, 92]. When the recent history of milkweed fibers is examined, it has been noted that during the Second World War, milkweed fibers were a very important material type from a tactical and economic point of view, and more than one million pounds (453.6 kg) of milkweed fibers were utilized to fill life-saving equipment and other flotation equipment [89, 93]. The damage to the trees from which kapok fiber was obtained during the war and the fact that these fibers became very difficult to find were very effective in increasing the demand for milkweed fiber. Milkweed fibers show similar properties to kapok fibers because they are fibers with wide lumen and very thin walls, just like kapok fibers. Milkweed fibers have a waxy layer that imparts hydrophobic properties like kapok fibers. Milkweed fibers can be an ideal kapok fiber substitute due to their light, hydrophobic, and insulating properties [94–96].

Milkweed fibers are evaluated in the category of seed fibers. However, fiber can also be obtained from the bark of the plant. Milkweed bark fibers are long, strong but brittle fibers. Although fibers from plant bark are akin to flax in many ways, fabrics from these fibers are very brittle and have little or no knitting qualities, so they are not widely available [73, 82]. Milkweed fibers have become the subject of many research from different fields to date [93, 97, 98]. In the studies conducted in the USA in the 1970s and 1980s, the use of milkweed as a biofuel was investigated, but unfortunately it was determined that it is an uneconomical source of biofuel [99–101]. In addition, there are research on the use of milkweed in the production of natural rubber, which can be an alternative to crude oil [89, 102, 103]. Milkweed fibers are still used as hypoallergenic fillers in pillows and quilts, and it is seen that oils obtained from milkweed seeds are used in body care products [95].

3   Structure of the *Asclepias Syriaca* (Milkweed) Plant

Milkweed plant, which can grow naturally in wild environments, can be grown in sandy and arid regions as well as in humid and swampy regions due to its easy adaptation to the environment [104]. This milkweed-like nature of the milkweed plant is very effective in making the milkweed plant have an affordable production cost. Milkweed, a perennial plant that could adapt to adverse soil conditions, has been thought as an alternative fiber resource in recent years. The plant, which used to belong to the *Asclepiadaceae* family, is currently classified in the subfamily *Asclepiadoideae* of the dogbane family *Apocynaceae* [94, 95, 99]. It is seen that farmers and scientists have cooperated together in the development of milkweed plant as an alternative natural fiber source since the late 1980s [76, 95, 104, 105].

The flowers of the milkweed plant are usually arranged in round flower clusters. The upper part of each flower, the corona, is comprised of five hoods. The corolla, which contains five fused petals, generates the lower part of the flower [100, 106, 107]. The shape of the hood differs by species, and some possess horn-like appendages. While flower color differs within the genus, it can be in numerous shades of white, yellow, green, purple, pink, orange, or red [107]. Milkweed plants generate their seeds in follicles. Arranged in overlapping rows, the seeds possess white silky filament-like fibers known as "pappus", "comas", or "floss". Follicles mature and break down over time. The seeds in the dried cocoon and the fibers around it are carried by the wind blowing [108]. Some milkweed species (native to the Americas, milkweed, Asclepias perennis) are aquatic species and have no floss around their seeds [109]. In India, it blooms between August and January each year, and the fruits ripen between October and February [90]. Milkweed plant has an upright and robust body and can grow up to 1.82 m in length. The elongation of the plant is varying depending on the species and, even within a species, plant height could differ depending on local genetics and climate. While the root morphology of milkweed plants varies between fleshy and woody structure, several species appear to be multi-rooted [110]. Along with the propagation by seed, some silkworms can remanufacture vegetatively by making new shoots from adventitious buds on their roots [111].

4 *Asclepias Syriaca* (Milkweed) Fibers

Milkweed fibers are single-celled cellulosic fibers obtained from the seed of the plant, just like cotton fibers. However, it does not have a crimped structure like cotton fibers [82, 112, 113]. Milkweed fibers have a smooth and glossy surface and a large internal space. Thanks to its hollow interior structure, it can exhibit insulation and float on water [88, 114, 115]. However, the smooth surface of milkweed fibers and the low durability against external loads cause spinnability problems [88, 104, 108, 114]. Milkweed fibers have a thin cell wall. The fiber wall consists of three regions as the outer wall, the inner wall, and the micro-fibers between these two walls [116]. In a study by Drean et al. [113], by analyzing milkweed fiber wall thicknesses, they noted from SEM images of the fibers that the wall thickness of milkweed fibers was around 1.27 μm, but there were large differences between the wall thicknesses of the fibers (CV: 30%-coefficient of variation).

The wall thickness of milkweed fibers also varies depending on whether the fibers are obtained from the seed or stem of the plant. While the wall thicknesses of the fibers obtained from the stem and seed of the plant vary between 6 and 7 μm, it is stated that the aspect ratio (ratio between length and diameter) of the fibers obtained from the stem of milkweed is around 67% higher than that of milkweed seed fibers [73]. Lignocellulosic fibers are mainly composed of a mixture of cellulose and hemi-cellulose which is called holocellulose and other chemical extractives such as lignin, ash, and wax. Milkweed fibers consist of 55% cellulose, 24% hemicellulose, 18% lignin, and other components such as minerals, waxes, and oils and are considered a lignocellulosic fiber type [87, 117]. Milkweed fibers have a lower density according to other cellulosic fibers as a result of its thin cell wall and hollow structure [73, 118]. The density of milkweed seed fibers is approximately 0.97 g/cm^3 [119].

5 *Asclepias Syriaca* (Milkweed) Fiber Properties

It is observed that different values are recorded in the literature on moisture recovery of milkweed fibers. In some studies, it is stated that milkweed fibers contain higher moisture and have higher moisture recovery compared to cotton fibers, since they can easily absorb water vapor thanks to the hollow channel structure that exists along the fiber length [113, 120, 121]. The fact that the moisture recovery of milkweed fibers is higher than that of cotton fiber can be associated with the high number of amorphous regions in the fiber structure as well as the hollow structure of the fiber [74, 82, 119]. In addition, since its hollow and porous structure is effective in absorbing moisture quickly, milkweed fibers exhibit high moisture permeability [121]. In this way, it has been noted that milkweed fibers could be utilized as an alternative raw material resource for applications requiring high moisture absorbency. It is observed that nonwovens made of milkweed fibers in particular exhibit high moisture absorbency potential [82, 119].

5.1 *Mechanical Properties of Milkweed Fibers*

It has been noted that the type of milkweed plants and various environmental conditions in which they are grown are effective in the different values of their chemical components in the fiber structure. This diversity also affects the mechanical properties of milkweed fibers [73]. When the mechanical characteristics of milkweed seed fibers and cotton fibers are compared, it was observed that the milkweed fibers exhibit lower strength and elongation values than cotton fibers. It has been suggested that the lower strength of milkweed seed fibers may be due mainly to the lower cellulose content, less percentage of crystalline according to cotton fibers, and very thin cell wall of milkweed fibers [112, 122].

5.2 *Thermal Behaviors of Milkweed Fiber*

Analyzing the thermal behaviors of milkweed fibers, Gu et al. found that various chemical components such as formic acid, acetic acid, and methanol were formed along with carbon dioxide (CO_2) and water (H_2O) during the pyrolysis decomposition process of the fibers (with no obvious synergistic interaction) [123]. It is stated that the susceptibility of milkweed fibers to thermal combustion will increase after the removal of the waxy layers from the fiber surfaces. Three reactions will occur in the thermal process (pyrolysis combustion) which is applied to milkweed fibers in two stages [123]:

(1)   An exothermic reaction resulted in a massive 58% weight loss at 100–300 °C;
(2)   A rapid combustion of about 38% sample mass loss at 300–450 °C;

(3)   Small weight loss between 450 and 600 °C (about 0.8%).

Also, in the study, it has been noted that the cellulose structure of milkweed fibers is more resistant to thermal degradation than the hemicellulose and lignin structures that exist in the milkweed fiber content.

5.3   Chemical Resistance of Milkweed Fibers

Like most cellulosic fibers, milkweed fibers are sensitive to acids. In a study by Sakthivel et al. [119], the resistance of milkweed fibers to weak and strong acids at different temperatures was investigated: milkweed fibers, like cotton fibers, are highly sensitive to strong acids such as sulfuric acid; milkweed fibers can be completely dissolved in strong acid solution. It has been noted that weak organic acids (acetic acid, formic acid, etc.) can be used in applications of milkweed fibers without damaging the fiber structure [119]. It was also stated that only color change is observed in the fibers after the controlled strong alkali treatments of milkweed fibers. However, uncontrolled processing may negatively influence the mechanical characteristics such as fiber strength and stiffness [82].

5.4   Dyeability of Milkweed Fibers

Milkweed fibers are natural cellulosic fibers, so their dyeability properties are struc-turally compatible with commonly used natural fibers such as cotton [87, 124]. As a result of dyeing processes of milkweed fibers with various dyestuff types, especially reactive dyestuffs that are widely used in the coloring of cotton fibers, the dyeing behaviors of the fibers such as dye uptake, color yield, and brightness were investi-gated [87, 124]. Milkweed fibers were subjected to various pre-treatment processes before dyeing to remove wax and oil from the fiber surface, and then, dyeing processes were carried out. In a study examining the dyeing properties of milkweed fibers with reactive dyestuffs, the dyeing procedure was carried out in a 1:20 liquor ratio according to the extrusion method. Dyeing processes were applied not only to the pre-treated milkweed fibers, but also to 100% cotton and 75–25% cotton/milkweed blended fibers (the milkweed fiber used in the blended yarn was included in the blend yarn structure after pre-treatment processes). As a result of dyeing processes, it has been reported that the highest K/S values are followed by 100% milkweed and 100% cotton yarns in cotton/milkweed blended yarns, respectively [87]. As an outcome of the results obtained, it was noted that better quality dyeing processes were obtained with the yarns using cotton and milkweed blend fibers.

5.5 Spinnability of Milkweed Fibers

Milkweed fibers face many problems and difficulties during the spinning processes, such as bobbin entanglement, frequent sliver breakage, and tissue condensation, due to their non-crimp structure and smooth surface [82, 113, 125–127]. The fact that the milkweed fibers are not attached to each other causes extra loosening of the fibers during the carding process. Also, the fibers are damaged considerably due to the fragility of the fibers [126]. Different methods have been investigated to minimize the mechanical damage that may occur during the spinning processes of milkweed fibers [126, 127]. There are some advised methods like blending fibers and fiber surface modification processes. It was thought that surface modifications could be an effective method in improving the spinning characteristics of the fibers by improving the adhesion of milkweed fibers and reducing the fiber loss during the carding process [87, 113, 121, 128]. It has also been stated that blending milkweed fibers effectively improves the spinnability of the fibers [126].

Sakthivel et al. [119] chemically modified the milkweed fiber surface with a liquid alkali solution containing 5% NaOH (5 min., 21 °C) to increase the spinnability of milkweed fibers. By using SEM imaging method, structural differences were detected between fiber surfaces nontreated and alkali treated. The folds that occur through the fiber length, which increase the friction of the fibers, were observed. In this way, it has been proven that by modifying the fiber surfaces, the spinnability of milkweed fibers can be increased. Another surface modification study was carried out by Gharehaghaji and Hayat-Davoodi [126]. In this research, it is aimed to increase the surface roughness of the fibers and to ensure that the fibers adhere to each other more by using the cold plasma method for the chemically modified milkweed fiber surfaces. Increasing the fiber roughness has been accepted as an efficient method for enhancing the spinning properties of milkweed fibers, as it will not only affect the surface friction but also reduce the fiber relaxation that occurs during the carding process. However, many cracks were detected on the cold plasma-treated milkweed fiber surfaces and it was predicted that these cracks that occurred after cold plasma treatment processes may cause the loss of strength and decrease in elongation percentages [126].

5.5.1 Milkweed Fiber Blends

Even though the surface modification applications provide effective conditions to increase the bonding of milkweed fibers, they cause various damage to the surface of the fibers, which can lead to a decrease in stamina. Louis and Andrews [82] had a research about the spinning properties of milkweed seed fibers by using rotor spinning system with various blending ratios cotton/milkweed fiber and determined that most of the obstacles during the carding process of milkweed seed fibers were due to fiber matching and lack of crimp. As a result, they advised that mixing milkweed fibers with different textile fibers (such as cotton) would improve the spinnability and reduce/prevent spinning problems and difficulties [82].

Dre'an et al. [113] investigated the yarn properties produced by ring spinning system containing milkweed/cotton fibers at different blending ratios. Relatively high ratios (up to 67%) of milkweed fibers were used as a mixture in the research. It has been noted that increasing milkweed fiber content in the mixture will occur more spinnability problems and difficulties during the process. In addition, they showed that the higher the cotton content in the blend, the lower the fiber loss during the spinning process due to higher fiber compatibility [113]. Sakthivel et al. [119] compared the characteristics of cotton/milkweed (25/75%) blend yarns and 100% cotton yarns. According to study, it has been stated that the evenness and the breaking strength of the 100% cotton yarns were higher than the cotton/milkweed blended yarns. It is also discussed in the study that the differences in fineness between blended fibers in the yarn structure will lead to different displacement behavior, resulting in spun yarn with uneven durability [119].

Bakhtiari et al. [121] investigated the spinnability and knitting properties of the fibers by producing knitted fabric samples in interlock structure with yarns spun with a cotton/silk cotton ring spinning system [121]. In order to avoid any damage to the fragility of the seed milkweed fibers, the fiber was subjected to very little mixing using a special opening machine. Roller speeds are precisely adjusted to reduce the possibility of fiber breakage and to minimize length shortening. In light of the data obtained as an outcome of this research, it has been determined that parameters such as relative humidity and temperature have very important effects on the production of cotton/milkweed mixed ring spinning system [121]. In addition to these studies, it has been noted that milkweed fibers can be utilized together with other fibers such as wool fibers in different blending ratios (25%/75%; 50%/50%; 75%/25%) [96].

6 Importance of *Asclepias Syriaca* (Milkweed) Fibers for the Sustainable Textile and Fashion Industry Application/Research Areas

Fibers obtained from the stem or seeds of the milkweed plant were found in prehistory and were used as raw materials for textile products in various regions of the world, especially in the south of the USA and Canada. These milkweed fibers have also utilized in food or medicinal purposes in addition to textile usages for centuries [129]. Milkweed fibers were the stuffing material of the life jackets during the Second World War due to their hydrophobic nature, light weight, and hollow structure [118, 130]. Today, the use of milkweed fibers is beyond being a stuffing material; functional products, composite structures, and technical textile products spread to a wide variety of areas. In more detail, milkweed fibers could be utilized in many different industrial purposes as technical textiles and other different application types such as filling materials, moisture-absorbing materials, sound and/or heat insulation materials, oil-cleaning materials, automotive textiles, personal care products, filtration equipment, geotextiles, and many kinds of composite structures. From the perspective of sustainable production, it is accepted that specialty fibers such as milkweed have a very valuable raw material potential in the future of the textile and fashion industry.

6.1 Daily Used Textile Products

Milkweed fibers are thought to have an important potential in the use of daily textile products as a result of their extraordinary properties, and many studies in the literature confirm this. Comfort, handling, breathability, water permeability, moisture absorption, and appearance properties are of great importance in the production of daily textile products. For this reason, it is estimated that thanks to the high moisture recovery feature of Milkweed fibers, its use in ready-to-wear products can provide a great advantage in terms of comfort and convenience. Studies examining the comfort properties of milkweed fibers at different mixing ratios also confirm this [82, 131]. Karthik and Murugan [125] compared 100% cotton fabrics with samples produced from cotton/milkweed (60/40) blended yarns, taking into account the handling and comfort characteristics of fabrics made from milkweed fibers.

According to the research, it is stated that the air permeability and moisture vapor permeability of milkweed/cotton blended fabrics were measured lower than the cotton fabric samples and it thought that the needle-like shape of milkweed fiber structures provides very high resistance to air flow [82, 122]. At first glance, these findings suggest that milkweed fibers are less comfortable. Bakhtiari et al. [121] have showed that the steam transmission rate in samples created from cotton/milkweed blended yarns was relatively higher than cotton counterparts. Karthik and Murugan [125] reported that the water absorption capability of the milkweed fibers is considerably higher than cotton fibers because of its hollow structure [125]. It is also determined that the thermal conductivity of the milkweed fabrics was considerably less than the cotton samples [122, 125].

This is also evidence of the thermal insulation properties that make milkweed fibers suitable for use as insulation materials. In addition, Crews et al. determined that fabrics made of milkweed are relatively thicker than cotton samples [104]. This also affects the insulating properties of the end products. Moreover, the water absorption ability and the comfort properties of milkweed fibers provide them properad features for being usable in personal care products such as towels, baby diapers, sanitary towels, tampons, bandage gauze, and cosmetic wipes [82, 131, 132].

6.2 Technical Textiles

Technical textiles are one of the most challenging application areas in sustainable fashion and textile production. Because the priority of these product groups is the preservation of their superior qualities, technical and functional properties. At this point, the light, hollow water-floating and insulating structure of milkweed fibers provides a great advantage for its use in technical textiles. Absorption properties of cellulosic fibers such as milkweed, comfort, skin-friendly properties, wipe-drying, and absorption performances make them suitable for use in many nonwoven products such as personal care and disposable products, medical and surgical textiles.

Rengasamy et al. [133] showed that unrubbed and unbleached natural fibers such as milkweed (with a waxy layer on the fiber surface) can be utilized as degreasing material instead of the most commonly used synthetic fibers in this field. Also, usage of natural milkweed fibers instead of synthetic sorbent fibers will provide some further benefits with regard to biodegradability and environmental advantages [134]. Karan et al. [134] had stated that 1 g of milkweed fiber has the potential to absorb 40 g of light crude oil at the ambient temperature. Milkweed fibers own this high oil-absorbency feature to their waxy coating on the fiber surface (due to the large amount of wax existence (3%) on the surface of the milkweed fiber. For instance, the amount of wax content in the cotton fiber is approximately 0.448%). Also, the wide and non-collapsing lumen of milkweed fibers is pointed to as other cause for the high-performance oil-absorbing capacity of the fibers.

To be clear, the greater absorbency of materials with higher porosity is usually attributed to capillary action that arises through hydrophobic interaction and capillary regions formed within fibrous structures [134–136]. Capillary action along the long hollow channel of the fibers, usually because of the hydrophobic interaction between the fiber surface and water molecules, was the key cause for the milkweed's capability to clean up oil spills. The smooth surface of milkweed fibers is thought to be other explanation why it can significantly clean up oil spills in comparison with other synthetic materials used [137–139]. Due to the low cellulose content, milkweed fibers can be expected to degrade slowly. This feature will provide novel potential applications in fisheries and shipping [133].

Evaluating the insulating characteristics of numerous materials, milkweed fibers were found to have significantly higher sound absorption because of its hollow structure [82, 108]. This result was also found in Hassanzadeh et al. [128, 140] and Hasani et al. It can also be seen in other studies by Hasani et al. [141]. Hassanzadeh et al. [128, 140] utilized numerous blends of polypropylene and seeded milkweed fibers to prepare sound-absorbing lightly needled nonwoven layers. The acoustic characteristics of the manufactured substrates were measured by the impedance tube technique. According to the results obtained, milkweed fiber ratio in the mixtures greatly influences the noise absorption coefficient (NAC). It was also reported that the sheets were composed of 100% milkweed (Persian, known as Estabragh) fibers, which provided the highest NAC value. They point out that the higher sound absorption compared to the same fibrous material consisting of solid fibers is due to air compression in the hollow channel of the milkweed fiber [141]. The attractive sound-absorbing properties of milkweed fibers make them suitable for use in the automotive and construction industries where sound pollution is a concern. According to another study, milkweed fibers can be used as soft and flexible felt insulation materials providing sound and heat insulation. In this research, a novel method has applied on the surfaces of the milkweed fibers aiming to increase the compatibility of the fibers and to minimize/prevent fiber loss during the carding process. This technique comprises a colloidal adhesive solution to mechanically bind the fibers, followed by straining and drying of the fibers [79].

Bahari et al. [142] investigated the uniaxial tensile and flexural characteristics of low-density thermally bonded nonwoven surface layers manufacture from milkweed/polypropylene fiber blends with different seeds [142]. According to the outcomes obtained, these fibers possess high prospective applicability to be utilized in technical textile manufactured goods. The resistance of milkweed resistance to outdoor environmental conditions makes it appropriate for the production of rope, carpet, fishing nets, and sewing thread [143]. Moreover, in the study of Eftekhari et al. [144] in 2019, it was stated that milkweed fibrous nonwoven surface structures can be successfully used in the separation of heavy metals such as Pb^{2+} and Ni^{2+} from aqueous solutions.

6.3 Composites

Composite structures have been appearing in all areas of our lives in recent years. It has a broad range of uses, from a simple bike to a complex space shuttle, from an airplane fuselage to a tennis racket to a speedboat. These composite structures, which are fiber-reinforced reinforced structures, are closely related to the textile industry. Textile fibers, textile structures, pre-treatments, post-processes, and coloring used in the textile industry all create different variations in composite production and constitute a large research and usage area. In line with the objectives such as sustainable production approach and ecofriendly manufacture, the use of natural fibers in composite structures has come to the fore and the production of bio composites has become a very large and important field. In this context, there are many studies in the literature on the use of milkweed fibers in composite structures [145–147]. Under this title, research on the usage of milkweed fibers in composite structures in the literature and their targeted usage areas is briefly summarized. The low density of milkweed fibers provides a great advantage for composite structures. Thanks to this feature, a higher amount of textile fibers per unit weight can make the coalescence during composite production. It ultimately results in a lighter weight composite structure including enhanced features [146]. Lightweight composites are very important in the transportation industry, especially in automotive applications through weight limitations and appropriate sound absorption. They can also be used for reinforcing cement composite structures [147].

6.4 Promising End-Uses

In a study conducted in 2021 on kapok fibers, which are very similar to milkweed fibers in terms of both structural and performance properties, kapok fibers were used as a raw material in research on controlled drug release. In the structure of kapok fibers, cellulose is chemically cross-linked with citric acid and loaded with chlorhexidine diacetate. And the results of the study determined that kapok fibers are a suitable material for the manufacturing of systems for drug release purposes. In this context, milkweed fibers have the potential to be used in similar drug delivery systems [148]. Also, it has been noted that the hollow structure of milkweed fibers is suitable for growing bacteria and this method has been researched [91].

In the above part of the book chapter, we discovered that milkweed fiber, a sustainable, biodegradable, renewable, and natural plant fiber, can be used in many different application types. The use and spread of milkweed fiber, which is a sustainable fiber, and similar sustainable textile materials in such different areas are promising for the sustainable textile sector. Obviously, sustainability in textile is not just about sustainable textile fibers. To sum up, there are many efforts and studies on many different types of sustainable textile production methods and technologies and a wide variety of sustainable textile materials [16, 23, 26, 28–30, 46, 62, 149–160] in order to enhance sustainable textile production and make these significant steps permanent. The increase in these steps for a more sustainable world is an important sustainability legacy that we will leave to future generations. It could be anticipated that these encouraging sustainable textile-oriented studies and efforts will increase and become more widespread around the world leading to more sustainable planet.

7 Conclusion

In recent years, we have often heard concepts such as "global warming", "climate crisis", and "sustainability" even in our daily lives, because the situation our planet is in has become so serious that the results of the deformation that has been going on for years have moved to dimensions that cannot be ignored. The textile and fashion industry, as is known, is one of the industries with a large environmental burden. Numerous studies have been conducted to reduce this burden, and these studies continue with great hope. The change in the advertisements and slogans of big brands is similarly promising. Concepts such as "sustainable production", "environmentally friendly production", "organic", "products made from recycled fibers", "carbon footprint", and "water footprint" are highlighted in the advertisements of their brands. In this way, it both contributes to the environmental awareness of the consumer and encourages other brands to produce with this awareness. One of the most common solution methods recommended for environmentally friendly production in the sustainable textile and fashion industry is to select the raw material to be used during product design from biodegradable, natural fibers that do not require the use of excess water, pesticides, chemicals, etc., during production.

Cotton fibers, which are the most widely used textile fibers, unfortunately, unless they are produced organically, are a natural type with a high environmental load and huge water consumption. At this point, each type that we have classified in the category of natural fibers is of huge importance for the sustainable textile and fashion industry. Although milkweed fiber is not a widely used fiber type today, it is a type that carries great potential in terms of fiber properties and fiber structure. By nature, milkweed fibers have a hollow channel running along the length of the fiber and thus possess high moisture content and moisture recovery properties. Thanks to these properties, they can be utilized as moisture-absorbing materials. Due to its high moisture permeability, milkweed fibers can be used as raw materials to produce fabrics with desired comfort characteristics. The hollow structure of milkweed also ensures that the fibers are lightweight and have better insulating characteristics than other commercially utilized natural fibers. Thus, natural milkweed fibers possess a great potential for being utilized as sound or heat insulation materials. Also, high oil-absorbent character of milkweed fibers makes them a perfect alternative for oil-cleaning materials compared to synthetic materials available in the market.

In addition to these, milkweed fibers can be used in many industrial purposes such as technical textiles and applications. Automotive textiles, personal care products, filtration equipment, geotextiles, and many kinds of composite structures can be pointed as just a few of their potential usages. However, milkweed fibers, which are not very suitable for spinning due to their structure, are generally used in the form of yarn production or nonwoven surface by mixing with other fibers. Therefore, further studies are needed to improve the spinnability of these fibers.

References

1. Fletcher K (2013) Sustainable fashion and textiles: design journeys. Routledge, New York
2. Muthu SS, Li Y, Hu JY, Mok PY (2012) Quantification of environmental impact and ecological sustainability for textile fibres. Ecol Ind 13(1):66–74
3. Fattahi FS, Khoddami A, Avinc O (2020) Sustainable, renewable, and biodegradable poly (lactic acid) fibers and their latest developments in the last decade. Sustain Text Apparel Ind, 173
4. Avinc O, Kalayci E, Yildirim FF, Yavas A (2015) Biologically inspired some textile protein fibers. In: International conference on life science and biological engineering, Tokyo, Japan
5. Yıldırım F, Avinç O, Yavaş A (2014) Soya Fasülyesi Protein Lifleri Bölüm 2: Soya Liflerinin Özelliklerinin Ve Kullanım Alanları. Uludağ Univ J Facul Eng 20(1):1–21
6. Kalaycı E, Avinç O, Yavaş A (2019) Usage of horse hair as a textile fiber and evaluation of color properties. Ann Univ Oradea Fascicle Text Leatherwork 2019(1):57–62
7. Kalaycı CB, Gündoğan M, Kalaycı E, Avinç O (2019) Color strength estimation Oyif Coir fibers bleached with peracetic acid. Ann Univ Oradea Fascicle Text Leatherwork 2019(2):65–70
8. Hasani H, Avinc O, Khoddami A (2017) Effects of different production processing stages on mechanical and surface characteristics of polylactic acid and PET fibre fabrics. Indian J Fibre Text Res (IJFTR) 42(1):31–37

9. Avinc O, Eren HA, Uysal P (2012) Ozone applications for after-clearing of disperse-dyed poly (lactic acid) fibres. Color Technol 128(6):479–487
10. Hasani H, Avinc O, Khoddami A (2013) Comparison of softened polylactic acid and polyethylene terephthalate fabrics using KES-FB. Fibres Text Eastern Euro 3(99):81–88
11. Avinc O, Eren HA, Uysal P, Wilding M (2012) The effects of ozone treatment on soybean fibers. Ozone Sci Eng 34(3):143–150
12. Avinc O, Day R, Carr C, Wilding M (2012) Effect of combined flame retardant, liquid repellent and softener finishes on poly (lactic acid)(PLA) fabric performance. Text Res J 82(10):975–984
13. Avinc O, Bone J, Owens H, Phillips D, Wilding M (2006) Preferred alkaline reduction-clearing conditions for use with dyed Ingeo poly (lactic acid) fibres. Color Technol 122(3):157–161
14. Avinc O, Phillips D, Wilding M (2009) Influence of different finishing conditions on the wet fastness of selected disperse dyes on polylactic acid fabrics. Color Technol 125(5):288–295
15. Avinc O, Khoddami A (2010) Overview of poly (lactic acid)(PLA) fibre. Fibre Chem 42(1):68–78
16. Avinc O (2011) Clearing of dyed poly (lactic acid) fabrics under acidic and alkaline conditions. Text Res J 81(10):1049–1074
17. Avinc O (2011) Maximizing the wash fastness of dyed poly (lactic acid) fabrics by adjusting the amount of air during conventional reduction clearing. Text Res J 81(11):1158–1170
18. Avinc O, Khoddami A, Hasani H (2011) A mathematical model to compare the handle of PLA and PET knitted fabrics after different finishing steps. Fibers Polym 12(3):405
19. Khoddami A, Avinc O, Ghahremanzadeh F (2011) Improvement in poly (lactic acid) fabric performance via hydrophilic coating. Prog Org Coat 72(3):299–304
20. Avinc O, Owens H, Bone J, Wilding M, Phillips D, Farrington D (2011) A colorimetric quantification of softened polylactic acid and polyester filament knitted fabrics to 'Water-spotting.' Fibers Polym 12(7):893
21. Fattahi F, Khodami A, Avinc O (2020) Nano-structure roughening on poly (Lactic Acid) PLA substrates: scanning electron microscopy (SEM) surface morphology characterization. J Nanostruct 10(2):206–216
22. Avinc O, Yavas A (2017) Soybean: for textile applications and its printing. In: Soybean—The basis of yield, biomass and productivity
23. Palamutcu S, Soydan AS, Avinc O, Günaydin GK, Yavas A, Kıvılcım MN, Demirtaş M (2019) Physical properties of different Turkish organic cotton fiber types depending on the cultivation area. Organic cotton. Springer, pp 25–39
24. Gedik G, Avinc O (2018) Bleaching of hemp (Cannabis sativa L.) fibers with peracetic acid for textiles industry purposes. Fibers Polym 19(1):82–93
25. Arık B, Avinc O, Yavas A (2018) Crease resistance improvement of hemp biofiber fabric via sol–gel and crosslinking methods. Cellulose 25(8):4841–4858
26. Günaydin GK, Avinc O, Palamutcu S, Yavas A, Soydan AS (2019) Naturally colored organic cotton and naturally colored cotton fiber production. Organic cotton. Springer, pp 81–99
27. Soydan AS, Yavas A, Günaydin GK, Palamutcu S, Avinc O, Kıvılcım MN, Demirtaş M (2019) Colorimetric and hydrophilicity properties of white and naturally colored organic cotton fibers before and after pretreatment processes. Organic Cotton. Springer, pp 1–23
28. Unal F, Avinc O, Yavas A (2020) Sustainable textile designs made from renewable biodegradable sustainable natural abaca fibers. Sustainability in the textile and apparel industries. Springer, Cham, pp 1–30
29. Unal F, Yavas A, Avinc O (2020) Contributions to sustainable textile design with natural raffia palm fibers. Sustainability in the textile and apparel industries, pp 67–86
30. Gedik G, Avinc O (2020) Hemp fiber as a sustainable raw material source for textile industry: can we use its potential for more eco-friendly production? In: Sustainability in the textile and apparel industries, pp 87–109
31. Gedik G, Avinç O, Yavaş A (2010) Kenevir lifinin özellikleri ve tekstil endüstrisinde kullanımıyla sağladığı avantajlar. Tekstil Teknolojileri Elektronik Dergisi 4(3):39–48

32. Kalaycı E, Avinc O, Yavaş A (2019) The effects of different alkali treatments with different temperatures on the colorimetric properties of lignocellulosic raffia fibers. Int J Adv Sci Eng Technol 7(1):15–19

33. Maia LC, Alves AC, Leão CP (2012) Sustainable work environment with lean production in textile and garment ındustry. In: Proceedings of ınternational conference on ındustrial engineering and operations management (ICIEOM2012)

34. Kalayci E, Avinc O, Yavas A, Coskun S (2019) Responsible textile design and manufacturing: environmentally conscious material selection. In: Alqahtani AY et al (eds) Responsible manufacturing: ıssues pertaining to sustainability. Taylor & Francis

35. Kumartasli S, Avinc O (2020) Important step in sustainability: polyethylene terephthalate recycling and the recent developments. In: Sustainability in the textile and apparel ındustries, p 1

36. Kumartasli S, Avinc O (2020) Recycling of marine litter and ocean plastics: a vital sustainable solution for ıncreasing ecology and health problem. In: Sustainability in the textile and apparel ındustries, p 117

37. Kumartasli S, Avinc O (2021) Recycled thermoplastics: textile fiber production, scientific and recent commercial developments. Recent developments in plastic recycling. Springer, pp 169–192

38. Eren HA, Yiğit I, Eren S, Avinc O (2020) Sustainable textile processing with zero water utilization using super critical carbon dioxide technology. In: Sustainability in the textile and apparel ındustries, p 179

39. Eren S, Avinc O, Saka Z, Eren HA (2018) Waterless bleaching of knitted cotton fabric using supercritical carbon dioxide fluid technology. Cellulose 25(10):6247–6267

40. Odabaşoğlu HY, Avinç OO, Yavaş A (2013) Susuz Boyama. Tekstil ve Mühendis 20(90):62–79

41. Yıldırım FF, Yavas A, Avinc O (2020) Bacteria working to create sustainable textile materials and textile colorants leading to sustainable textile design. Sustainability in the textile and apparel industries. Springer, Cham, pp 109–126

42. Eren HA, Yiğit I, Eren S, Avinc O (2020) Ozone: an alternative oxidant for textile applications. In: Sustainability in the textile and apparel ındustries, p 81

43. Avinc O, Celik A, Gedik G, Yavas A (2013) Natural dye extraction from waste barks of Turkish red pine (Pinus brutia Ten.) timber and eco-friendly natural dyeing of various textile fibers. Fibers Polym 14(5):866–873

44. Gedik G, Yavaş A, Avinç OO, Şimşek Ö (2013) Cationized natural dyeing of cotton fabrics with corn poppy (Papaver rhoeas) and investigation of antibacterial activity

45. Eren HA, Avinc O, Uysal P, Wilding M (2011) The effects of ozone treatment on polylactic acid (PLA) fibres. Text Res J 81(11):1091–1099

46. Yavaş A, Avinc O, Gedik G (2017) Ultrasound and microwave aided natural dyeing of nettle biofibre (Urtica dioica L.) with madder (Rubia tinctorum L.). Fibres Text Eastern Euro

47. Yıldırım FF, Avinc O, Yavas A, Sevgisunar G (2020) Sustainable antifungal and antibacterial textiles using natural resources. Sustainability in the textile and apparel industries. Springer, Cham, pp 111–179

48. Yıldırım FF, Yavas A, Avinc O (2020) Printing with sustainable natural dyes and pigments. In: Sustainability in the textile and apparel industries–production process sustainability, pp 1–35

49. Avinc O, Erişmiş B, Eren HA, Eren S (2016) Treatment of cotton with a laccase enzyme and ultrasound. De Redactie, p 55

50. Gedik G, Avinc O, Yavas A, Khoddami A (2014) A novel eco-friendly colorant and dyeing method for poly (ethylene terephthalate) substrate. Fibers Polym 15(2):261–272

51. Yildirim F, Sevgisunar H, Yavaş A, Avinç O, Çelik A (2014) UV Korumada Ekolojik Çözümler. Tekstil ve Mühendis 21(96):36–51

52. Gedik G, Avinc O, Yavaş A (2011) Bromus Tectorum Bitkisinin Tekstilde Doğal Boyarmadde Kaynağı Olarak Kullanımı. Tekstil Teknolojileri Elektronik Dergisi 5(1):40–47

53. Karakan Günaydin G, Palamutcu S, Soydan AS, Yavas A, Avinc O, Demirtaş M (2020) Evaluation of fiber, yarn, and woven fabric properties of naturally colored and white Turkish organic cotton. J Text Inst 111(10):1436–1453
54. Arık B, Yavaş A, Avinc O (2017) Antibacterial and wrinkle resistance improvement of nettle biofibre using Chitosan and BTCA. Fibres Text Eastern Euro
55. Unal F, Avinc O, Yavas A, Eren HA, Eren S (2020) Contribution of UV technology to sustainable textile production and design. Sustainability in the textile and apparel industries. Springer, Cham, pp 163–187
56. Unal F, Yavas A, Avinc O (2020) Sustainability in textile design with laser technology. Sustainability in the textile and apparel industries. Springer, Cham, pp 263–287
57. Rahmatinejad J, Khoddami A, Mazrouei-Sebdani Z, Avinc O (2016) Polyester hydrophobicity enhancement via UV-Ozone irradiation, chemical pre-treatment and fluorocarbon finishing combination. Prog Org Coat 101:51–58
58. Rahmatinejad J, Khoddami A, Avinc O (2015) Innovative hybrid fluorocarbon coating on UV/ozone surface modified wool substrate. Fibers Polym 16(11):2416–2425
59. Davulcu A, Eren HA, Avinc O, Erişmiş B (2014) Ultrasound assisted biobleaching of cotton. Cellulose 21(4):2973–2981
60. Eren HA, Avinc O, Erişmiş B, Eren S (2014) Ultrasound-assisted ozone bleaching of cotton. Cellulose 21(6):4643–4658
61. Tungtriratanakul S, Setthayanond J, Avinç OO, Suwanruji P, Sae-Bae P (2016) Investigation of UV protection, self-cleaning and dyeing properties of nano TiO2-treated poly (lactic acid) fabric
62. Kurban M, Yavas A, Avinc O, EREN HA (2016) Nettle biofibre bleaching with ozonation. DE REDACTIE, p 45
63. Waite M (2009) Sustainable Textiles: the role of bamboo and a comparion of bamboo textile properties-part 1. J Text Apparel Technol Manage 6(2)
64. Shangnan Shui AP (2013) World apparel fibre consumption survey. International Cotton Advisory Committee, Washington
65. Sharma A (2013) Eco-friendly textiles: a boost to sustainability. Asian J Home Sci 8(2):768–771
66. Petry F (2008) Environmental protection and sustainability in the textile industry. Text Finish 7–8:86–88
67. Hayes LL (2001) Synthetic textile innovations: polyester fiber-to-fiber recycling for the advancement of sustainability. AATCC Rev 11(4):37–41
68. Karthik T, Murugan R (2013) Milkweed fibers-properties and potential applications. Melliand China 7:012
69. Kalaycı E, Avinç OO, Bozkurt A, Yavaş A (2016) Tarımsal atıklardan elde edilen sürdürülebilir tekstil lifleri: Ananas yaprağı lifleri. Sakarya Üniversitesi Fen Bilimleri Enstitüsü Dergisi 20(2):203–221
70. Kozlowski R, Wladyka–Przybylak M (2001) Natural polymers, wood and lignocellulosic materials. In: Horrocks AR, Price D (eds) Fire retardant materials. Woodhead Publishing Limited, Cambridge
71. Satyanarayana K, Guimarães J, Wypych F (2007)Studies on lignocellulosic fibers of Brazil. Part I: source, production, morphology, properties and applications. Compos Part A Appl Sci Manuf 38(7):1694–1709
72. Konak S (2014) Bamya bitkisinden suda çürütme yöntem ile lif elde edilmesi ve elde edilen lifin çeşitli fiziksel kimyasal ve mekanik özelliklerinin ölçümü, in Tekstil Mühendisliği Bölümü. Pamukkale Üniversitesi Fen Bilimleri Enstitüsü: Denizli, Türkiye.
73. Ashori A, Bahreini Z (2009) Evaluation of Calotropis gigantea as a promising raw material for fiber-reinforced composite. J Compos Mater 43(11):1297–1304
74. Karthik T, Murugan R (2013) Characterization and analysis of ligno-cellulosic seed fiber from Pergularia daemia plant for textile applications. Fibers Polym 14(3):465–472
75. Wikipedia, t.f.e. *Asclepias*. 2018 [cited 2018 October]; Available from: https://en.wikipedia.org/wiki/Asclepias

76. Witt MD, Nelson LA (1992) Milkweed as a new cultivated row crop. J Prod Agric 5(1):167–171
77. Hartzler RG, Buhler DD (2000) Occurrence of common milkweed (Asclepias syriaca) in cropland and adjacent areas. Crop Prot 19(5):363–366
78. Yeargan KV, Allard CM (2005) Comparison of common milkweed and honeyvine milkweed (Asclepiadaceae) as host plants for monarch larvae (Lepidoptera: Nymphalidae). J Kansas Entomol Soc 78(3):247–251
79. Hassanzadeh S, Hasani H (2017) A review on milkweed fiber properties as a high-potential raw material in textile applications. J Ind Text 46(6):1412–1436
80. Kalayci E, Yildirim FF, Avinc OO, Yavas A (2015) Textile fibers used in products floating on the water. Textile science and economy VII. Zrenjanin, Serbia, pp 85–90
81. Von Bargen K, Jones D, Zeller R, Knudsen P (1994) Equipment for milkweed floss-fiber recovery. Ind Crops Prod 2(3):201–210
82. Louis GL, Andrews BK (1987) Cotton/milkweed blends: a novel textile product. Text Res J 57(6):339–345
83. Varshney A, Bhoi K (1987) Some possible industrial properties of Calotropis procera (Aak) floss fibre. Biological wastes
84. Turkoglu KB, Kalayci E, Avinc O, Yavas A (2019) Oleofilik Buoyans Özellikli Kapok Lifleri ve Yenilikçi Yaklaşımlar. Düzce Üniversitesi Bilim ve Teknoloji Dergisi 7(1):61–89
85. Martin AC, Zim HS, Nelson AL (1961) American wildlife & plants: a guide to wildlife food habits: the use of trees, shrubs, weeds, and herbs by birds and mammals of the United States. 1961: Courier Corporation.
86. Muenscher WC (1939) Poisonous plants of the United States. Poisonous plants of the United States
87. Bahreini Z, Kiumarsi A (2008) A comparative study on the dyeability of stabraq (milkweed) fibers with reactive dyes. Prog Color Color Coat 1(1):19–26
88. Prasad P (2006) Physio-chemical properties of a nonconventional fiber: aak (calotropis procera). J Text Assoc 67(2):63–66
89. Gaertner EE (1979) The history and use of milkweed (Asclepias syriaca L.). Econ Botany 33(2):119–123
90. Parrotta JA (2001) Healing plants of peninsular India. CABI publishing
91. Karthik T, Murugan R (2016) Milkweed—a potential sustainable natural fibre crop. Sustainable fibres for fashion industry. Springer, pp 111–146
92. Ulbricht H (1940) Die bastfasern von asclepias syriaca L. Forscbung 14(4):232–237
93. Berkman B (1949) Milkweed—a war strategic material and a potential industrial crop for sub-marginal lands in the United States. Econ Bot 3(3):223–239
94. Witt MD, Knudsen HD (1993) Milkweed cultivation for floss production. In: Janick J, Simon JE (eds) New crops. Wiley, New York, pp 428–431
95. Knudsen HD, Zeller RD (1993) The milkweed business. In: Janick J, Simon JE (eds) New crops. New York, Wiley
96. Nehring J (2014) The potential of Milkweed Floss as a natural fiber in the textile industry. J Undergraduate Res, 63–68
97. Moore RJ (1946) Investigations on rubber-bearing plants: IV. Cytogenetic studies in asclepias (Tourn.) L. Can J Res 24(3): 66–73
98. Stevens O (1945) Asclepias syriaca and A. speciosa, distribution and mass collections in North Dakota. Am Midland Nat, 368–374
99. Adams RP, Balandrin MF, Martineau JR (1984) The showy milkweed, Asclepias speciosa: a potential new semi-arid land crop for energy and chemicals. Biomass 4(2):81–104
100. Borders B, Lee-Mäder E (2014) Milkweeds: a conservation practitioner's guide. Portland, OR: xerces society for invertebrate conservation. Proc R Soc B 282(20141734):9
101. Phippen WB (2007) Production variables affecting follicle and biomass development in common milkweed. In: Janick J, Whipkey A (eds) Issues in new crops and new uses. ASHS Press Alexandria Virginia USA, pp 82–88

102. Roşu A, Danaila-Guidea S, Dobrinoiu R, Toma F, Roşu DT, Sava N, Manolache C (2011) Asclepias syriaca L.–an underexploited industrial crop for energy and chemical feedstock. Rom Biotechnol Lett 16(6)
103. Beckett RE, Stitt RS (1935) The desert milkweed (Asclepias subulata) as a possible source of rubber. United States Department of Agriculture, Economic Research Service
104. Cox Crews P, Rich W (1995) Influence of milkweed fiber length on textile product performance. Cloth Text Res J 13(4):213–219
105. Flynn P, Vidaver AK (1995) Xanthomonas campestris pv. asclepiadis, pv. nov., causative agent of bacterial blight of milkweed (Asclepias spp.). Plant disease (USA)
106. Luna T, Dumroese RK (2013) Monarchs (Danaus plexippus) and milkweeds (Asclepias species) the current situation and methods for propagating milkweeds. Native Plants J 14(1):5–16
107. USDA (2018) Common Milkweed Asclepias syriaca L. 2003 [cited 2018 October]; Available from: https://plants.usda.gov/plantguide/pdf/cs_assy.pdf
108. Crews PC, Sievert SA, Woeppel LT, McCullough EA (1991) Evaluation of milkweed floss as an insulative fill material. Text Res J 61(4):203–210
109. Woodson RE (1954) The North American species of Asclepias L. Ann Mo Bot Gard 41(1):1–211
110. CHEM J (1944) Milkweed helps solve fiber problem. J Chem Educ 21(2):54
111. Evetts L, Burnside O (1974) Root distribution and vegetative propagation of Asclepias syriaca L. Weed Res 14(5):283–288
112. Reddy N, Yang Y (2009) Extraction and characterization of natural cellulose fibers from common milkweed stems. Polym Eng Sci 49(11):2212–2217
113. Drean J-YF, Patry JJ, Lombard GF, Weltrowski M (1993) Mechanical characterization and behavior in spinning processing of milkweed fibers. Text Res J 63(8):443–450
114. Jones D, Von Bargen K (1992) Some physical properties of milkweed pods. Trans ASAE 35(1):243–246
115. Andrews BK, Kimmel LB, Bertoniere NR, Hebert J (1989) A comparison of the response of cotton and milkweed to selected swelling and crosslinking treatments. Text Res J 59(11):675–679
116. Shaikhzadeh Najar S, Haghighat-Kish M (1998) Structure and properties of a natural celulosic hollow fiber. Int J Eng 11(2):101–108
117. Timell T, Snyder J (1955) Molecular properties of milkweed cellulose. Text Res J 25(10):870–874
118. Knudsen HD (1990) Milkweed floss fiber for improving nonwoven products. In: TAPPI Nonwovens conference, TAPPI Press, Atlanta
119. Sakthivel J, Mukhopadhyay S, Palanisamy N (2005) Some studies on Mudar fibers. J Ind Text 35(1):63–76
120. Karthik T, Murugan R (2013) Analysis of comfort properties of cotton/milkweed blended rotor yarn fabrics. Melliand Int 19(4)
121. Bakhtiari M, Hasani H, Zarrebini M, Hassanzadeh S (2015) Investigation of the thermal comfort properties of knitted fabric produced from Estabragh (Milkweed)/cotton-blended yarns. J Text Inst 106(1):47–56
122. Karthik T (2014) Studies on the spinnability of milkweed fibre blends and its influence on ring compact and rotor yarn characteristics
123. Gu P, Hessley RK, Pan W-P (1992) Thermal characterization analysis of milkweed flos. J Anal Appl Pyrol 24(2):147–161
124. Shakyawar D, Dagur R, Gupta N (1999) Studies on milkweed fibres
125. Karthik T, Murugan R (2013) Influence of spinning parameters on milkweed/cotton DREF-3 yarn properties. J Text Inst 104(9):938–949
126. Gharehaghaji AA, Davoodi SH (2008) Mechanical damage to estabragh fibers in the production of thermobonded layers. J Appl Polym Sci 109(5):3062–3069
127. Bahl M, Arora C, Rao PJV (2013) Surface modification of milkweed fibres to manufacture yarns. Text Potpouri, 33–35

128. Hassanzadeh S, Hasani H, Zarrebini M (2014) Analysis and prediction of the noise reduction coefficient of lightly-needled Estabragh/polypropylene nonwovens using simplex lattice design. J Text Inst 105(3):256–263
129. Jeeshna M, Manorama S, Paulsamy S (2009) Antimicrobial property of the medicinal shrub Glycosmis pentaphylla. J Basic Appl Biol 3(1&2):25–27
130. Srinivas CA, Babu GD (2013) Mechanical and machining characteristics of calotropis gigentea fruit fiber reinforced plastics. Int J Eng Res Tech 2:1524–1530
131. Zarehshi A, Ghane M (2021) Study of the water vapor permeability of multiple layer fabrics containing the milkweed fibers as the middle layer. J Text Inst, 1–7
132. Rajesh Kumar C, Raja D, Kumar SKS, Prakash C (2020) Study on moisture behavior properties of milkweed and milkweed/cotton blended sanitary napkins. J Nat Fibers, 1–12
133. Rengasamy RS, Das D, Praba Karan C (2011) Study of oil sorption behavior of filled and structured fiber assemblies made from polypropylene, kapok and milkweed fibers. J Hazard Mater 186(1):526–532
134. Karan CP, Rengasamy R, Das D (2011) Oil spill cleanup by structured fibre assembly. Indian J Fibre Text Res 36:190–200
135. Choi HM, Cloud RM (1992) Natural sorbents in oil spill cleanup. Environ Sci Technol 26(4):772–776
136. Choi HM, Moreau JP (1993) Oil sorption behavior of various sorbents studied by sorption capacity measurement and environmental scanning electron microscopy. Microsc Res Tech 25(5–6):447–455
137. Hindi SS (2013) Characteristics of some natural fibrous assemblies for efficient oil spill cleanup. Int J Sci Eng Invest 2(16):10
138. Subramoniapillai V, Thilagavathi G (2021) Oil spill cleanup by natural fibers: a review. Res J Text Apparel
139. Panahi S, Moghaddam MK, Moezzi M (2020) Assessment of milkweed floss as a natural hollow oleophilic fibrous sorbent for oil spill cleanup. J Environ Manage 268:110688
140. Hassanzadeh S, Zarrebini M, Hasani H (2014) An investigation into acoustic properties of lightly needled Estabragh nonwovens using the Taguchi method. J Eng Fibers Fabr 9(3):155892501400900300
141. Hasani H, Zarrebini M, Zare M, Hassanzadeh S (2014) Evaluating the acoustic properties of Estabragh (milkweed)/hollow-polyester nonwovens for automotive applications. Text Sci Eng 4:1–6
142. Bahari N, Hasani H, Zarrebini M, Hassanzadeh S (2016) Investigating the effects of material and process variables on the mechanical properties of low-density thermally bonded nonwovens produced from Estabragh (milkweed) natural fibers. J Ind Text 46(3):719–736
143. Oudhia P (2002) Allelopathic potential of useful weed Calotropis gigentea R. Br: A review. in Abstracts. Third World Congress on Allelopathy: Challenges for the New Millennium, National Institute for Agro-Environmental Sciences (NIAES), Tsukuba, Japan
144. Eftekhari E, Hasani H, Fashandi H (2019) Removal of heavy metal ions (Pb2+ and Ni2+) from aqueous solution using nonwovens produced from lignocellulosic milkweed fibers. J Indus Text, 1528083719888931
145. Ovlaque P, Bayart M, Elkoun S, Robert M (2021) Milkweed floss-reinforced thermoplastics: interfacial adhesion and related mechanical properties. Compos Interfaces, 1–21
146. Reddy N, Yang Y (2010) Non-traditional lightweight polypropylene composites reinforced with milkweed floss. Polym Int 59(7):884–890
147. Merati A (2014) Reinforcing of cement composites by estabragh fibres. J Inst Eng (India): Ser E, 95(1):27–32
148. Peraza-Ku SA, Escobar-Morales B, Rodríguez-Fuentes N, Cervantes-Uc JM, Uribe-Calderon JA (2021) Ceiba pentandra cellulose crosslinked with citric acid for drug release systems. Carbohydr Res 504:108334
149. Bakan E, Avinc O (2021) Sustainable carpet and rug hand weaving in Uşak province of Turkey. Handloom sustainability and culture. Springer, pp 41–93

150. Avinc O, Bakan E, Demirçalı A, Gedik G, Karcı F (2020) Dyeing of poly (lactic acid) fibres with synthesised novel heterocyclic disazo disperse dyes. Color Technol 136(4):356–369
151. Setthayanond J, Sodsangchan C, Suwanruji P, Tooptompong P, Avinc O (2017) Influence of MCT-β-cyclodextrin treatment on strength, reactive dyeing and third-hand cigarette smoke odor release properties of cotton fabric. Cellulose 24(11):5233–5250
152. Avinc O, Wilding M, Phillips D, Farrington D (2010) Investigation of the influence of different commercial softeners on the stability of poly (lactic acid) fabrics during storage. Polym Degrad Stab 95(2):214–224
153. Khoddami A, Avinc O, Mallakpour S (2010) A novel durable hydrophobic surface coating of poly (lactic acid) fabric by pulsed plasma polymerization. Prog Org Coat 67(3):311–316
154. Avinc O, Khoddami A (2010) Overview of poly (lactic acid)(PLA) fibre: Part II: Wet processing; pretreatment, dyeing, clearing, finishing, and washing properties of poly (lactic acid) fibres
155. Avinc O, Wilding M, Bone J, Phillips D, Farrington D (2010) Colorfastness properties of dyed, reduction cleared, and softened poly (lactic acid) Fabrics. AATCC Rev 10(5)
156. Avinc O, Wilding M, Bone J, Phillips D, Farrington D (2010) Evaluation of colour fastness and thermal migration in softened polylactic acid fabrics dyed with disperse dyes of differing hydrophobicity. Color Technol 126(6):353–364
157. Avinc O, Wilding M, Gong H, Farrington D (2010) Effects of softeners and laundering on the handle of knitted PLA filament fabrics. Fibers Polym 11(6):924–931
158. Günaydin GK, Yavas A, Avinc O, Soydan AS, Palamutcu S, Şimşek MK, Dündar H, Demirtaş M, Özkan N, Kıvılcım MN (2019) Organic cotton and cotton fiber production in Turkey, recent developments. Organic cotton. Springer, pp 101–125
159. Soydan AS, Yavaş A, Avinç OO, Günaydın GK, Kivilcim MN, Demirtaş M, Palamutcu S (2019) The effects of hydrogen peroxide and sodium hypochlorite oxidizing treatments on the color properties of naturally colored green cotton. Euro J Eng Nat Sci 3(2):1–10
160. Kurban M, Yavaş A, Avinç O (2011) Isırgan Otu Lifi ve Özellikleri. Tekstil Teknolojileri Elektronik Dergisi 5(1):84–106

Extracellular Polymeric Substances in Textile Industry

Murat Topal and E. Işıl Arslan Topal

Abstract Different microorganisms secrete extracellular polymeric substances that are also known as exopolymeric substances, exopolymers, extracellular polysaccharides, and exopolysaccharides. Extracellular polymeric substances that protect the microorganisms from the environmental stresses are mainly composed of lipids, proteins, polysaccharides, and nucleic acids. They have unique characteristics with functional properties such as addhesion, binding activity, water retention, sorption and so on. The extracellular polymeric substances are valuable products because of their various applications in different industries about textile, pharmaceutical, food, agriculture, cosmetic, and environment. The extracellular polymeric substances are valuable for textile industry as being sustainable resource as well as environmental friendly properties. In this chapter, we have evaluated extracellular polymeric substances followed by the main functions and applications of the extracellular polymeric substances. Moreover, we have emphasized the applications for various extracellular polymeric substances in textile.

Keywords Xanthan · Pullulan · Alginate · Cellulose · Chitosan

1 Introduction

Textile industry, whose origins date back to prehistory, has become one of the most common mass production industries today [1–5]. Textile industry positively affects the economic development worldwide [6]. In most of the countries, textile industries are accountable for creating an important number of jobs worldwide. Countries that develops may join in worldwide production networks and worldwide value chains

M. Topal (✉)
Department of Chem. and Chem. Processing Technologies, Tunceli Vocation School, Munzur Uni., Tunceli, Turkey

E. I. Arslan Topal
Department of Env. Eng., Eng. Fac., Firat Uni., Elazig, Turkey

through textile sector while developing technology and abilities of textile industry [5].

Polysaccharides have structures of linear and highly branched macromolecules. The polysaccharides are highly important biopolymers that play significant role in the development and growth of livings. Polysaccharides exist in most of livings, including in tissues of plants' leaves, seeds, and stems, animals' body fluids, cell wall, and extracellular fluids of yeast, bacteria, and fungi [7–9]. Microbial polysaccharides are exopolymers either insoluble or soluble [10]. Extracellular polymeric substances that are metabolic by products of microbes [10, 11] are exreted out of cell as loosely attached slime layer or tightly bound capsule to carry out functions of source of nutrients during shortage and cell protection against external stress [10, 12, 13]. Extracellular polymeric substances are synthesized by eukaryotes and prokaryotes [14]. Exopolysaccharides are composed of phospholipids, carbohydrates, DNA, proteins, and non carbohydrate substituents [10, 15–17]. Extracellular polymeric substances have high water binding ability, capacity of water retention, extensive gelation and swelling potential [10, 18].

Microbial polysaccharides are value added substances and are preferred for various aims [10, 18, 19]. Some of the applications of extracellular polymeric substances include decolorization of dye and heavy metal removal from effluents, management of leachate, soil remediation and reclamation, treatment of water and wastewater [20]. Alginate, starch, chitosan, cellulose, pectin, gelatin, etc. are widely used extensively for diverse applications [21]. Alginate is preferred in biomedical applications because of low toxicity, mild gelation, and relatively low cost of it [22, 23]. Chitosan is widely applied in multifunctional finishing of textile materials because of the safety of it [24]. Various polysaccharides are preferred as additives in medical and food applications as well as as flocculant and treatment agent in industrial applications. They are environmentally friendly [25–28]. Pullulan has adhesive property and foam preservation feature while drying and dissolution in water, respectively [29–34]. Dextran is preferred in biomedical applications due to high cytocompatibility, chemical functionality and variability of it [35–37]. Welan obtained from Alcaligenes sp. is applied in textile industry [38]. The carrageenan and xanthan are highly produced [13]. In summary, the extracellular polymeric substances find application in food, textile, cosmetic, therapeutics, and bioremediation [10].

In this chapter, we have evaluated extracellular polymeric substances followed by the main functions of the extracellular polymeric substances. Moreover, we have emphasized the applications for various extracellular polymeric substances in textile.

2   Extracellular Polymeric Substances

The exopolysaccharides are high molecular weight polysaccharides produced by microbes [12, 13]. Exopolysaccharides are classified as soluble exopolysaccharides and bound exopolysaccharides. Extracellular polysaccharides are soluble in form of slime, soluble macromolecules, and colloids in growing liquid media that completely dissolved in solution or dispersed with outside wall [39]. The bound extracellular polymeric substances are packed closely and structure is layered into 2 with inner layer of tightly bound extracellular polymeric substances and outer layer of loosely bound extracellular polymeric substances [20, 40]. Extracellular polymeric substances like cellulose, sphingan, xanthan, and alginate promote formation of biofilm on surfaces of cell as protecting barrier [10, 41]. Extracellular polymeric substances play most evident role against cell recognition, phagocytosis, desiccation, toxic compounds, osmotic stress, and antibiotics [10]. The extracellular polymeric substances are classified into heteropolysaccharides and homopolysaccharides. The homopolysaccharides are unbranched or branched and composed of fructose or glucose and are classified into alpha-d-glucans (mutan, dextran, reuteran, and alternan), beta-d-glucans, fructans (levan and inulin) and polygalactans [10, 42]. The heteropolysaccharides are consist of d-glucose, rare sugars, and d-galactose as well as N-acetylgalactosamine, N-acetylglucosamine, or glucuronic acid may exist. They are called xanthan, kefiran and gellan [10, 43]. Some of the extracellular polymeric substances (alginate, cellulose, xanthan, pullulan, chitosan, dextran, and curdlan) are detailed as follows.

The alginates are biocompatible, biodegradable, nontoxic, and nonimmunogenic biopolymer polyelectrolytes [44, 45] that are obtained from algal and bacterial sources (exopolysaccharide of bacteria including *Pseudomonas aeruginosa*). They are unbranched polysaccharides [46, 47]. Commercial alginate is obtained from *Phaeophyceae* that includes A. nodosum, and M. pyrifera [23, 48]. Alginate obtained from bacteria can be produced from Azotobacter and Pseudomonas [23, 49]. Alginate obtained from bacteria may have applications in biomedical applications [23]. The alginates with natural origin are researched due to decolourization and flocculation abilities of them [44, 45, 50–54].

Cellulose has a macromolecule polysaccharide matrix [55–57]. It is insoluble in water and solvents with organic origin because of the rigid structure [55, 57]. It is abundant in natural fibers and plants [57, 58]. Cellulose may be obtained from algae, sea animals, and bacteria [57, 59]. The properties of non toxicity, biodegradability, and high nanoporosity of it make cellulose good industrial compound [60–63].

Xanthan gum is obtained from *Xanthomonas campestris* [64]. Production of xanthan from various agroindustrial residues has been reported by different researchers [63]. Stredansky and Conti [65] and Moshaf et al. [66] reported xanthan gum production from *Xanthomonas campestris* by utilizing agroindustry wastes. Rončević et al. [67] reported biosynthesis of xanthan from *Xanthomonas campestris* in bioreactor by using wastewater of winery. Gondim et al. [68] reported xanthan gum production by *X. axonopodis* pv. Xanthan gum is used as matrix component for drug delivery systems. It forms suspensions having stability in aqueous media [64, 69, 70]. It is used in food industry because of its valuable characteristics [27, 71]. There are many researchs using xanthan gum because of adequate economic feasibility and its high soil strengthening [27, 72–83].

Pullulan is a linear and unbranched extracellular polymeric substance [34, 84]. Bauer [85] firstly noticed pullulan during fermentation of *A. pullulans* [34]. Pullulan is an odourless, tasteless, nonmutagenic and nontoxic edible natural polymer [34,

86, 87]. Pullulan is an important industrial polymer. Because it finds applications in industries of food, pharmaceutical and biomedical [88]. Pullulan is produced as amorphous slime on exterior of cells of microbes [89–93]. The pullulan production has been studied by various authors [63]. The pullulan was produced by *A. pullulans* in tank reactor [94]. High pullulan amount from *A. pullulans* by usage of waste of cassava was obtained [95]. The pullulan secretion by *A. pullulans* by usage of waste of potato starch was reported [63, 96]. Studies dealing with pullulan production focused on *A. pullulans* because of the high production efficiency of it [90, 97].

The chitosan is obtained from chitin contained in insects, squid bones, and exoskeletons of crabs and shrimps like crustaceans. The chitosan is the 2nd most abundant amino polysaccharide on earth [24, 27, 98–101]. It is a polycationic polysaccharide with high biocompatibile, biodegradabile, antibacterial properties and with no toxicity [57, 102–108]. It is a biopolymer with applications of textile, food, pharmaceutical and chemical [63, 109]. Chitosan has become an agent for biomaterials and food products because of the compatibility of it with the human cells. Chitosan is widely used as environmentally friendly agricultural products because of its biodegradability [27, 110, 111]. Chitosan is preferred as coagulant for removal of pollutants from wastewater and groundwater [27, 112–114].

Dextran was noticed by Pasteur from slime producing bacteria [115]. It is secreted by various gram positive, facultatively anaerobe cocci [36, 116]. It is one of the extracellular microbial polymers used industrially first. It is utilized as an emulsifier, as oil drilling mud additive, and as soil stabilizer [27, 117, 118].

Curdlan is a polysaccharide that is obtained from *A. faecalis* [119, 120]. Curdlan has been approved as additive in food that could be used safely. It is highly preferred in food industry because of valuable features of it [120, 121]. There is attention to the researches on biomedical applications of curdlan [122–124]. It is also commonly used in pharmaceutical industry. In civil and geotechnical engineering, there are attempts to use as pore clogging agent for reduction of hydraulic conductivity of soil and adsorbent for treatment of contaminated soil [27, 119, 125, 126].

3   Main Functions and Applications

Extracellular polymeric substance is renewable, biodegradable, ecofriendly, nontoxic, and economically valued product. Some of the functional properties of extracellular polymeric substances include adhesion, aggregation of bacterial cells, binding activity, cell–cell recognition, cohesion, electron acceptor or donor, energy and nutrient source, enzymatic activity, export of cell components, protective barrier, water retention, sorption, transfer of genetic information [20].

There is an increase in use of natural polymers in various industries. Natural polysaccharides are non toxic, biodegradabile and bioactive. Thus, natural polysaccharides have applied in different industries (e.g. oil industry, textile industry, and food industry) [64]. Extracellular polymeric substances produced by microbes have advantages with regard to industrial applications when compared to other agents with natural origin [10]. Production time of extracellular polymeric substances is much shorter than the other natural materials. Furthermore, extraction method of extracellular polymeric substances is simple [14, 127]. Extracellular polymeric substances have great potential applications in adhesives, textile, cosmetology, treatment of wastewater, food additives, and pharmacology [128–132].

The exopolysaccharides have antiviral effects [14, 133]. Extracellular polymeric substances hinder viral absorption and penetration into host cell. They hinder reverse transcriptase activity of different retroviruses. There are some studies about the antiviral activity of exopolysaccharides [14]. Arena et al. [134] and Gugliandolo et al. [135] noticed that exopolysaccharides secreted by *B. teriformis* and *Bacillus licheniformis* T14 strain had antiviral activity, respectively [14]. Chitosan has anti microbial activity because of the polycationic nature of it [24, 100, 136, 137]. The nonionic pullulan is nonimmunogenic, nonmutagenic, noncarcinogenic, and nontoxic [64, 138] that confering it important role in different uses in various industries [139]. Therefore, it is one of the most preferred polysaccharides in industries [43, 64, 138, 140]. Levan that is one of the extracellular polymeric substances having properties of water solubility, antitumor, nontoxicity, cholesterol lowering properties and ecofriendly adhesives finds different applications [132, 141, 142]. Kefiran benefits includes tumor growth retardation, wound healing and antimicrobial properties, and cholesterol lowering [132, 141, 142].

Extracellular polymeric substances are used as absorbent and adsorbent to remove dye and heavy metal from effluents, respectively. Extracellular polymeric substance is used as bioflocculant in wastewater treatment [20]. The chitosan features make it the most probable material for different industries as cosmetics, treatment of wastewater, agriculture, pharmaceutical, textile dyeing and finishing, fibre formation, and biomedical fields [24, 101, 110, 143–153].

Extracellular polymeric substance is preferred for remediation of soil and control of erosion [20]. Polysaccharides (e.g. xanthan gum, sodium alginate, and beta glucan) and new biopolymers (*R. tropici* extracellular polymeric substance, *L. mesenteroides* extracellular polymeric substance) are used in soil stabilization [28]. The production of cover material from extracellular polymeric substance as an ecocover and replacement for conventional materials (e.g. vegetative cover and geomembrane) for improving soil stability is suggested by Siddharth et al. [20].

Extracellular polymeric substances find use area in food industry. Pullulan is able to form thin films that are oxygen impermeable, transparent, and oil resistant. The films that have oxygen impermeability, formed with pullulan may be preferred for preservation of food [34, 154] because of safety of pullulan for human consumption [64]. Gellan that is the important microbial origin polysaccharide obtained commercially by *S. elodea* is popular as gelatinizing agent in food items. Gellan is also preferred as stabilizer, thickener, and emulsifier in food industry [64, 155].

4 Applications in Textile

The categorization of textile industries depends on type of fabrics produced by them. They include cellulosic fabrics obtained from plants, protein fabrics obtained from animals, and synthetic fabrics produced artificially [6]. Industrial textile products are classified as follows: industrial textiles, medical textile, geo textile, agro textile, construction textile, protective clothing, packaging textile, sports textile, aerospace and automotive textile [156–158].

Various polymers with natural origin have electrospun for applications such as wound dressings [159–161]. One of the polymers generally used in wound care are fibres manufactured from naturally available polymers as alginate, chitosan, and cellulose. The chitosan is one of the most commonly preferred polymers with natural origin along with textile materials as wound dressings. It offers dressing which would show scar prevention [162]. The novel chitosan membrane as dressing consisting of skin surface on top layer has been used by Fwu-Long et al. [163]. Fwu-Long et al. [163] notified that the chitosan membrane was good as wound dressing. The wound dressings obtained from alginate are important as wound management aids. Alginate provides satisfactory dressing for lightly contaminated wound and cavity [162]. Bacterial cellulose is preferred in wound dressing material preperation due to high porosity of it [38, 164, 165].

Extracellular polymeric substances are used as efficient binding agents with hydrogels and color dyes in textile industry due to properties as stabilizer, viscosity, cross linking ability with fabrics [38]. Natural polymers have electrospun for military protective clothing [160, 161]. The bacterial cellulose is used in special clothing for burn patients due to high porosity [38, 164, 165]. Alginate fibers are preferred in sports clothing preparation due to hydrophilic property of it. The calcium alginate fibers are preferred in carpet, firefighter clothing, and military clothing. The bacterial cellulose, chitosan, and pullulan find applications in textile industry as thinning agent, combining agent, and water resistant covers [38]. The cellulosic fibers will be indispensable on the current textile market because of specific properties of them (e.g. absorbency and moisture management) [166–170]. The cellulosic fibers have original properties in moisture management that enable the body for the regulation of the temperature and thereby give comfortable feel to textiles in skin contact directly [169–171]. The characteristic is good for clothing applications [170].

The natural polymers of polysaccharides are used in textile printing [172]. Sodium alginate is the most widely preferred thickener in textile printing. Because, it has penetrability and screenability [173, 174]. Calcium alginate is used to make textile print thickeners and other applications requiring barrier, biocompatible, and ion chelating features [175]. Neumann et al. [176], Fijan et al. [177] and Bu et al. [178] studied interactions of sodium alginate and charged ionic surfactants, polysaccharide thickeners, and aqueous mixtures of modified alginate with various surfactants, respectively [174]. Kobašlija and McQuade [175] prepared non toxic and biodegradable calcium alginate coating. Colored calcium alginate coatings may be used in textile due to ease of application and removal, and biocompatibility. Lacoste et al. [179]

reported usage of sodium alginate as adhesive binder for wood fibres/textile waste fibres biocomposites. The authors reported that promising biocomposites based on sodium alginate were manufactured. They suggested that the alginate appeared as possible candidate for green biopolymer adhesives market. In the field of textile, also chitosan may be used as dye fixing agent, in pigment printing of cellulosic fabric as binder and thickener [101, 180–182].

The textile dyeing is one of the application areas of exopolysaccharides. The xanthan solutions have high viscosity at low concentrations. The xanthan solutions are pseudoplastic [102, 183–185]. Xanthan displays high shear stability [186]. Due to the valuable properties it has application area in textiles [187]. The textile dyeing is one of the applications of xanthan for its pseudoplasticity [187]. Moreover, chitosan improves the fastness features of dyed fabrics [101, 180–182].

The exopolysaccharides are used in textile sizing. Pullulan is used in textile industry as textile sizing agent [93]. This extracellular polymeric substance is one of the commercially important microbial extracellular polymeric substances and has been used in textiles [34, 188]. Pullulan is nonmutagenic, nontoxic, nonhygroscopic, water soluble, odorless, tasteless, and edible polymer. The pullulan is used in different industrial applications including textile due to perfect film forming features and typical rheological features of pullulan [34].

Textile industry are looking for textiles anti microbial finishing process based on sustainable biopolymers from the point of view of industrialization, environmental friendliness, and economic concern [138]. The extracellular polymeric substances has been broadly used in textile finishing [14, 127]. Antiviral materials may be applied to antiviral textiles [24]. Islam et al. [152] reviewed role of various biopolymers in antimicrobial textile development. Zhang et al. [14] reviewed methods of combining antiviral materials with textiles and proposed the prospects for antiviral textile research. In textile finishing, chitosan provides supplementary alternatives. Chitosan has been preferred as functional agent for the development of insect repellent, deodorant, aroma, fire retardant, water resistant, UV blocking, and anti microbial textile surfaces [24, 189]. Original properties of it have been explored for application in textile industry [24]. The properties of it including total biodegradability, biocompatibility with tissues of plants and animals, anti microbial and anti fungal activity, nontoxicity make it the most prospective material for various fields as textile and fabrics engineering for products of hygiene and healthcare (e.g. bandages) [101, 145, 146, 148–150]. The chitosan may an additive when spinning antimicrobial fibers for providing antimicrobial agent for textiles [190]. Several researchers studied antibacterial functional finishing of various textile materials and grafting of chitosan for enhancement of antibacterial activity onto textiles [24, 172, 191–207]. The alginate role in development of antimicrobial textiles and different processing routes for antibacterial finishing of textiles are also reported [138, 174]. Some researchers have reported improvement of cotton fabric properties by nanochitosan, anti microbial finishing of textiles by chitosan in durable press finishing of cotton [101, 208–210]. There are various researches on application of chitosan nanoparticles in textile industry [108]. Ali et al. [202] reported usage of chitosan nanoparticles in textile industry as finishing agents able to provide antimicrobial activity. Lu et al. [211] and

Yang et al. [212] studied usage of chitosan nanoparticles to enhance properties of breaking strength, shrinkproof and wrinkle resistance. Cheung et al. [213] reported usage of chitosan nanoparticles as enhancers of textile dyeing. The chitosan nanoparticles are excellent candidate for textile dyeing. The chitosan nanoparticles interact with different fabrics [108]. Costa et al. [108] reported that dying textile fabrics using textile dye loaded chitosan nanoparticles was possible. Authors reported that good entrapment efficiency was obtained by the chitosan tripolyphosphate ionic gelation method. Furthermore, nanoencapsulated dyes were reported as non cytotoxic towards human keratinocyte cell line cells. They claimed that nanoencapsulated dyes based textile dyeing procedure bypassed conventional dye fabric specific. Islam and Butola [24] reported that results may improve additional studies to tap chitosan as new anti microbial agent that can minimize unwanted activities of anti microbials available in market place. Approved usefuly of both preparation technology and nano sized cellulose material application in different fields (e.g. tissue stents and wound dressing) has also been reported [55, 57].

The various flame retardant agents have been applied onto textiles as solutions and additives for enabling textile products to meet fire safety regulatory standards [214]. The extracellular polymeric substances based flame retardant materials may are one of the high performance materials [215]. The worldwide consumption of flame retardants is covered by halogenated flame retardants that have been associated with harmful effects in environment and human Furthermore, usage of them in textile products is progressively banned. Design of biobased flame retardant materials is desirable as a replacement for synthetic counterparts [216, 217]. Kim et al. [217] hypothesised that exopolysaccharides obtained from aerobic granular sludge resulting from wastewater treatment which is widely used could be used for the development of biobased flame retardant materials. The exopolysaccharides obtained from granular sludge and activated sludge have fire resistant property that make them suitable flame retardant materials for coatings. The flax fabrics coated with exopolysaccharides obtained from sludge meet requirements of flame retardancy in US Federal Aviation Regulation standards [217].

5 Conclusions and Future Perspectives

In this chapter, extracellular polymeric substances have evaluated followed by the main functions and applications of the extracellular polymeric substances. Moreover, the applications for various extracellular polymeric substances in textile industry have emphasized. The extracellular polymeric substances are preferred by various industries including textile industry for different applications because of their unmatched properties. These biomacromolecules are also environmentally sustainable. There is huge potential for more intensive use of extracellular polymeric materials in textile industry, as extracellular polymeric materials are both environmentally friendly and economical. The future perspective involve the new applications of extracellular polymeric substances obtained from various sources by various ways in textile industry.

References

1. Smelser NJ (2013) Social change in the industrial revolution: n application of theory to the british cotton industry. Routledge
2. Nayak R, Padhye R (2015) Introduction: the apparel industry. Woodhead Publishing, Garment Manufacturing Technology, pp 1–17
3. Amaral MCD, Zonatti WF, Silva KLD, Karam Junior D, Amato Neto J, Baruque-Ramos J (2018) Industrial textile recycling and reuse in Brazil: case study and considerations concerning the circular economy. Gest Prod 25:431–443
4. Leal Filho W, Ellams D, Han S, Tyler D, Boiten VJ, PaΩ CÇco A, Moora H, Balogun A-L (2019) A review of the socio-economic advantages of textile recycling. J Cleaner Prod 218:10–20
5. Ranasinghe L, Jayasooriya VM (2021) Ecolabelling in textile industry: a review. Resour Environ Sustain, 100037
6. Yaseen DA, Scholz M (2019) Textile dye wastewater characteristics and constituents of synthetic effluents: a critical review. Int J Environ Sci Technol 16:1193–1226
7. Xie J-H, Jin M-L, Morris GA, Zha X-Q, Chen H-Q, Yi Y, Li J-E, Wang Z-J, Gao J, Nie S-P, Shang P, Xie M-Y (2016) Advances on bioactive polysaccharides from medicinal plants. Crit Rev Food Sci Nutri 56:S60–S84
8. Yu Y, Shen M, Song Q, Xie J (2018) Biological activities and pharmaceutical applications of polysaccharide from natural resources: a review. Carbohydr Polym 183(235):91–101
9. Albuquerque PBS, Oliveira WF, Silva PMS, Correira MTS, Kennedy JF, Coelho LCBB (2020) Epiphanies of well-known and newly discovered macromolecular carbohydrates—a review. Int J Biol Macromol 156:51–66
10. Angelin J, Kavitha M, Exopolysaccharides from probiotic bacteria and their health potential. Int J Biol Macromole 162:853–865
11. Ziadi M, Bouzaiene T, M'Hir S, Zaafouri K, Mokhtar F, Hamdi M, Boisset-Helbert C (2018) Evaluation of the efficiency of ethanol precipitation and ultrafiltration on the purification and characteristics of exopolysaccharides produced by three lactic acid bacteria. Biomed Res Int 2018:1896240
12. Suresh Kumar A, Mody K, Jha B (2007) Bacterial exopolysaccharides—a perception. J Basic Microbiol 47:103–117
13. Soumya MP, Sasikumar K, Pandey A, Nampoothiri KM (2019) Cassava starch hydrolysate as sustainable carbon source for exopolysaccharide production by Lactobacillus plantarum. Bioresour Technol Rep 6:85–88
14. Zhang Y, Fan W, Sun Y, Chen W, Zhang Y (2021) Application of antiviral materials in textiles: a review. Nanotechnol Rev
15. De Vuyst L, Degeest B (1999) Heteropolysaccharides from lactic acid bacteria. FEMS Microbiol Rev 23:153–177
16. Freitas F, Alves VD, Reis MAM (2011) Advances in bacterial exopolysaccharides: from production to biotechnological applications. Trends Biotechnol 29:388–398
17. Nwodo UU, Green E, Okoh AI (2012) Bacterial Exopolysaccharides: functionality and prospects, 14002–14015
18. Majee SB, Avlani D, Biswas GR (2017) Rheol Behav Pharm Appl Bacterial Exopolysaccharides 7:224–232
19. Dilna SV, Surya H, Aswathy RG, Varsha KK, Sakthikumar DN, Pandey A, Nampoothiri KM (2015) Characterization of an exopolysaccharide with potential health-benefit properties from a probiotic Lactobacillus plantarum RJF4. LWT Food Sci Technol 64:1179–1186

20. Siddharth T, Sridhar P, Vinila V, Tyagi RD (2021) Environmental applications of microbial extracellular polymeric substance (EPS): a review. J Environ Manage 287:112307
21. Venkatachalam D, Kaliappa S (2021) Superabsorbent polymers: a state-of-art review on their classification, synthesis, physicochemical properties, and applications. Rev Chem Eng
22. Gombotz WR, Wee SF (1998) Protein release from alginate matrices. AdvDrug Delivery Rev 31:267–285
23. Lee KY, Mooneya DJ (2012) Alginate: Properties and biomedical applications. Prog Polym Sci 37(2012):106–126
24. Islam S, Butola BS (2019) Recent advances in chitosan polysaccharide and its derivatives in antimicrobial modification of textile materials. Int J Biol Macromol 121:905–912
25. Mudgil D, Barak S, Khatkar BS (2014) Guar gum: processing, properties and food applications-A Review. J Food Sci Technol 51(3):409–418
26. Osmalek T, Froelich A, Tasarek S (2014) Application of gellan gum in pharmacy and medicine. Int J Pharm 466(1–2):328–340
27. Chang I, Lee M, Tran ATP, Lee S, Kwon YM, Im J, Cho GC (2020) Review on biopolymer-based soil treatment (BPST) technology in geotechnical engineering practices. Transp. Geotech. 24:100385
28. Huang J, Kogbara RB, Hariharan N, Masad EA, Little DN (2021) A state-of-the-art review of polymers used in soil stabilization. Constr Build Mater 305:124685
29. Kato T, Katsuki T, Takahashi A (1984) Static and dynamic solution properties of pullulan in a dilute-solution. Macromolecules 17:1726–1730
30. Kato T, Okamoto T, Tokuya T, Takahashi A (1982) Solution properties and chain flexibility of pullulan in aqueous solution. Biopolymers 21:1623–1633
31. Nishinari K, Kohyama K, Williams PA, Philips GO, Burchard W, Ogino K (1984) Solution properties of pullulan. Macromolecules 24:5590–5593
32. Kawahara K, Ohta K, Miyamoto H, Nakamura S (1984) Preparation and solution properties of pullulan fractions as standard samples for water-soluble polymers. Carbohyd Polym 4:335–356
33. Nordmeier E (1993) Static and dynamic light-scattering solution behavior of pullulan and dextran in comparison. J Phys Chem 97:5770–5785
34. Sugumaran KR, Ponnusami V (2017) Review on production, downstream processing and characterization of microbial pullulan. Carboh Polym 173:573–591
35. Bachelder EM, Pino EN, Ainslie KM (2017) Acetalated dextran: a tunable and acid-labile biopolymer with facile synthesis and a range of applications. Chem Rev 117(3):1915–1926
36. Hu Q, Lu Y, Luo Y (2021) Recent advances in dextran-based drug delivery systems: from fabrication strategies to applications. Carbohyd Polym 264:117999
37. Lee H, Han Y, Park JH (2022) Enhanced deposition of Fe(III)-tannic acid complex nanofilm by Fe(III)-embedded dextran nanocoating. Appl Surf Sci 573:151598
38. Shukla A, Mehta K, Parmar J, Pandya J, Saraf M (2019) Depicting the exemplary knowledge of microbial exopolysaccharides in a nutshell. Eur Polymer J 119:298–310
39. Laspidou CS, Rittmann BE (2002) A unified theory for extracellular polymeric substances, soluble microbial products, and active and inert biomass. Water Res 36:2711–2720
40. Wingender J, Th R, Neu HC (1999) Flemming, microbial extracellular Polymeric Substances: Characterization. Struct Funct 279
41. Kodali VP, Das S, Sen R (2009) An exopolysaccharide from a probiotic: biosynthesis dynamics, composition and emulsifying activity food research international an exopolysaccharide from a probiotic: biosynthesis dynamics, composition and emulsifying activity. Food Res Int 42:695–699
42. Torino MI, De Valdez GF, Mozzi F (2015) Biopolymers from lactic acid bacteria. Novel Appl Foods Beverages 6:1–16
43. Prajapati VD, Jani GK, Khanda SM (2013) Pullulan: an exopolysaccharide and its various applications. Carbohydr Polym 95:540–549
44. Yang JS, Xie YJ, He W (2011) Research progress on chemical modification of alginate: a review. Carbohydr Polym 84:33–39

45. Salehizadeh H, Yan N, Ramin Farnood R (2018) Recent advances in polysaccharide bio-based flocculants. Biotechnol Adv 36(1):92–119
46. Draget KI (2009) Alginates. In: Phillips GO, Williams PA (eds) Handbook of hydrocolloids, p 379e95
47. Pawar SN, Edgar KJ (2012) Alginate derivatization: a review of chemistry, properties and applications. Biomaterials 33:3279–3305
48. Smidsrod O, Skjak-Bræk G (1990) Alginate as immobilization matrix for cells. Trend Biotechnol 8:71–78
49. Remminghorst U, Rehm BHA (2006) Bacterial alginates: from biosynthesis to applications. Biotechnol Lett 28:1701–1712
50. Sand A, Yadav M, Mishra DK, Behari K (2010) Modification of alginate by grafting of N-vinyl-2-pyrrolidone and studies of physicochemical properties in terms of swelling capacity, metal-ion uptake and flocculation. Carbohydr Polym 80:1147–1154
51. Zhao YX, Gao BY, Wang Y, Shon HK, Bo XW, Yue QY (2012) Coagulation performance and floc characteristics with polyaluminum chloride using sodium alginate as coagulant aid: a preliminary assessment. Chem Eng J 183:387–394
52. Yuan YH, Jia DM, Yuan YH (2013) Chitosan/sodium alginate, a complex flocculating agent for sewage water treatment. Adv Mater Res 641–642:101–114
53. Rani P, Mishra S, Sen G (2013) Microwave based synthesis of polymethyl methacrylate grafted sodium alginate: its application as flocculant. Carbohydr Polym 91:686–692
54. Diaz-Barrera A, Gutierrez J, Martinez F, Altamirano C (2014) Production of alginate by Azoto-bacter vinelandii grown at two bioreactor scales under oxygen-limited conditions. Bioprocess Biosyst Eng 37:1133–1140
55. Sun D, Zhou L, Wu Q, Yang S (2007) Preliminary research on structure and properties of nano-cellulose. J Wuhan Univ Technol Mater Sci Ed 22(4):677–680
56. Ying W, Huifang L, Xiaomin F, Yuanqing X, Yanrong R, Tao D (2013) Preparation and characterization of a biodegradable poly (malic-co-butanediol) elastomer. China Elastomerics (1):4
57. Ambaye TG, Vaccari M, Prasad S, van Hullebusch ED, Rtimi S (2022) Preparation and applications of chitosan and cellulose composite materials. J Environ Manage 301:113850
58. Prasad S, Singh A, Korres NE, Rathore D, Sevda S, Pant D (2020) Sustainable utilization of crop residues for energy generation: a life cycle assessment (LCA) perspective. Bioresour Technol 303:122964
59. Alavi M (2019) Modifications of microcrystalline cellulose (MCC), nanofibrillated cellulose (NFC), and nanocrystalline cellulose (NCC) for antimicrobial and wound healing applications. E-Polymers 19(1):103–119
60. Bielecki S, Krystynowicz A, Turkiewicz M, Kalinowska H (2005) Bacterial cellulose. In: Steinbüchel A, Doi Y (eds), Biotechnology of polymer: from Synthesis to patents. Wiley-VCH, Weinheim, Germany, pp 381–434
61. Castro C, Zuluaga R, Putaux JL, Caro G, Mondragon I, Gañán P (2011) Structural charac-terization of bacterial cellulose produced by Gluconacetobacter swingsii sp. from Colombian agroindustrial wastes. Carbohydr Polym 84:96–102
62. Hussain Z, Sajjad W, Khan T, Wahid F (2019) Production of bacterial cellulose from industrial wastes: a review. Cellulose 26:2895–2911
63. Sodhi AS, Sharma N, Bhatia S, Verma A, Soni S, Batra N (2022) Insights on sustain-able approaches for production and applications of value added products. Chemosphere 286(1):131623
64. Kalia S, Choudhury AR (2019) Synthesis and rheological studies of a novel composite hydrogel of xanthan, gellan and pullulan. Int J Biol Macromol 137:475–482
65. Stredansky M, Conti E (1999) Xanthan production by solid state fermentation. Process Biochem 34:581–587
66. Moshaf S, Hamidi-Esfahani Z, Azizi MH (2011) Optimization of conditions for xanthan gum production from waste date in submerged fermentation. World Acad Sci Eng Technol 57:521–524

67. Rončević Z, Grahovac DS, Vučurović DJ, Dodić J (2019) Utilisation of winery wastewater for xanthan production in stirred tank bioreactor: bioprocess modelling and optimization. Food Bioprod Process 117:113–125
68. Gondim TS, Pereira RG, Fiaux SB (2019) Xanthan gum production by Xanthomonas axonopodispv. Mangiferae indicae from glycerin of biodiesel in different media and addition of glucose. Acta Sci Biol Sci 41:e43661–e43661
69. Eren NM, Santos PHS, Campanella O (2015) Mechanically modified xanthan gum: rheology and polydispersity aspects. Carbohydr Polym 134:475–484
70. Kang M, Oderinde O, Liu S, Huang Q, Ma W, Yao F, Fu G (2019) Characterization of xanthan gum-based hydrogel with Fe3+ ions coordination and its reversible sol-gel conversion. Carbohydr Polym 203:139–147
71. García-Ochoa F, Santos VE, Casas JA, Gómez E (2000) Xanthan gum: production, recovery, and properties. Biotechnol Adv 18:549–579
72. Khachatoorian R, Petrisor IG, Kwan C-C, Yen TF (2003) Biopolymer plugging effect: laboratory-pressurized pumping flow studies. J Petrol Sci Eng 38:13–21
73. Bouazza A, Gates W, Ranjith P (2009) Hydraulic conductivity of biopolymer-treated silty sand. Géotechnique 59:71–72
74. Chang I, Im J, Prasidhi AK, Cho G-C (2015) Effects of xanthan gum biopolymer on soil strengthening. Constr Build Mater 74:65–72
75. Ayeldeen MK, Negm AM, El Sawwaf MA (2016) Evaluating the physical characteristics of biopolymer/soil mixtures. Arabian J Geosci 9:1–13
76. Latifi N, Horpibulsuk S, Meehan CL, Majid MZA, Tahir MM, Mohamad ET (2017) Improvement of problematic soils with biopolymer—an environmentally friendly soil stabilizer. J Mater Civ Eng 29:04016204
77. Qureshi MU, Chang I, Al-Sadarani K (2017) Strength and durability characteristics of biopolymer-treated desert sand. Geomech Eng 12:785–801
78. Cabalar A, Wiszniewski M, Skutnik Z (2017) Effects of Xanthan Gum biopolymer on the permeability odometer, unconfined compressive and triaxial shear behavior of a sand. Soil Mech Found Eng 54:356–361
79. Lee S, Chang I, Chung M-K, Kim Y, Kee J (2017) Geotechnical shear behavior of xanthan gum biopolymer treated sand from direct shear testing. Geomech Eng 12:831–847
80. Im J, Tran ATP, Chang I, Cho G-C (2017) Dynamic properties of gel-type biopolymer-treated sands evaluated by Resonant Column (RC) tests. Geomech Eng 12:815–830
81. Lee S, Im J, Cho G-C, Chang I (2019) Laboratory triaxial test behavior of xanthan gum biopolymer-treated sands. Geomech Eng 17:445–452
82. Chang I, Kwon Y-M, Im J, Cho G-C (2019) Soil consistency and interparticle characteristics of xanthan gum biopolymer–containing soils with pore-fluid variation. Can Geotech J 56:1206–1213
83. Kwon Y-M, Ham S-M, Kwon T-H, Cho G-C, Chang I (2020) Surface-erosion behaviour of biopolymer-treated soils assessed by EFA. Géotech Lett 10:1–7
84. Carolan G, Catley BJ, McDougal FG (1983) The location of tetrasaccharide units in pullulan. Carbohydr Res 114:237–242
85. Bauer R (1938) Physiology of dematium pullulans de bary. Zentralbl Bacteriol Parasitenkd Infektionskr Hyg Abt 2(98):133–167
86. Okada K, Yoneyama M, Mandai T, Aga H, Sakai S, Ichikawa T (1990) Digestion and fermentation of pullulan. J Japan Soc Nutr Food Sci 43:23–29
87. Raychaudhuri R, Naik S, Shreya AB, Kandpal N, Pandey A, Kalthur G, Mutalik S (2020) Pullulan based stimuli responsive and sub cellular targeted nanoplatforms for biomedical application: synthesis, nanoformulations and toxicological perspective. Int J Biol Macromol 161:1189–1205
88. Singh RS, Kaur N, Kennedy JF (2015) Pullulan and pullulan derivatives as promising biomolecules for drug and gene targeting. Carbohyd Polym 123:190–207
89. Sutherland IW (1993) Xanthan. In: Swings JG, Civerolo EL (eds) Xanthomonas. Chapman & Hall, London, pp 363–388

90. Singh RS, Saini GK, Kennedy JF (2008) Pullulan: microbial sources, production and applications. Carbohyd Polym 73(4):515–531
91. Singh RS, Saini GK, Kennedy JF (2009) Downstream processing and characterization of pullulan from a novel colour variant strain of Aureobasidium pullulans FB-1. Carbohyd Polym 78(1):89–94
92. Singh RS, Kaur N, Rana V, Kennedy JF (2016) Recent insights on applications of pullulan in tissue engineering. Carbohyd Polym 153:455–462
93. Ganie SA, Rather LJ, Li Q (2021) A review on anticancer applications of pullulan and pullulan derivative nanoparticles. Carbohydr Polym Technol Appl 2:100115
94. Lazaridou A, Roukas T, Biliaderis CG, Vaikous H (2002) Characterization of pullulan produced from beet molasses by Aureobasidium pullulans in a stirred tank reactor under varying agitation. Enzyme Microb Technol 31(1–2):122–132
95. Viveka R, Varjani S, Ekambaram E (2021) Valorization of cassava waste for pullulan production by Aureobasidium pullulans MTCC 1991. Energy Environ 32(6):1086–1102
96. Barnett C, Smith A, Scanlon B, Israilides CJ (1999) Pullulan production by Aureobasidium pullulans growing on hydrolysed potato starch waste. Carbohydr Polym 38:203–209
97. Cheng KC, Demirci A, Catchmark JM (2011) Pullulan: biosynthesis, production, and applications. Appl Microbiol Biotechnol 92:29–44
98. Duarte ML, Ferreira MC, Marvão MR, Rocha J (2002) An optimised method to determine the degree of acetylation of chitin and chitosan by FTIR spectroscopy. Int J Biol Macromol 31(1):1–8
99. Zuo D, Tao Y, Chen Y (2009) Preparation and characterization of blend membranes of polyurethane and superfine chitosan powder. Polym Bull 62:713–725
100. Dina R, Hans-Georg S (2009) Chitosan and its antimicrobial potential—a critical literature survey. Microb Biotechnol 2(2):186–201
101. Muzaffar S, Bhatti IA, Zuber M, Bhatti HN, Shahid M (2016) Synthesis, characterization and efficiency evaluation of chitosan-polyurethane based textile finishes. Int J Biol Macromole Part A 93:145–155
102. Rinaudo M, Milas M (1978) Polyelectrolyte behaviour of a bacterial polysaccharide from Xanthomonas campestris: comparison with carboxymethylcellulose. Biopolymers 17:2663–2678
103. Li Q, Zhou JP, Zhang LN (2009) Structure and properties of the nano-composite films of chitosan reinforced with cellulose whiskers. J Polym.Sci Part B Polym Phys 47:1069–1077
104. Kumar A, Vimal A, Kumar A (2016) Why Chitosan? From properties to perspective of mucosal drug delivery. Int J Biol Macromol 91:615–622
105. Khalil HA, Davoudpour Y, Saurabh CK, Hossain MS, Adnan AS, Dungani R, Paridah MT, Sarker MZI, Fazita MN, Syakir MI, Haafiz MKM (2016) A review on nanocellulosic fibres as new material for sustainable packaging: process and applications. Renew Sustain Energy Rev 64:823–836
106. Alavi M, Rai M (2019) Recent progress in nanoformulations of silver nanoparticles with cellulose, chitosan, and alginic acid biopolymers for antibacterial applications. Appl Microbiol Biotechnol 103(21):8669–8676
107. Yang J, Dahlström C, Edlund H, Lindman B, Norgren M (2019) pH-responsive cellulose–chitosan nanocomposite films with slow release of chitosan. Cellulose 26(6):3763–3776
108. Costa EM, Silva S, MarianaVeiga PatriciaBaptista, Tavaria FK, Pintado ME (2021) Textile dyes loaded chitosan nanoparticles: characterization, biocompatibility and staining capacity. Carbohyd Polym 251:117120
109. Ravindran R, Jaiswal AK (2016) Exploitation of food industry waste for high-value products. Trends Biotechnol 34:58–69
110. Hirano S, Kitaura S, Sasaki N, Sakaguchi H, Sugiyama M, Hashimoto K et al (1996) Chitin biodegradation and wound healing in tree bark tissues. J Environ Polym Degrad 4:261–265
111. Linden JC, Stoner RJ, Knutson KW, Gardner-Hughes CA (2000) Organic disease control elicitors. Agro Food Ind Hi Tech 11:32–34

112. Liu XD, Tokura S, Haruki M, Nishi N, Sakairi N (2002) Surface modification of nonporous glass beads with chitosan and their adsorption property for transition metal ions. Carbohydr Polym 49:103–108
113. Cheung WH, Ng JCY, McKay G (2003) Kinetic analysis of the sorption of copper(II) ions on chitosan. J Chem Technol Biotechnol 78:562–571
114. Renault F, Sancey B, Badot PM, Crini G (2009) Chitosan for coagulation/flocculation processes—an eco-friendly approach. Eur Polym J 45:1337–1348
115. Pasteur L (1861) On the viscous fermentation and the butyrous fermentation. Bull Soc Chim Fr 11:30–31
116. Jeanes A, Haynes WC, Wilham C, Rankin JC, Melvin E, Austin MJ, Cluskey J, Fisher B, Tsuchiya H, Rist C (1954) Characterization and classification of dextrans from ninety-six strains of bacteria1b. J Am Chem Soc 76(20):5041–5052
117. Pini R, Canarutto S, Guidi GV (1994) Soil microaggregation as influenced by uncharged organic conditioners. Commun Soil Sci Plant Anal 25:2215–2229
118. Fechner A, Knoth A, Scherze I, Muschiolik G (2007) Stability and release properties of double-emulsions stabilised by caseinate–dextran conjugates. Food Hydrocolloids 21:943–952
119. Zhan XB, Lin CC, Zhang HT (2012) Recent advances in curdlan biosynthesis, biotechnological production, and applications. Appl Microbiol Biotechnol 93(2):525–531
120. Li H, Yang H, Xua J, Gaoa Z, Wua J, Zhub L, Zhan X (2022) Novel amphiphilic carboxymethyl curdlan-based pH responsive micelles for curcumin delivery. LWT2, 153:112419
121. Popescu I, Pelin IM, Ailiesei GL, Ichim DL, Suflet DM (2019) Amphiphilic polysaccharide based on curdlan: synthesis and behaviour in aqueous solution. Carbohyd Polym 224:115157
122. Cai Z, Zhang H (2021) The effect of carboxymethylation on the macromolecular conformation of the $(1 \to 3)$-β -D-glucan of curdlan in water. Carbohyd Polym 272:118456
123. Lehtovaara BC, Gu FX (2011) Pharmacological, structural, and drug delivery properties and applications of 1,3-beta-glucans. J Agric Food Chem 59(13):6813–6828
124. Cai Z, Zhang H (2017) Recent progress on curdlan provided by functionalization strategies. Food Hydrocolloids 68:128–135
125. D'Cunha NJ, Misra D, Thompson AM (2009) Experimental investigation of the applications of natural freezing and curdlan biopolymer for permeability modification to remediate DNAPL contaminated aquifers in Alaska. Cold Reg Sci Technol 59:42–50
126. Ivanov V, Stabnikov V (2017) Bioclogging and biogrouts. In: Construction biotechnology. Springer, Singapore, pp 139–178
127. Yildiz H, Karatas N (2018) Microbial exopolysaccharides: resources and bioactive properties. Process Biochem 72:41–46
128. Deepak V, Ram Kumar Pandian S, Sivasubramaniam SD, Nellaiah H, Sundar K (2016a) Optimization of anticancer exopolysaccharide production from probiotic Lactobacillus acidophilus by response surface methodology. Prep Biochem Biotechnol 46(3):288–297
129. Deepak V, Ramachandran S, Balahmar RM, Pandian SRK, Sivasubramaniam SD, Nellaiah H, Sundar K (2016b) In vitro evaluation of anticancer properties of exopolysaccharides from Lactobacillus acidophilus in colon cancer cell lines. In Vitro Cellular Dev Biol Anim 52(2):163–173
130. Patel AK, Michaud P, Singhania RR, Soccol CR, Pandey A (2010) Polysaccharides from probiotics: new developments as food additives. Food Technol Biotechnol 48(4):451–463
131. Rehm BH (2010) Bacterial polymers: Biosynthesis, modifications and applications. Nat Rev Microbiol 8(8):578
132. Saadat YR, Khosroushahi AY, Gargari BP, A comprehensive review of anticancer, immunomodulatory and health beneficial effects of the lactic acid bacteria exopolysaccharides. Carbohydr Polym 217:79–89
133. Baba M, Snoeck R, Pauwels R, de Clercq E (1988) Sulfated polysaccharides are potent and selective inhibitors of various enveloped viruses, including herpes simplex virus, cytomegalovirus, vesicular stomatitis virus, and human immunodeficiency virus. Antimicrob Agents Ch 32(11):1742–1745

134. Arena A, Maugeri TL, Pavone B, Iannello D, Gugliandolo C, Bisignano G (2006) Antiviral and immunoregulatory effect of a novel exopolysaccharide from a marine thermotolerant Bacillus licheniformis. Int Immunopharmacol 6(1):8–13

135. Gugliandolo C, Spanò A, Lentini V, Arena A, Maugeri TL (2014) Antiviral and immunomodulatory effects of a novel bacterial exopolysaccharide of shallow marine vent origin. J Appl Microbiol 116(4):1028–1034

136. Rabea EI, Badawy MET, Stevens CV, Smagghe G, Steurbaut W (2003) Chitosan as antimicrobial agent: applications and mode of action. Biomacromol 4(6):1457–1465

137. Kong M, Chen XG, Xing K, Park HJ (2010) Antimicrobial properties of chitosan and mode of action: a state of the art review. Int J Food Microbiol 144(1):51–63

138. Li J, He J, Huang Y (2017) Role of alginate in antibacterial finishing of textiles. Int J Biol Macromol

139. Ghimici L, Constantin M (2020) A review of the use of pullulan derivatives in wastewater purification. React Funct Polym 149:104510

140. Choudhury AR, Saluja P, Prasad GS (2011) Pullulan production by an osmotolerant Aureobasidium pullulans RBF-4A3 isolated from flowers of Caesulia axillaris. Carbohydr Polym 83:1547–1552

141. Patel S, Majumder A, Goyal A (2012) Potentials of exopolysaccharides from lactic acid bacteria. Indian J Microbiol 52(1):3–12

142. Zannini E, Waters DM, Coffey A, Arendt EK (2016) Production, properties, and industrial food application of lactic acid bacteria-derived exopolysaccharides. Appl Microbiol Biotechnol 100(3):1121–1135

143. Dodane V, Vilivalam VD (1998) Pharmaceutical applications of chitosan. Pharm Sci Technol Today 1(6):246–253

144. Lalov IG, Guerginov II, Krysteva MA, Fartsov K (2000) Treatment of waste water from distilleries with chitosan. Water Res 34(5):1503–1506

145. Shao J, Yang Y, Zhong Q (2003) Studies on preparation of oligoglucosamine by oxidative degradation under microwave irradiation. Polym Degrad Stab 82:395–398

146. Giri Dev VR, Neelkandan R, Sudha N, Shamugasundaram OL, Nadaraj RN (2005) Chitosan—a polymer with wider applications. Text Mag 7:83–86

147. Ham-Pichavant F, Sèbe G, Pardon P, Coma V (2005) Fat resistance properties of chitosan-based paper packaging for food applications. Carbohyd Polym 61(3):259–265

148. Kean T, Roth S, Thanou M (2005) Trimethylated chitosans as non-viral gene delivery vectors: cytotoxicity and transfection efficiency. J Controlled Release 103:643–653

149. Li J, Gong Y, Zhao N (2005) Preparation of N-butyl chitosan and study of its physical and biological properties. J Appl Polym Sci 98:1016–1024

150. Harish Prashanth KV, Tharanathan RN (2007) Chitin/chitosan: modifications and their unlimited application potential—an overview. Trends Food Sci Technol 18:117–131

151. Pillai CKS, Paul W, Sharma CP (2009) Chitin and chitosan polymers: chemistry, solubility and fiber formation. Prog Polym Sci 34(7):641–678

152. Islam S, Shahid M, Mohammad F (2013) Green chemistry approaches to develop antimicrobial textiles based on sustainable biopolymers—a review. Ind Eng Chem Res 52(15):5245–5260

153. Anitha A, Sowmya S, Kumar PTS, Deepthi S, Chennazhi KP, Ehrlich H, Tsurkan M, Jayakumar R (2014) Chitin and chitosan in selected biomedical applications. Prog Polym Sci 39(9):1644–1667

154. Yuen S (1974) Pullulan and its applications. Process Biochem 9:7–22

155. Coutinho DF, Sant SV, Shin H, Oliveira JT, Gomes ME, Neves NM, Khademhosseini A, Reis RL (2010) Modified Gellan gum hydrogels with tunable physical and mechanical properties. Biomaterials 31:7494–7502

156. Sun X, Zhou M, Sun Y (2015) Classification of textile fabrics by use of spectroscopy-based pattern recognition methods. Spectrosc Lett 49(2):1–31

157. Founda A, Hassan SED, Salem SS, Shaheen TI (2018) In-Vitro cytotoxicity, antibacterial, and UV protection properties of the biosynthesized Zinc oxide nanoparticles for medical textile applications. Microb Pathog 125:252–261

158. Roy M, Sen P, Pal P (2020) An integrated green management model to improve environmental performance of textile industry towards sustainability. J Clean Prod 271:122656
159. Wang X, Ding B, Li B (2013) Biomimetic electrospun nanofibrous structures for tissue engineering. Mater Today 16(6):229–241
160. Hu X, Liu S, Zhou G, Huang Y, Xie Z, Jing X (2014) Electrospinning of polymeric nanofibers for drug delivery applications. J Control Release 185:12–21
161. Shah TV, Vasava DV (2019) A glimpse of biodegradable polymers and their biomedical applications. e-Polymers
162. Gupta B, Agarwal R, Alam MS (2010) Textile-based smart wound dressings Indian. J Fibre Text Res 35:174–187
163. Mi F-L, Shyu S-S, Wu Y-B, Lee S-T, Shyong J-Y, Huang R-N (2001) Biomaterials, 22:165
164. Retegi A, Gabilondo N, Pena C, Zuluaga R (2010) Bacterial cellulose films with controlled microstructure–mechanical property relationships. Cellulose 17:661–669
165. Babu RP, O'connor K, Seeram R (2013) Current progress on bio-based polymers and their future trends. Prog Biomater 2:8. https://doi.org/10.1186/2194-0517-2-8
166. Eichhorn S, Hearle JWS, Jaffe M et al (2009) Handbook of textile fibre structure: natural, regenerated, inorganic and specialist fibres. Woodhead Publishing Limited, UK
167. Woodings C (2001) Regenerated cellulose fibres. The Textile Institute CRC, UK
168. Shen L, Patel MK (2010) Life cycle assessment of manmade cellulose fibres. Lenzinger Ber 88:1–59
169. TFY (2013) The Fibre Year 2013—World Survey on Textiles & Nonwovens. Issue 14, 2014
170. Michud A, Tanttu M, Asaadi S, Ma YB, Netti E, Kaariainen P, Persson A, Berntsson A, Hummel M, Sixta H (2016) Ionic liquid-based cellulosic textile fibers as an alternative to viscose and lyocell. Textil Res J 86:543–552
171. Hämmerle FM (2011) The cellulose gap (the future of cellulose fibres). Lenzinger Ber 89:12–21
172. Abdel-Mohsen AM, Aly AS, Hrdina R, Montaser AS, Hebeish A (2012) Biomedical textiles through multifunctioalization of cotton fabrics using innovative methoxypolyethylene glycol-N-chitosan graft copolymer. J Polym Environ 20(1):104–116
173. Sostar Turk S, Schneider R (2000) Printing properties of a high substituted guar gum and its mixture with alginate. Dyes Pigments 47:269–275
174. Chen F, Long J-J (2018) Influences of process parameters on the apparent diffusion of an acid dye in sodium alginate paste for textile printing. J Clean Prod 205:1139–1147
175. Kobašlija M, McQuade DT (2006) Removable colored coatings based on calcium alginate hydrogels. Biomacromol 7(8):2357–2361
176. Neumann MG, Schmitt CC, Iamazaki ET (2003) A fluorescence study of the interactions between sodium alginate and surfactants. Carbohydr Res 338:1109–1113
177. Fijan R, Sostar-Turk S, Lapasin R (2007) Rheological study of interactions between non-ionic surfactants and polysaccharide thickeners used in textile printing. Carbohydr Polym 68:708–717
178. Bu H, Kjoniksen A-L, Knudsen KD, Nystrom B (2007) Characterization of interactions in aqueous mixtures of hydrophobically modified alginate and different types of surfactant. Colloids Surf A 293:105–113
179. Lacoste C, Hage RE, Bergeret A, Corn S, Lacroix P (2018) Sodium alginate adhesives as binders in wood fibers/textile waste fibers biocomposites for building insulation. Carbohyd Polym 184:1–8
180. Bahmani SA, East GC, Holme I (2000) The application of chitosan in pigment printing. Color Technol 116:94–99
181. Öktem T (2003) Surface treatment of cotton fabrics with chitosan. Color Technol 119:241–246
182. Achwal WB (2003) Chitosan and its derivatives for textile finishing. Colourage 50(8):51–76
183. Milas M, Rinaudo M (1986) Properties of xanthan gum in aqueous solutions: role of the conformational transition. Carbohydr Res 158:191–204
184. Richardson RK, Ross-Murphy SB (1987) Nonlinear viscoelasticity of polysaccharide solutions. 2. Xanthan polysaccharide solutions. Int J Biol Macromol 9:257–264

185. Nolte H, John S, Smidsrùd O, Stokke B (1992) Gelation of xanthan with trivalent metal ions. Carbohydr Polym 18:243–251
186. Chen CSH, Sheppard EW (1980) Conformation and shear stability of xanthan gum in solution. Polym Eng Sci 20:512–516
187. Becker A, Katzen F, Pühler A, Ielpie L (1998) Xanthan Gum biosynthesis and application: a biochemical/genetic perspective. Appl Microbiol Biotechnol 50:145–152
188. Xue S-J, Jiang H, Chi Z (2019) Over-expression of Vitreoscilla hemoglobin (VHb) and flavohemoglobin (FHb) genes greatly enhances pullulan production. Int J Biol Macromol
189. Lim S-H, Hudson SM (2003) Review of chitosan and its derivatives as antimicrobial agents and their uses as textile chemicals. J Macromol Sci C Polym Rev J 43(2):223–269
190. Simoncic B, Tomsic B (2010) Structures of novel antimicrobial agents for textiles—a review text. Res J 80:1721–1737
191. Younsook S, Il YD, Kyunghye M (1999) Antimicrobial finishing of polypropylene nonwoven fabric by treatment with chitosan oligomer. J Appl Polym Sci 74(12):2911–2916
192. Zhang Z, Chen L, Ji J, Huang Y, Chen D (2003) Antibacterial properties of cotton fabrics treated with chitosan. Text Res J 73(12):1103–1106
193. Ye W, Leung MF, Xin J, Kwong TL, Lee DKL, Li P (2005) Novel core-shell particles with poly(n-butyl acrylate) cores and chitosan shells as an antibacterial coating for textiles. Polymer 46(23):10538–10543
194. El-tahlawy KF, El-bendary MA, Elhendawy AG, Hudson SM (2005) The antimicrobial activity of cotton fabrics treated with different crosslinking agents and chitosan. Carbohydr Polym 60(4):421–430
195. Jung KH, Man-Woo H, Wan M, Jiang Y, Hee HS, Jung-Sook B, Hudson SM, Inn-Kyu K (2007) Preparation and antibacterial activity of PET/chitosan nanofibrous mats using an electrospinning technique. J Appl Polym Sci 105(5):2816–2823
196. Tseng H-J, Hsu S-H, Wu M-W, Hsueh T-H, Tu P-C (2009) Nylon textiles grafted with chitosan by open air plasma and their antimicrobial effect. Fibers Polym. 10(1):53–59
197. Janjic S, Kostic M, Vucinic V, Dimitrijevic S, Popovic K, Ristic M, Skundric P (2009) Biologically active fibers based on chitosan-coated lyocell fibers. Carbohydr Polym 78(2):240–246
198. Alonso D, Gimeno M, Olayo R, Vázquez-Torres H, Sepúlveda-Sánchez JD, Shirai K (2009) Cross-linking chitosan into UV-irradiated cellulose fibers for the preparation of antimicrobial-finished textiles. Carbohydr Polym 77(3):536–543
199. Dev VRG, Venugopal J, Sudha S, Deepika G, Ramakrishna S (2009) Dyeing and antimicrobial characteristics of chitosan treated wool fabrics with henna dye. Carbohydr Polym 75(4):646–650
200. Alonso D, Gimeno M, Sepúlveda-Sánchez JD, Shirai K (2010) Chitosan-based microcapsules containing grapefruit seed extract grafted onto cellulose fibers by a non-toxic procedure. Carbohydr Res 345(6):854–859
201. Tayel AA, Moussa SH, El-Tras WF, Elguindy NM, Opwis K (2011) Antimicrobial textile treated with chitosan from Aspergillus niger mycelial waste. Int J Biol Macromol 49(2):241–245
202. Ali SW, Rajendran S, Joshi M (2011) Synthesis and characterization of chitosan and silver loaded chitosan nanoparticles for bioactive polyester. Carbohydr Polym 83(2):438–446
203. Joshi M, Khanna R, Shekhar R, Jha K (2011) Chitosan nanocoating on cotton textile substrate using layer-by-layer self-assembly technique. J Appl Polym Sci 119(5):2793–2799
204. Ivanova NA, Philipchenko AB (2012) Superhydrophobic chitosan-based coatings for textile processing. Appl Surf Sci 263:783–787
205. Teli MD, Sheikh J (2012) Extraction of chitosan from shrimp shells waste and application in antibacterial finishing of bamboo rayon. Int J Biol Macromol 50(5):1195–1200
206. Liu J, Liu C, Liu Y, Chen M, Hu Y, Yang Z (2013) Study on the grafting of chitosan–gelatin microcapsules onto cotton fabrics and its antibacterial effect. Colloids Surf B Biointerfaces 109:103–108
207. Sheikh J, Bramhecha I (2018) Multifunctional modification of linen fabric using chitosan-based formulations. Int J Biol Macromol 118:896–902

208. Huang K, Wu WJ, Chen JB, Lian HS (2008) Application of low-molecular-weight chitosan in durable press finishing. Carbohydr Polym 73:254–260
209. Joshi M, Ali SW, Purwar R (2009) Ecofriendly antimicrobial finishing of textiles using bioactive agents based on natural products. Indian J Fiber Text. Res 34:295–304
210. Chattopadhyay D, Inamdar MS (2014) Improvement in properties of cotton fabric through synthesized nano-chitosan. Indian J Fiber Text Res 38:14–21
211. Lu Y-H, Chen Y-Y, Lin H, Wang C, Yang Z-D (2010) Preparation of chitosan nanoparticles and their application to Antheraea pernyi silk. J Appl Polym Sci 117(6):3362–3369
212. Yang H-C, Wang W-H, Huang K-S, Hon M-H (2010) Preparation and application of nanochitosan to finishing treatment with anti-microbial and anti-shrinking properties. Carbohyd Polym 79(1):176–179
213. Cheung WH, Szeto YS, McKay G (2009) Enhancing the adsorption capacities of acid dyes by chitosan nano particles. Biores Technol 100(3):1143–1148
214. Horrock AR (2011) Flame retardant challenges for textiles and fibres: new chemistry versus innovatory solutions. Polym Degrad Stab 96:377–392
215. Innes A, Innes J (2011) Flame retardants. In: Applied plastics engineering handbook
216. IHS Markit (2017) The flame retardants market, Flameretrdant-online, https://www.flameretardants-online.com/flame-retardants/market
217. Kim NK, Mao N, Lin R, Bhattacharyya D, van Loosdrecht MCM, Lin Y (2020) Flame retardant property of flax fabrics coated by extracellular polymeric substances recovered from both activated sludge and aerobic granular sludge. Water Res 170:115344

Sustainable Agrotextile: Jute Needle-Punched Nonwoven Preparation, Properties and Use in Indian Perspective

Surajit Sengupta, Sanjoy Debnath, and Manik Bhowmick

Abstract Jute is one of the important cash crops in India and Bangladesh. Previously, jute conventional products like hessian, sacking and carpet backing were secured by mandatory packaging act of Government of India, and 40 lakhs families were earning their bread and butter directly or indirectly by jute sector. With the slow abolishment of mandatory packaging and capture of jute Indian as well as export market by synthetic one, jute cultivation and use are in threat. India is an agriculture-based country and produces different crops like grains, vegetables, fruits, flowers, medicinal crops, etc., in different agro-climatic zones basically in an unorganized way. It was felt that if jute can be introduced as geotextiles for eco- or green cultivation, a major share of jute will be utilized in this area saving the life of lakhs of families in future. Needle-punched nonwoven is a potential alternative system, very much suited to the industry because of its low wage proportion and high productivity resulting in low cost of production. The needled fabric has high bulk and possesses good hydraulic, thermal insulation and impact resistance property with sufficient strength. With the above-mentioned aim, some trials have been made to use the jute needle-punched nonwoven in the area of mulching in the rain-fed humid zone and semi-arid zone. The trial has also been made on horticultural pot, grass mat, artificial seed germination medium at dry, hard and stone ground. It was found that there is a great potential of jute needle-punched nonwoven in agrotextile with enhanced performance. This chapter deals with the production and properties of jute needle-punched nonwoven and some case studies for using such fabric in agrotextiles. Scope, limitations and economy are also discussed. In the end, suggestions and further researchable points have been mentioned. It will help both industry and academia, especially teachers, students and technologists. It will also help the textile as well as agricultural practitioners.

Keywords Agriculture · Artificial soil substitute · Horticultural pot · Jute · Mulching · Nonwoven · Grass mat · Properties

S. Sengupta (✉) · S. Debnath · M. Bhowmick
Mechanical Processing Division, ICAR-National Institute of Natural Fibre Engineering and Technology, 12 Regent Park, Kolkata 700040, India
e-mail: drssengupta42@gmail.com

S. S. Muthu (ed.), *Sustainable Approaches in Textiles and Fashion*, Sustainable Textiles:
Production, Processing, Manufacturing & Chemistry,
https://doi.org/10.1007/978-981-19-0878-1_3

1 Introduction

India is an agriculture-based country and produces different crops like grains, vegetables, fruits, flowers, medicinal crops, etc., in different agro-climatic zones basically in an unorganized way. For the years, plastic material and synthetic nets are used in agriculture for different purposes like mulching, weed control, sapling/horticultural bag, seedbed cover, shading for nurseries, carry seedling, wrapping or covering of plant parts and products, etc. Therefore, plastics or synthetic has a great role in agriculture, but removal or disposal of synthetic material after use is difficult and it is harmful to the environment. Presently, eco-friendly and sustainable farming is coming up in Western countries in a great way. In India, the big farmers and organized farming sectors are also encouraging organic farming as they are earning more profit out of these agricultural produces.

Mankind has widely accepted the materials developed from cotton, wool and silk, and they are still used abundantly recovering the initial fall due to the impact of synthetic fibres and proving its sustainability. But the manufacture of rayon and synthetic fibres has attained the status of major world industry, and output is increasing due to process friendliness, tailor-making opportunity, cheapness and attractiveness. With these parameters, natural fibre cannot be sustained in competition with synthetic fibres. Therefore, to regain the lost reign of natural fibres, more knowledge and education regarding natural fibres highlighting their positive qualities, i.e. eco-friendliness, renewability, biodegradability, sustainability, hygroscopic nature, dyeability, low carbon footprint, etc., to the consumers, are essential. Moreover, the new area of available, under-exploited or unexploited natural fibres is important for exploring their property-application fusion with the help of research and development. The globe is blessed with plenty of renewable natural fibre resources obtained from plants and animals. Hundreds of different natural fibres have been examined as raw materials for cloth. It was found that many of these fibres along with their by-products have immense potential for utilization as raw material for various applications including textiles [44]. For several decades, jute is being used in agriculture as packaging. At present, jute packaging is facing stiff competition from synthetic products [40]. It has become imperative, therefore, to diversify the uses of jute on the one hand and make changes in the production system on the other.

Textile materials can be produced by different technologies such as woven, nonwoven, knitting and braided. These technologies are used by humankind for the development of different technical textiles having a lot of different variable properties. These technologies can be used to impart different suitable properties and engineer the material to the required specification. All the concepts of fabric manufacture start with the formation of yarn from the fibres available except with the nonwoven technologies. Woven structures are manufactured by interlacing two sets of yarns in a perpendicular direction to each other. Knitted fabrics are manufactured by interloping one set of yarn. Braided structure is manufactured by interlacing two sets of yarns at less than perpendicular position.

The 'nonwoven' technologies developed half a century ago, when these fabrics were often regarded as low-price alternates for conventional textiles and generally made from carded fibres on converted textile processing machinery [45]. The spinning or weaving process is absent in the nonwoven processing. In this system, bonding of the carded fibre web is done by chemical, mechanical or thermal methods replacing the conventional textiles production system of weaving or knitting. Needle-punched nonwoven is a potential alternative system, very much suited to the industry because of its low wage proportion and high productivity resulting in low cost of production [3]. The needled fabric has high bulk and possesses good hydraulic, thermal insulation and impact resistance properties with sufficient strength [40]. Such fabric, made out of annually renewable fibre like jute, is eco-friendly and biodegradable.

Recently, several trials have been conducted on agricultural practices using jute needle-punched nonwoven. The fabric structure has been optimized for some specific uses. The performance of optimized fabric has been compared with control and existing materials. The trials show a great potential of nonwoven in such applications. Mulching is a common practice in agriculture, where the cultivating field is covered usually with a synthetic sheet for weed control and moisture retention of the soil. Several trials have been made using woven and needle-punched jute nonwoven fabric in different crops and agro-climatic zones. The jute fabric is also compared with conventional methods (using straw, wood chips, banana leaf, etc.) and plastic materials. Moreover, trials have also been made and discussed as horticultural pot, prefabricated grass mat and artificial soil substitute. The nonwoven agrotextile shows better biomass and growth, better moisture absorption and retention, higher yield, better weed suppression, control in soil temperature and improved nutrient enrichment due to biodegradation of nonwoven [40]. It also reduces soil erosion and nutrient loss during watering or rain. Scope, limitations and economy have also been discussed. In the end, suggestions and further researchable points have been mentioned.

2 Nonwovens

2.1 What is Nonwoven?

Nonwoven is a fabric made by an unconventional method of fabric preparation. In this production system, the fabric is prepared directly from fibres by bonding them mechanically, chemically or some other way with the help of shorter machinery line up, eliminating conventional methods of spinning, weaving and knitting. According to the American Society for Testing Materials (ASTM D1117.80), the definition is as follows: 'A nonwoven is a textile structure produced by the bonding or interlocking of fibres, or both, accomplished by mechanical, chemical, thermal or solvent means and combinations thereof. The term does not include paper or fabrics that are woven, knitted or tufted'.

In fibre processing, it is common to make first a web (a thin semi-transparent layer of fibres where fibres are attached by surface cohesion only) and then to lay several webs on top of each other to form a butt, which goes directly to bonding by mechanical, chemical, thermal or solvent means [43].

2.2 Why Nonwoven Fabric?'

Nonwovens have many fold advantages over other types of textiles. They are as follows: [43]

- The rate of production is very high.
- The manufacturing line is shorter than the other textile processes.
- It is a continuous process directly from the raw material to the finished fabric.
- The labour cost of manufacture is low because there is no need for material handling as there is in other textile processes.
- It produces cheaper fabric for the same area density.
- This industry can produce a very wide range of fabric properties from open weddings suitable for insulation containing only 2–3% fibres by volume to stiff reinforcing fabrics where the fibre content may be over 80% by volume.
- In many instances, the amount of raw materials per unit of production is decreased, or lower-quality materials can be used to achieve an effect similar to that obtainable with superior material with traditional equipment.

The nonwoven fabric has proved its potential in the synthetic arena. Nonwoven machinery has reached a high level of engineering quality and design. The continued development of the process and its product has allowed the nonwoven fabric to become widely used in both domestic and industrial situations. Blankets and floor coverings are probably the most common domestic application, whereas filtration media, civil engineering substrates and papermaking felt dominate the industrial market [43].

2.3 Nonwoven Versus Woven

These two types of fabrics have distinctly different structures and properties. Therefore, their uses are different. In woven fabric, a strong and compact fabric can be made with the help of a series of steps being yarn as intermittent product. It is produced interlacing two series of yarns using the weaving process, whereas needle-punched nonwoven is made by needling or entanglement of fibres. Woven fabrics are much stronger than nonwoven fabrics. Nonwoven fabrics are used where hydraulic, insulation and compression properties are important than strength, durability, abrasion, etc.

2.4 Fibres for Nonwoven

Any fibre of vegetable or animal origin, regenerated or manmade type which can be converted into fibre web, is suitable for nonwoven preparation [44]. The fibre wastes can also be used as raw material. Here, jute is converted to the web after two carding and cross-lapping.

2.5 Why Jute? [21, 43]

- Unique natural colour; smooth and glossy fibre surface.
- High strength, modulus and dimension stability.
- Moderate draping.
- Good moisture absorption and breathability.
- Good bleach ability, dyeability and printability.
- Low cost.
- Annually renewable and abundantly grown in India.
- Biodegradability and eco-friendliness, vegetation and bio-technical support with the enhancement of organic matters and nutrient levels to the soil after degradation.
- Resistance to weather and microbial attack is better than cotton (Fig. 1).

Fig. 1 Jute fibre reed

3   Needle-Punching Nonwoven [20, 26]

3.1   Web Formation

Needle punching is a process for converting webs of fibre into coherent fabric structures, normally using barbed needles, which produce mechanical bonds within the web [43]. The jute reed is softened and subsequently carded to make a web. This web is laid manifold to get a batt. In the needle-punching machine, the batt is mechanically entangled to make a fabric. Thus, before needling can commence, a web of fibres is required. Webs are described by the orientation of fibres contained within them.

Parallel-laid web

Thus, a parallel-laid web has fibres lying along its length and is formed by adding together the individual webs from a series of cards placed inline or folding the output from a single card on top of itself to the required weight. Here, the web width is restricted to that of the card.

Cross-laid web

A cross lapper is used to produce cross-laid webs. In this process, an inclined conveyor collects the carded web and carries the undisturbed web through a double vertical conveyor, which traverses it across a moving bottom lattice and thus produces a multiple and thick web in which fibres lie diagonally across the width. In this case, the web can be produced to a width limited only by the cross lapper capability.

Random laid web

'Random webs' are those in which the fibre orientation is not biased either along or across the web length but is more evenly balanced between these two and other directions. Air-laying techniques are necessary to produce such webs, and there is at present a width restriction to 5.6 m. One advantage of air-laying systems is increased production speeds above those possible with cross-lapping (Fig. 2).

Methods of Batt Production

Batt production using a carding machine

A nonwoven machine usually consists of automatic fibre blending and opening unit feeding automatically to one or more roller and clearer cards. The output of a card is a thin sheet of fibres called a web, where fibres are adhered by frictional contact only. The mass per unit area of card web is too low (about 40 g/m^2) to be used directly in a nonwoven machine to make fabric. The uniformity and required mass/unit area can be increased by laying several card webs over each other to form the batt. The fibre orientation in the batt is of three types: parallel laying, cross laying and random laying (Fig. 3) [43]

The different ways of batt formation are described below.

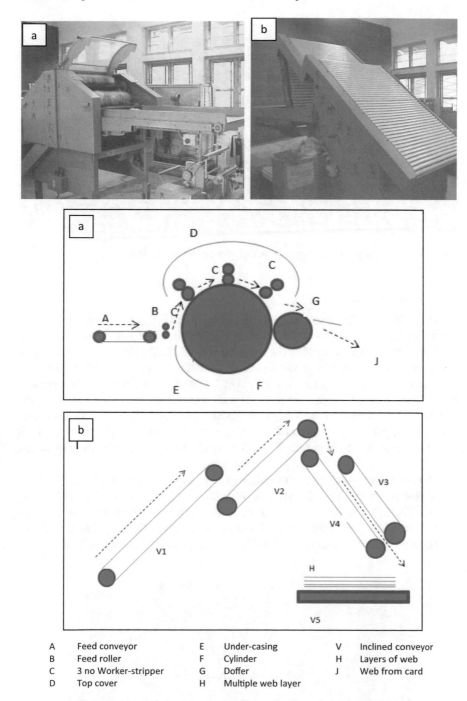

A	Feed conveyor	E	Under-casing	V	Inclined conveyor
B	Feed roller	F	Cylinder	H	Layers of web
C	3 no Worker-stripper	G	Doffer	J	Web from card
D	Top cover	H	Multiple web layer		

Fig. 2 **a** Roller and clearer card for making web; **b** cross lapper for crossed batt

(a)   (b)   (c)

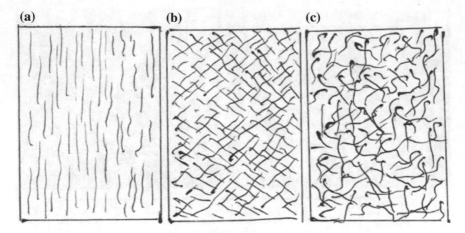

Fig. 3 Web types, **a** parallel laying, **b** cross laying and **c** random laying

(i) *Parallel laying*

A parallel-laid web has fibres lying along its length and is formed by adding together the individual webs from a series of cards placed inline or folding the output from a single card on top of itself to the required weight. Batt formation in this way is simple and cheap. Figure 4 shows three cards and a long conveyor lattice. The webs from each card fall onto the lattice forming a batt with three times the mass per unit area. Since in parallel laying all the card webs are parallel to each other, it follows that most of the fibres will lie along with the batt (Fig. 1a) [27]. The fabric formation direction (lengthwise) is called the machine direction and the right angle to this is called the cross direction (widthwise). This is the simplest and cheapest way of batt formation.

It has two disadvantages or limitations. The strength of parallel-laid fabric in the cross direction is much weaker than the fibre because it depends mainly on bonding strength (Fig. 5). On the contrary, the fabric strength in machine direction depends on fibre strength. The weakness of the fabric in the cross direction has a profound effect on the possible uses of the fabric. Briefly, it can be used when high strength

Card 1   Card 2   Card 3

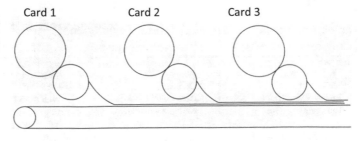

Fig. 4 Method of parallel laying

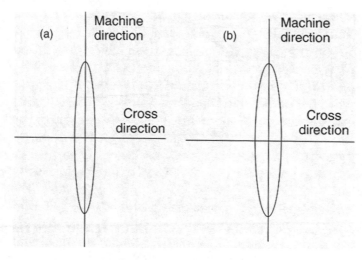

Fig. 5 Polar diagram of **a** fibres lying and **b** strength of the fabric

is not required in all directions. In such laying, the width of fabric is limited by the card width [14].

(ii)   *Cross laying*

Cross-laid webs are produced using a cross lapper, which takes web from the card and traverses it across a moving bottom lattice and thus builds a web in which fibres lie diagonally. In cross laying, the cards are placed at right angles to the main conveyor and the web is laid in a zig-zag way on the feed conveyor of the needle loom (Fig. 6)

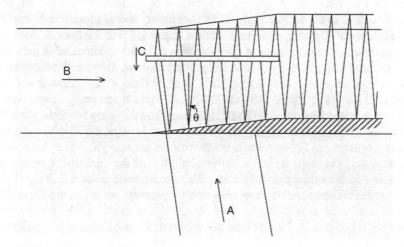

Fig. 6 Production of cross-laid butt, **a** web from card; **b** conveyor; **c** Zig-zag web laying mechanism; θ: angle of webs laying direction

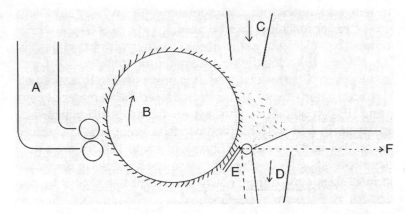

Fig. 7 Air-laying mechanism, **a** feed; **b** opening roller; **c** air blow; **e** stripping rail; **f** batt conveyor

[27]. Several layers of the web are formed on slow-moving conveyor B where fibre orientation is like Fig. 3b.

In such a mechanism, the butt is thicker in the edges. It can be eliminated by the faster movement of the traversing conveyor at the edges. Moreover, the card speed is slower which is limited by the speed of the cross-laying system. In this case, the majority of fibres are oriented towards the cross direction. This also produces anisotropy in the batt. Stretching the butt in the machine direction is a solution to that. This is the most used system in the industry.

(iii)   *Random laying*

Figure 7 is a schematic diagram describing the principle of an air-lay machine. Fibres are first fed in the hopper opening/ blending section. In this section, fibres are opened to a small fluffy tuft, blended thoroughly and produce a uniform thick sheet of material. Opened fibre from this section is fed into the back of hopper A, which delivers a uniform sheet of fibres to the feed rollers. The fibre is then taken by the toothed roller B, which is revolving at high speed. There may or may not be worker and stripper rollers set to roller B to improve the opening power. A strong air stream C dislodges the fibres from the surface of roller B and carries them onto the permeable conveyor on which the batt is formed. The stripping rail E prevents fibre from re-circulating round cylinder B. The airflow at D helps the fibre to stabilize in the formation zone [14]. Figure 8 shows the formation zone where fibre falls onto an inclined plane. The angle depends on the width, w, and the thickness, t, of the batt.

The air-laying method produces the final batt in one stage without making a lighter weight web. It is also capable of running at high production speeds, but the width of the final batt is the same as the width of the air-laying machine. The bonded fabric from air-laid batt obtains better machine direction to cross direction strength ratio.

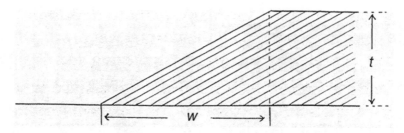

Fig. 8 Formation zone

3.2 Nonwoven Preparation

Needle looms

The most common mode of operation is one in which the needles oscillate vertically on a fixed stroke through web material supported between two plates. A schematic diagram of the loom design is shown in Fig. 9, together with a photograph of the needling zone in a laboratory loom (Fig. 10). The plates are drilled with holes to match the pattern of needles in the needle board. The web must be supported until it passes into the inlet gap between the two plates, and nip rollers or a compression roll may help pass the web into the needling zone, particularly if the loom is the first of a series or is working alone. In these cases, the top plate (the stripper) will be inclined so that there is more room at the entry side for the passage of bulky web before needling. The bottom plate (the bed) supports the web during needling, and the gap between these plates can be adjusted depending on the thickness of the web being processed. The needled fabric is taken away by take-up rollers. The action of these rollers can be either intermittent or continuous. The widths of most needle looms lie in the range 2.2–4.4 m, but looms up to 16 m wide are available for papermaking felts [20].

Extent of needling

The extent to which a web is needled is specified by two factors, depth of needle penetration and punch density [10, 41].

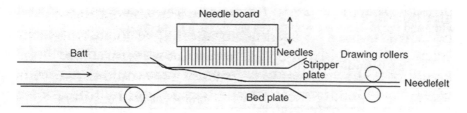

Fig. 9 Schematic diagram of loom design

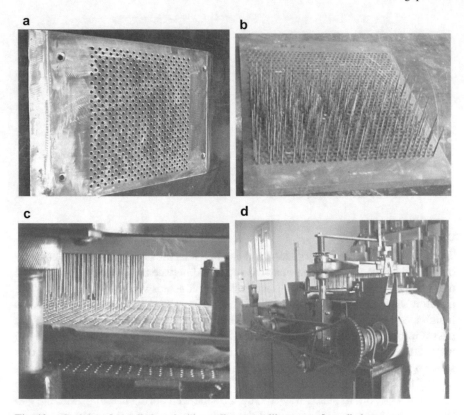

Fig. 10 **a** Bedplate; **b** needle board with needles; **c** needling zone; **d** needle loom

(a)   Depth of needle penetration

The depth of needle penetration is defined as the distance the needlepoint passes below the top surface of the bedplate [23]. This amounts to the length of the needle protruding through the web and is normally in the range of 3–17 mm, depending on the thickness and the composition of the web for needling.

(b)   Punch density

The punch density per unit area is a function of the number of needles per unit board width and the distance the web moves during each loom cycle.

$$\text{Punch density/unit area}(\text{cm}^2) = \frac{\text{needles/unit width(cm) of board}}{\text{web movement/loom cycle(cm)}}$$

$$\text{web movement/loom cycle (cm)} = \frac{\text{throughput speed(cm/min)}}{\text{loom punches/min}}$$

Needle board

The needles are held in a board (Fig. 10B), which itself is securely attached to the needle arm of the loom. Two important needle board parameters are needle density and needle arrangement. The needle density is normally defined in terms of needles per unit board width since this is relevant to the calculation of punch density. The number of needles per unit width is determined by the character of the web being needled. The art of needle board design is to reduce as far as possible the superposition of punches so that the fabric surface is uniform and has no definite features, such as large craters or tracks along its length. Thus, the needle board design affects the process and the product. The density of the needles governs how rapidly a given punch density can be achieved, and their arrangement plays an important part in the final fabric appearance.

Needle:

The needle is an important part of the needle loom because it creates a loop inside the fabric by which a needled fabric is formed. Hence, its parameters are also very important. These parameters are [15]:

- Size and shape: length, single or double reduction, type of point, type of finish.
- Blade gauge: gauge, cross-sectional shape.
- Barb arrangement: spacing, no of edges, no of barbs in the edge.
- Barb shape.

Though needle specifications depend on many factors, the suitable specifications for jute may be 15 × 18 × 21 × R/SP × 3 × 1/4 × 9; that is, the needle should be a double reduction (15 shank gauge, 18 intermediate gauge), coarse (21 gauge), regular barb, standard protrusion, 3-inch length, 1/4 inch barb spacing and 9 number of barbs [9]. Figure 3.4 shows these specifications (Fig. 11).

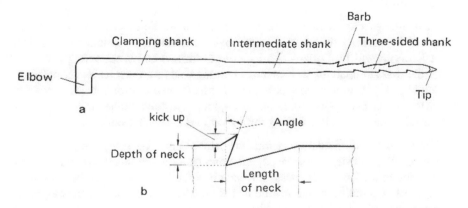

Fig. 11 Barbed needle

Process Flow Diagram of Cross-laid Jute Nonwoven Preparation

Jute reed
⇩
Sprayed with batching oil emulsion
⇩
Processed in softener
⇩
Storing for 72 hours
⇩
Sliver in breaker card
⇩
Sliver in finisher card
⇩
Laying of sliver in roller and clearer card conveyor
⇩
Card web-fed in a cross lapper
⇩
Pre-needled jute fabric in needle loom
⇩
Needle punched fabric preparation in a loom with required parameters.

3.3   Structure

The structural properties of jute needle-punched fabrics are very complex since the special inherent nature of the jute fibre is of a wide variation in fibre length, fibre fineness, mushiness and reed tapering.

Vertical structure

The enlarged picture of the nonwoven fabric cross-section clearly shows vertical structure. It consists of fibre pegs, fibre entanglement and fibre orientation in a vertical direction along with the horizontal orientation of fibres. Fibrous peg or loop produces as needles pass through the web as fibres become hooked around barbs during penetration (Fig. 12). This is so when either low depth of penetration or fine gauge needles are in operation, but, under other conditions, loops are either hidden in large pegs or destroyed as they are pulled further through at higher penetration [20].

- The amount of reorientation or interlacement of the fibres increases gradually as the needling density is increased, and consequently, the fabric becomes denser with the reduction in thickness. The length of the loops increases for the higher depth of penetration of the needle and simultaneously it changes the degree of entanglement of the fibre assembly.

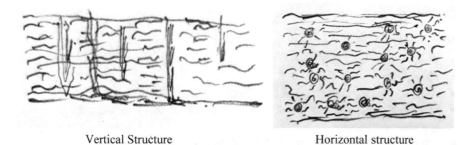

Vertical Structure   Horizontal structure

Fig. 12 Structure of needle-punched nonwoven

- A coarse gauge needle has a greater influence on the reorientation and interlacement of a very large number of fibres resulting in the formation of thicker loops in comparison with a fine gauge of the needle.
- High barb protrusion shows that the fibrous mass is not bound as fairly and firmly as a standard protrusion barb.
- Interlacements of the fibres are associated with pegs and fibre reversals around the greater neighbouring fibres, and as a result, a large number of fibres are closely associated with multiple entanglement centres.
- The number of fibres of the loops in the vertical structures of the fabrics is increased with the higher web weight [42].

The extent of fibre transfer from one layer to another brought about by the barbs of the penetrating needles is important for understanding the structural integrity of a needle-punched fabric [40]. The transfer index of jute-reinforced needle-punched nonwoven with 14.3 mm depth of needle penetration is 40%. As needling density increases, the fibre transfer also increases [42].

Horizontal structure

The magnified surface view shows the horizontal structure which is the sideway disturbance of the fibres in the web that do not come into contact with needle barbs (Fig. 12). It reveals that:

- Fibres in the needle-punched fabric form a layered structure and their arrangement in the fabric depending on the type of fibres and mode of needling.
- Horizontal structure of the fabric changes with a higher amount of needling.
- Longitudinal paths of the parallel-laid fibre on the surface of the web tends to change to a circular area of the transferred fibres and gauges.
- The degree of smoothness and surface uniformity of the jute needle-punched fabric gradually increases if the needling density is distributed in a higher number of passages of the web movement [42].

These two structures are not necessarily distinct. Many fibres that are reoriented remain partly in the horizontal plane. It is thought that such behaviour is important in the realization of maximum fabric strength [40].

Factors affecting structure

Web variables [22]:

- Weight (i.e. mass/ unit area).
- Thickness.
- Density, which is the result of mass/unit area and thickness.
- Fibre type, i.e. whether jute has been used alone or in blends with natural or synthetic fibres,
- Fibre dimensions, i.e. fineness and length distribution.
- Fibre orientation, i.e. fibres can be oriented in various ways. The most common orientations are parallel laying, cross laying and random laying.

Machine variables:

- Needle blade gauge: Coarse type needle is most suitable for jute and jute-like fibres, i.e. 21–25 gauge needle; in the case of jute blended with finer synthetic fibres, the finer gauge may be used.
- Barb shape, barb spacing: Regular barb, standard protrusion barbs may be used for jute and jute blends.
- Number of barbs.
- Depth of penetration.
- Web movement per loom cycle.
- Punching speed.
- Punch density.

3.4 Properties

Tensile Property

The mechanism of tensile behaviour of needle-punched nonwoven fabric and ultimate breakdown can be explained in the following way [12, 42]. The fibre pegs formed by penetration of needle and existence of same fibre in both the surfaces of the fabric are responsible to generate resistance to deformation. Hearle et al. [13, 25] pointed out that due to extension, the fibre pegs act to build up the transverse forces within the structure. The load-extension curve of a needle-punched nonwoven fabric can be divided into two deformation regions. The initial region shows negligible resistance to deformation, and this is followed by a jammed, stiff region. In the initial stage of extension, fibres are pulled straight, helped by slippage at crossover points, with only a nominal development of tension. This is followed by the building up of transverse forces from the fibres under tension passing round and into fibre pegs, resulting in a steep rise in load with a small increase in extension. During initial extension, fibre alignment and straightening take place. Adjacent pegs are pushed close together. Eventually, fabric density increases due to stretching. So, fibres are locked together in frictional contact and substantial resistance results in the steep rise stress with small

Table 1 Physical properties of jute needle-punched nonwoven textiles with or without reinforcement [42]

Jute needle-punched nonwoven	Fabric weight (g/m^2)	Thickness (cm)	Density (g/cm^2)	Breaking load (kg)	Tenacity (g/tex)	Thermal insulation(Marsh %)
Without reinforcement	1000–1500	0.90–0.14	0.10–0.14	2.5–6.5	0.06–0.12	–
Hessian reinforcement	270–570	0.35–0.70	0.07–0.08	18.0–32.5	0.70–1.70	29.36
Polyethylene reinforcement	210–990	0.35–0.90	0.04–0.13	1.3–7.1	1.30–7.10	–

increments of extension. As frictional forces are overcome, slippage and breakage of fibres occur, the ultimate result being fabric breakdown.

The factors responsible for self-complicated structural imperfections of needled jute fabric were analysed theoretically by Debnath [4, 22]. The jute web had very poor strength and dimensional stability, but the tensile strength of jute nonwovens improved with a higher fibre-loop length [22, 42]. A detailed study was carried out by Debnath [5, 25] on the effect of various parameters on jute needle-punched nonwoven fabric. It was reported that better entanglement of fibres could be achieved either by an increase in the needling density up to an optimum level with a constant depth of penetration of the needle or vice versa. The breaking load of the fabric, made with several numbers of needle loom passages of the web, expected to be greater than that of the fabric made with its one passage only. In general, the fabric made with a low protrusion, finer gauge, 9-barbed needle or a needle of closer barb spacing showed higher values of both stress and strain. The value of tenacity of the fabric made by needling on one surface only was higher than that of the fabric made by needling on both top and bottom surfaces [42].

Some physical properties of jute nonwoven fabric have been shown in Table 1. The stress–strain diagram of such fabric is demonstrated in Fig. 13.

It was also reported [3] that the tenacity values of the needled fabrics increased with an increase in the batching emulsion or binning period (both for web and for fabric) up to a certain level, and the optimum conditions were found to be (10%, 30%) and (48 h, 72 h), respectively [42].

Woven fabric reinforced nonwoven

Debnath (1979b) [3] investigated the effect of jute hessian reinforcement in needle-punched nonwoven varying the web weight and quality of hessian used as a base or in the centre. The tensile properties of jute needled fabric with reinforcement are mainly influenced by the tensile properties of the reinforcement fabric. Longer and finer fibres were more suitable for the formation of better interlocked fibrous structures by the needling process, causing a higher value of stress and strain of the fabric. A stress–strain diagram of jute nonwoven reinforced with different reinforcing materials has been shown in Fig. 14 [42].

58

S. Sengupta et al.

Fig. 13 Stress–strain
diagram [31] of **a** reinforcing
material, **b** needled web
without reinforcing material,
c needled web with
reinforcing material

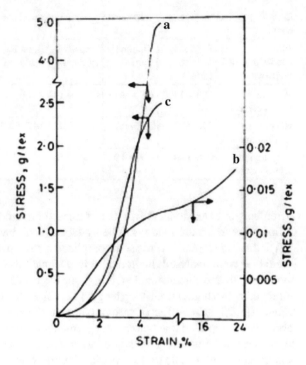

Fig. 14 Stress–strain
diagram of nonwoven
reinforced with different
material, **a** jute hessian, **b**
cotton bandage cloth, **c**
cotton gauge cloth, **d**
polyethylene film [31]

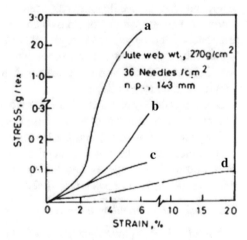

In another work [31], it was reported that batching oil emulsion treatment on jute
fibres and woollenized jute was found to give improved processability and tensile
properties to nonwovens. Reinforcing fabric placed at the centre shows better tensile
property than that at the base (Fig. 15). Figure 16 shows that woollenization in
the fabric form gives much higher extensibility than nonwoven fabric made from
woollenized jute fibre.

Fig. 15 Stress–strain
diagram of hessian
reinforced jute nonwoven,
x-x-x: Hessian at centre,
_____: Hessian at base [31]

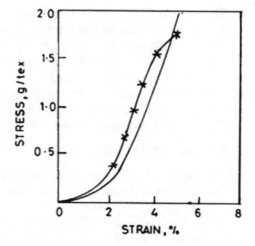

Fig. 16 Stress–strain
diagram of reinforced
nonwoven, **a** jute, **b**
woollenized jute, (3)
woollenized in fabric form
[31]

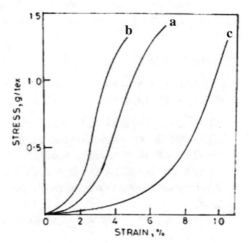

Thermal Insulation

The thermal conductivity of needle-punched nonwovens can be measured by the Lees
disc method [31]. This property of parallel-laid and randomly laid needle-punched
nonwoven fabrics using jute, jute blends with synthetic fibres, jute caddis (mill waste)
and woollenized (alkali-treated) jute was investigated by Debnath [6]. It was found
that:

- Parallel-laid needle-punched nonwoven has shown better thermal insulation
  behaviour than random laid nonwoven.
- Nonwoven from jute caddies can also be used as a thermal insulator due to its low
  conductivity.

- The blending of woollenized jute improves the thermal insulation property when blended with pineapple leaf and ramie fibre.
- Needle-punched nonwoven, made out of woollenized jute/wool blend, shows better thermal insulation than those made out of woollenized jute/polypropylene or woollenized jute/acrylic blend.
- The sandwich (layered) blending of polypropylene or acrylic with woollenized jute appears to be a better insulator than homogeneous blending. Higher crimp in fibres shows better insulation by the fabric [42].

Sengupta et al. (1985d) showed that:

- The thermal conductivity of jute nonwoven woollenized in fabric form is much higher than nonwoven fabric, made from jute or woollenized jute.
- Thermal conductivity decreases with the decrease in the ratio of the reinforcing material weight to web weight and the increase in the thickness of the fabric.
- It is increased with the increase of punch density and depth of needle penetration.

These studies explored the possibility of utilization of needle-punched jute-based nonwovens as thermal insulation medium with an advantageous cost–benefit ratio [42].

Air Permeability

Air permeability of jute needle-punched nonwoven with reinforcing jute cloth was studied by [31]. It was observed [32] that:

- At lower needle penetration, the air permeability is higher and it is reduced with higher needle penetration up to a certain limit.
- Very high needle penetration results in fibre rupture and damage to the reinforcing fabric which increases the air permeability.
- Batching oil emulsion treatment did not affect the air permeability of nonwoven fabric.
- Air permeability is increased with the increase in the reinforcing material weight by web weight ratio.
- Air permeability is higher when the reinforcing material is used at the centre of the web instead of a surface.
- Open the construction of the reinforcing fabric, the more is the air permeability.
- An increase in the needling density increases the fabric consolidation and thereby reduces sectional air permeability.
- Woollenized jute nonwovens show lower air permeability.

Compression and Recovery

Sengupta et al. [36] were ranked different important parameters using a statistical method and found that needling density, depth of needle penetration and fabric area density are the three most significant parameters for compressional behaviour of needle-punched nonwoven. Taking these parameters, they proposed the statistical model [35, 37]. From the contour diagrams, it was found that 15–16 mm depth of needle penetration, 170 punches/cm^2 needling density and 650–700 g/m^2 area density are process variables for minimum compressibility [42].

A good correlation (Table 2) is observed [35] between compressional or recovery parameters (α or β) and the properties of jute needle-punched nonwoven fabrics. It suggests that α and β can help to predict the other properties of the needle-punched nonwoven.

Sengupta [35] observed the nature of the repeated compression recovery cycles with the following conclusions [36, 37, 42, 43],

Table 2 Correlation coefficient between compressional behaviour and properties of jute nonwoven fabrics [35]

Compressional behaviour	Properties of jute nonwoven fabrics				
	tenacity (cN/tex)	extension at break (%)	energy to break (J)	Air permeability (cc/s/cm^2)	Thermal insulation value (%)
Compressional parameter (α)	−0.7540	0.7696	−0.8028	0.8028	0.9327
Recovery parameter (β)	−0.7603	0.7478	−0.8000	0.7921	0.9047

Table 3 Properties of 500 g/m^2 jute needle-punched nonwoven sample [42, 43]

• Bulk Density, g/cc	0.11
• Thickness, cm	0.52
• Tenacity, g/tex	0.43
• Breaking strain, %	20
• Work of rupture, cN	1.145
• Thermal insulation, Marsh %	29.36
• Thermal conductivity, Cal. deg^{-1} cm^{-1} s^{-1}	2.56 × 10^{-4}
• Sectional air permeability, ml s^{-1} cm^{-1}	51.2
• Fibre shredding, mg m^{-2}	258
• Average cumulative thickness loss in dynamic loading for 250 impacts, %	51.5
• Abrasion resistance, cycles	35.8
• Compression and recovery	
• Compressional parameter (α)	0.1109
• Recovery parameter (β)	0.1537
• Energy loss, %	76.73
• Thickness loss, %	49.52
• Wetting, times of water absorbed	6
• Volume swelling, %	62

Table 4 Percent improvement of biomass and other characteristics of cultivation [42]

	On control	On synthetic sheet	On straw
Plant dry weight (g)	33.58	13.66	−1.12
Shoot dry weight (g)	36.80	8.97	−1.74
Root dry weight (g)	36.37	29.00	−3.34
Plant height (cm)	25.95	8.87	0.19
Root depth (cm)	25.81	8.82	−0.46
No. of leaves. $plant^{-1}$	30.80	13.00	−9.06
Leaf area (cm^2)	26.64	9.41	2.06
Leaves dry weight. $plant^{-1}$ (g)	23.76	7.07	−1.21
Number of fruits. $plant^{-1}$	43.38	25.40	−4.88
Average fruit fresh weight (g)	32.10	13.99	−4.41
Fruit length (cm)	28.25	10.96	1.54
Water after 3 days of irrigation (%)	47.48	4.00	−6.43
Water after 7 days of irrigation (%)	64.17	5.02	−8.61
Water after 14 days of irrigation (%)	83.26	9.64	−12.71
Soil temperature, °F	−20.83	−28.75	9.62
Yield, t/ha	28.54	11.03	−3.50

- Most of the changes in the compressional properties take place between the first and second compression cycles.
- After that, the compressional behaviour of the fabric remained all most unchanged.
- As the rate of compressional deformation increases, the compressional and recovery parameters including energy loss decrease.
- There is no significant effect of ultimate compressional pressure.
- Compressional and related parameters increase with an increase in the number of plies.
- There was an instantaneous compression on the application of compressive load, and subsequently, thickness loss increases with time at a diminishing rate. The thickness loss becomes almost unchanged after reaching the maximum. Recovery for that load also follows a similar trend.
- Jute nonwoven takes about 30–40 s, whereas polypropylene and jute-polypropylene blend needed 120–140 s time for stabilization in a thickness value [12].

Roy et al. [24] reported that the compressibility of jute/viscose blended fabric is initially reduced with the increase in fabric weight, punch density and depth of penetration; and after attaining a minimum value, thickness increases.

Relation of compressibility with other properties [35].

With the increase in compressional pressure on jute needle-punched nonwoven, tenacity and elongation-at-break cv%, tensile modulus decrease sharply, whereas fabric tenacity increases initially and after optimum, it decreases. This optimum tenacity increases on the addition of polypropylene fibre with jute or wetting of jute needle-punched nonwoven. It increases initially with the increase in needling density or depth of needle penetration or area density, and beyond a certain value, it decreases [42].

Abrasion Resistance [42]

Sengupta et al. observed [31] that:

- An increase in the ratio of the reinforcing material to web weight increases the abrasion resistance of the fabric.
- Woollenization of jute decreases and batching oil emulsion treatment increases the abrasion resistance of nonwoven compared to untreated jute.
- Jute nonwoven, woollenized in fabric form shows much higher abrasion resistance than untreated jute nonwoven and nonwoven made from treated jute.
- Jute hessian reinforcement shows the highest abrasion resistance followed by bandage and gauge cloth, respectively.
- When the reinforcing material is used at the centre of the web, the abrasion resistance shows a significantly higher value.
- It decreases with an increase in needling density and increases with an increase in depth of needle penetration.

Wetting

Samajpati et al. [28] studied the absorbency of needle-punched jute nonwoven fabrics with different area densities and punch densities. It was observed [42] that:

- Sorptive capacity and rate of sorption increase with an increase in area density.
- Fabric with an area density of 300 g/m^2 and punch density of 150 punches/cm^2 shows the highest absorptive capacity as well as the rate of sorption.
- The sorptive capacity of the fabric without oil is higher than that of the fabric with oil. For the rate of sorption, the trend was reversed.
- Woollenized jute and bleached jute nonwoven absorb 2.7 times and 6.08 times water respectively of its weight.

Studies [3] showed that the tensile behaviour of wet jute and jute blended fabrics is much higher than those in normal conditions. The stick–slip effect of the load elongation curve during extension reduces remarkably on wetting.

4 Jute Needle-Punched Nonwoven in Agrotextiles

Textiles are now widely used for different agricultural end uses like sunscreens, windshields, harvesting, protecting the crop from birds, etc. These are mainly poly-olefin, polyester, and polyethylene nets/sheets/cloth. Some more applications of the synthetic sheet have been given in Fig. 17.

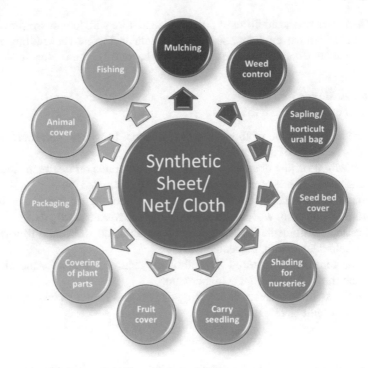

Fig. 17 Application of synthetic textile in agriculture

Jute is used in agriculture as packaging for centuries. Recently, in the search of diversified uses, jute fabric with low areal density in the form of woven, knitted or nonwoven is slowly becoming popular in the following application areas.

- Helps to prevent soil erosion.
- Adds organic matter to the soil.
- Feeds soil life and improves soil structure.
- Decreases water loss due to evaporation.
- Works against weeds.
- Affects soil temperature control.
- Adds a look to the landscape.

In this chapter, the use of jute needle-punched nonwoven mat in the areas of agriculture and horticulture will be discussed with case studies. *From the sustainability point of view, jute needle-punched nonwoven is more sustainable not only for eco-friendly material but also low use of man, machine, power and low eco-friendly waste generation resulting low carbon footprint. Moreover, whenever it is used in agrotextiles, it adds nutrients to the soil and reduces soil erosion which adds points in favour of sustainability.* Recently, several trials have been conducted on these areas using jute needle-punched nonwoven (Fig. 18).

Fig. 18 Jute needle-punched nonwoven fabric

From a manufacturing point of view, one advantage of needle-punched nonwoven is that by changing three main process parameters, e.g. areal density, punch density and depth of needle penetration, most of the required properties of nonwoven fabric, e.g. bulk density, porosity, thickness, strength, elongation, drape, rigidity, permeability, moisture retentivity, thermal insulation, etc., can be controlled easily [16]. These properties are key factors for use and better sustainable agriculture which can be customized depending on requirements for crop and agro-climatic zone. So, to use the jute nonwoven in agricultural practices, there is a need for optimization of fabric specification for different crops and agro-climatic zones. Some efforts have been reported here in comparison with control and existing natural as well as synthetic materials. The trials show a great potential of nonwoven in such applications. Figure 18 shows photographs of jute needle-punched fabric.

The potential of jute and its needle-punched nonwoven in favour of agrotextiles has been shown in Fig. 19.

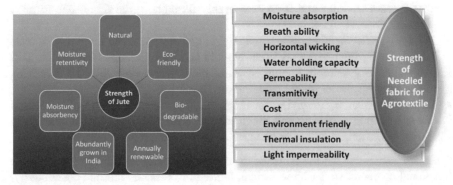

Fig. 19 Strength of jute fibre and needled nonwoven from jute on agrotextile

4.1 Agricultural Mulch Fabric: Case Studies

Optimization of Nonwoven Fabric for Mulching on Strawberry Cultivation

Mulching is a common practice in agriculture, where the cultivating field is covered usually with a synthetic sheet for weed control and moisture retention of the soil [17]. In a study [40, 42], the plastic sheet mulching has been compared with jute woven and mechanically entangled sheet (nonwoven) of mesta (similar to jute) in strawberry cultivation. Seven samples were prepared varying process variables of mesta entangled sheet as follows: areal density from 300 to 700 g/m^2, punch density from 160 to 220 punches/cm^2 and depth of needle penetration from 10 to 14 mm. One sample was prepared from mesta waste.

It was observed that mesta nonwoven from 500 g/m^2 areal density, 190 punches/cm^2 punch density and 10 mm depth of needle penetration produces (4.62 mm thickness) the best mulching product in comparison with other fabrics including plastic and jute woven, resulting in better biomass and growth, better moisture absorption and retention, higher yield, better weed suppression, control in soil temperature and improved nutrient enrichment due to biodegradation of nonwoven. The initially higher cost of raw material is justified by the higher return at the end. Nonwoven from waste mesta fibre shows further improvement in terms of yield, plant and fruit quality [33, 40].

Other observations are as follows [40]:

- Mesta nonwoven fabric loses strength after about two months but makes a solid coating over the surface which performs well up to four to five months depending on area density, structure and soil characteristics.
- Mesta fabric holds the loose soil particles and resists the shifting of soil particles during rain/thunderstorm/blowing of strong air/watering resulting in a decrease of soil erosion.
- Water is not accumulated on mesta fabric mulching like a synthetic sheet. Mesta absorbs excess water and it spreads evenly throughout the fabric due to good permeability and transmitivity.
- The low-quality synthetic sheet material is generally used in mulching. It tears into pieces during use due to direct sunlight and a temperature change. Then, its removal is very difficult. Labour costs and much effort are involved in removing all the synthetic sheet pieces from the soil [38].
- Mesta nonwoven fabric laying for mulching is much easier than synthetic sheet due to higher area density and rough surface. Synthetic sheet mulch tends to be shifted in the action of strong air current if sides are not fixed with the ground.

In another study [16] the effect of the needling density and needle gauge on the functional characteristics (fabric weight, fabric thickness and thermal conductivity) of jute-based nonwovens was studied for the frost protection of crops. The results show that:

- The effect of needling density is higher on the fabric weight than the needle gauge. For the same needle gauge, an increase in needling density also increases the thermal conductivity.
- Needling density and needle gauge have opposite effects on fabric thickness.
- Response surfaces show that:
- The higher the needling density, the smaller is the fabric thickness.
- The thermal conductivity increases with a decrease in thickness, resulting in lower thermal insulation.
- With the increase in needling density, thermal insulation decreases.
- Thus, their frost protection functionality can be enhanced to ensure an extended plant growing period. The critical points of the needle-punched nonwoven characteristics have also been suggested.

Effect on green gram and groundnut

Sarkar et al. [29] studied the jute agrotextiles of 400–1000 gsm on yield and moisture efficiency of green gram (VignaradiataL.). It was observed that:

- The yield was increased with an increase in g/m^2 of the nonwoven fabric.
- Moisture efficiency of the crop was increased significantly by 70.27% for mulched area over control.

Sarkar et al. [30] studied the jute agrotextiles of 400–1000 gsm yield and yield attributes, dry matter accumulation, water use efficiency as well as accumulated agrometeorological indices and thermal utilization of groundnut (variety J L-24).

- Response of pod yield over control showed increment over control.
- The water use efficiency increased by 69.12% due to the various treatments over control.
- Heat unit and thermal unit's use efficiencies showed the most accurate utilization of thermal indices.

So, it can be concluded that jute agrotextile has the potential to improve crop yield as well as soil fertility and productivity status [16]

Comparisons of jute needle-punched fabric with conventional materials

A uniform, porous and bulky mechanically entangled fibrous sheet (nonwoven) from eco-friendly and natural, low-grade/process waste jute fibre has been designed for strawberry cultivation in the semi-arid zone [40]. The agricultural performance of this engineered textile material has been compared with the commonly used polythene sheet and traditionally used sawdust, straw and banana leaf in the same strawberry field under the same conditions using random complete block design with three replicas [34]

It was observed [40]that:

- Waste jute nonwoven mulch shows improvement over the polythene sheet.

- Performance of straw is comparable to jute nonwoven as mulching material considering biomass/ growth characteristics, fruit characteristics, soil moisture retention, soil temperature and yield.
- Jute nonwoven reduces weed by 68% which is much better than other biodegradable mulches.
- Jute nonwoven fabric as mulch is economically viable for high-valued crops.
- Jute fabric reduces erosion and nutrient loss during storms or rain or blowing of strong air and watering.
- Waste jute needle-punched engineered nonwoven of 500 g/m^2 is a promising alternative to existing polythene sheet or other traditional materials in strawberry cultivation.
- The application of jute waste in nonwoven also promotes bio-waste utilization for sustainable and green agriculture and is a promising alternative as a soil cover (Fig. 20).

Cultivation of summer variety tomato in Gangetic alluvial soil

It has been observed [8, 38] that the 500 g/m^2 Jute nonwoven was found superior in terms of plant growth and productivity as the mulch of summer variety tomato (which suffers from a crisis of water) than 250 g/m^2 jute nonwoven, sawdust, banana leaf, white and black plastics. Dry banana leaves and straw are dried/ destroyed/removed from the ground due to strong wind as these are lighter when dried. It has been found that there is a lot of weed germination in the control (no mulching sheet) and under the white sheet due to the penetration of sunlight. Plastics are getting damaged and brittle on continuous exposure to sunlight [38]. Using jute needle-punched nonwoven mulching, production of tomato increased by 12%, plant height growth improved 15%, and leaf dry weight increased by 23% compared to plastic mulching. Table 5 presents the optimized jute nonwoven fabric for tomato cultivation in the alluvial soil (Fig. 21).

Fig. 20 Strawberry cultivation on jute nonwoven mulch

Table 5 Fabric for a summer variety tomato at alluvial soil [38]

Type of nonwoven	Needle punched
Area density, g/m^2	500
Density, g/cc	0.16
Fibre grade	TD6/waste
Needling density, punchs/cm^2	160
Depth of needle penetration, mm	10
Breaking tenacity, cN/tex	0.49
Breaking strain (%)	14
Sectional air permeability, cc/s/cm	76
Thermal insulation, tog	0.65

Fig. 21 Tomato cultivation **a** control, **b** plastic, **c** jute

Broccoli and cauliflower in dry lateritic soil

Manna et al. [18] have shown that jute mulch significantly improves moisture content, organic C, available N, P and K contents and microbial population of soil, reducing weed population and thereby enhancing the growth and yield of broccoli (*Brassica oleracea L.*) in dry lateritic soil. They concluded that 350 gsm jute nonwoven mulch provides the most favourable soil condition compared to other mulches to broccoli and resulted in the highest growth and yield of the crop and subsequent improvement in soil health. ICAR-NINFET has also carried out many trials on jute agrotextile mulches at different agro-climatic zones to study the effect of jute mulch on weed control and soil moisture preservation on broccoli and cauliflower. All the results were very encouraging in favour of jute nonwoven mulching. Some of the photographs of the trials are given in Figs. 22 and 23.

Effect on capsicum pointed gourd, sweet lime and turmeric

Nag et al. [19] demonstrated efficient management of soil moisture on sweet lime and turmeric cultivation in a red lateritic zone by jute nonwoven compared to control where no mulching materials were used. In another study [7], the application of jute agrotextile mulches increased the yield of capsicum and pointed gourd.

Fig. 22 Trials on broccoli crop

Fig. 23 Trials on coloured cauliflower crop

4.2 Application on Horticultural/Nursery Bag [17]

It has been found that 350 g/m^2 needle-punched nonwoven made from waste jute
with scrim cloth reinforcement is suitable for a horticultural bag for germination
and transportation of seedlings. A trial of germination of the flower using above-
mentioned fabric shows that leaf dry weight increases by 24%, soil loss reduces by
17%, soil nutrient increases by 6%, and plant height growth increases by 27% after
one month in comparison with a plastic bag [38]. It can be planted directly into the
soil within approximately one and a half months without removing the bag which
is not possible for plastic. Moreover, the removal of plastic may damage the roots
which are harmful to plant. Jute nonwoven allows the hindrance free growth of roots
as it can easily penetrate the jute nonwoven.

Table 6 shows plant growth and different leaf parameters at various horticultural
pots under trial. The optimized properties of needle-punched jute nonwoven fabric
for horticultural pots are depicted in Table 7 and Fig. 24.

Table 6 Plant and leaf parameters on different horticultural pots [38]

Treatment	Plant height, cm	No. of leaf	Leaf length, mm	Leaf width, mm	Leaf dry weight, g
Mud tub	11.65	8.9	51.3	46.3	2.06
Plastic bag	8.12	6.4	47.2	38.7	1.43
Jute bag	10.74	8.3	50.7	44.5	1.97

Table 7 Nonwoven fabric properties optimized for horticultural pots [11, 38]

Type of nonwoven	Needle punched with scrim backing
Area density, g/m^2	350
Density, g/cc	0.18
Fibre grade	TD6 / waste
Needling density, punchs/cm^2	200
Depth of needle penetration, mm	12
Breaking tenacity, cN/tex	0.63
Breaking strain (%)	10
Sectional air permeability, cc/s/cm	105
Thermal insulation, tog	0.46

Fig. 24 Jute nonwoven horticultural pot

4.3 Prefabricated Grass Mat from Jute Nonwoven

The garden beautification includes a lawn where the grass is grown as a green carpet for a landscape. Nowadays, landscaping can be done within a small time by laying readymade grass carpets part by part over the barren land. These carpets are prepared in some nurseries on a plastic sheet and thick soil layer which is prepared manually with required fertilizer and other requirements to grow grass over it. It needs a huge volume of soil to be transported from one place to another. In this process, grass has been grown inside the nurseries and then shifted in the roll form to lay in the lawns, roof, terrace, play arena for beautification, drainage and soil erosion control.

An experiment [38] has been conducted to make grass carpet using jute needle-punched nonwoven replacing about 85% soil by volume as grass growing medium. Different types of nonwoven, as stated below, have been prepared for this purpose.

- Needle-punched nonwoven from TD_3 jute (500 g/m^2).
- Needle-punched nonwoven from waste jute (500 g/m^2).
- Needle-punched nonwoven from waste jute with 25-micron plastic at the back.
- Needle-punched nonwoven from waste jute with coir net at the back.
- Needle-punched nonwoven with a double layer of waste jute.
- Needle-punched nonwoven from waste jute with jute woven fabric at the back.

The flat ground or slope of a nursery is cleaned and levelled. Thin plastic/back material is laid over the ground for ease of withdrawal. Needle-punched nonwoven is laid over it. Soil dust is prepared with fertilizer as per the requirement of the growth of grass. These soil particles are spread, manually and as evenly as possible, on nonwoven fabric as about 3–4 mm layer. Grasses are planted at equal distances. Now, it is ready for carpet preparation. Water, fertilized or other ingredients for better growth are applied from time to time as required. After 30–45 days, the carpet is ready for application on the lawn. Then, the grass carpets are removed by cutting in small pieces (4ft x 4ft or some other suitable size), rolled, transported and laid over the clean ground which will be converted to a grass lawn. Before laying only a plastic sheet is to be removed. Other backing layers will be applied on the lawn along with the grass mat 2.

On comparing, 500 g/m^2 waste jute nonwoven with double layer along with approximately 4 mm soil layer shows the best result. Withdrawal time is very important so that the grass-root should not reach the under soil.

It has also been found that:

- A nonwoven-based mat is four times lighter and lower in density than a soil mat which helps in transportation and laying.
- Removal of the mat from the ground in the nursery is easier for plastic, but it has to be removed before application and is also difficult to dispose of.
- Coir net backing is the second-best for removal. But late removal gives trouble due to penetration of root in the ground soil.
- Jute nonwoven replaces soil for about 85% by volume.
- Grass production on nonwoven increases by about 12% by weight.
- Appearance is much better in health and with attractive deep green colour for the nonwoven mat.
- The amount of water requirement was reduced by about 19% in the case of the nonwoven mat (Fig. 25).

Fig. 25 Fabric laying, grass plantation and prefabricated grass mat

5　Artificial Soil Substitute

Needle-punched nonwoven fabrics from low-grade jute with different process parameters (punch density, areal density and depth of needle penetration) have been prepared and examined for soil substitute.

- The best quality of germination was observed with 0.116 g/cm^3 bulk density in the case of a bed having multiple layers of 500 g/m^2 areal density jute needle-punched nonwoven.
- The required bulk density has been achieved using 160 punches/cm^2 and 13 mm depth of penetration. It may be a good alternative as a medium of cultivation.
- The design of the bed is also proposed in the article.
- In this artificial system, cultivation can be made in hostile conditions and plant growth is better than soil.
- Its moisture-holding capacity and temperature control of medium help in better agriculture.

This is highly applicable in the regions where soil is either not available or not suitable for cultivation [39]

This system has the following advantages [39]

Fig. 26 Jute nonwoven bed preparation and germination of lentil

- The process is simple and farmers friendly.
- No chance of erosion.
- Loss of water and fertilizer reduced considerably.
- On degradation, jute increases nutrients.
- The bed can be reused.
- The bed can be disposed of in the field directly, after controlled germination.
- The bed is lighter than the soil. Hence, transportation or shifting is easier (Fig. 26).

6 Use of Jute Nonwoven as Protective Cloth in Agriculture

Jute woven sacks after several uses as packaging reused as animal cover to protect from cold was an age-old practice in the Indian context, especially in hilly areas. Jute is a very good thermal insulator due to its fine structure. It is quite obvious that when the needle-punched fabric will be used in place of woven sacks, it will give better warmth to the animal with lighter weight due to its pore size distribution and low density. Therefore, it can be comfortably used as a protective cover for pet animals, e.g. cow, dog, horse, etc. (Fig. 27).

Fruits can be protected by covering them with jute needle-punched nonwoven during transportation. It protects the soft and delicate fruits from abrasion and mild shock due to its rough surface as well as compressive resiliency. In place of synthetic protective cover, jute needle-punched nonwoven can be effectively used

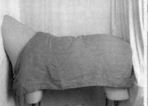

Fig. 27 Animal cover and fruit cover

for apple, orange, pear, guava, etc. The breathability and thermal insulation property of nonwoven keep the fruit healthy for a longer time.

6.1 Survey on Potential

An all India survey was conducted by M/s the Nielsen Company, Kolkata, under a project of Jute Technology Mission [38] Random probable stakeholders were selected from different states of India and enquired about the potential of using jute nonwoven in agrotextiles. In their report, the following points were highlighted:

- Mulching enhances weed control.
- Presently, mulching practised in India are of two types—organic mulch and inorganic mulch. Plastic mulch is most popular in the inorganic category and widely used by nurseries as well as cultivators/farmers.
- In high-valued crops, plastic mulch is mainly used.
- Different states use different mulching materials in different crops.
- In India, mulching is practised mainly in strawberry, capsicum, tomatoes, brinjal, water melon, musk melon and chilli.
- The crops for which mulching is carried out in most parts of India is tomatoes, water melons and brinjal.
- Mulching is most popular in the states of Maharashtra, Tamil Nadu and Andhra Pradesh.
- The Compound Annual Growth rate for the sale of mulching products for the last three years has been 5.8%.
- About 85% of nurseries are most willing to use jute nonwoven as mulch if it is available.
- About 5% of nurseries are not aware of jute mulching, so they are keen to know.
- About 15% of nurseries are not willing to use jute mulching due to higher costs.
- The nurseries have suggested that jute nonwoven can be used to make nursery bags. They are well aware that plants grown in nursery bags are 15% better than plants grown in pots. Moreover, transportation is easier. They believe that this is a potential area where jute nonwoven can be introduced instead of plastic.
- Around 80% of farmers of Punjab, Haryana, Tamil Nadu, Andhra Pradesh, Maharashtra and Gujarat are willing to use jute nonwoven in mulching and weed suppression, whereas around 20% of farmers are not willing, though 100% are not aware of the use of jute nonwoven. 50% of farmers of Chhattisgarh who are poor and have low landholding are not willing to use nonwoven. Farmers of Punjab, Haryana and UP are interested in the trial basis first if they are provided fabric samples.
- Farmers in the states of Tamil Nadu, Andhra Pradesh, Punjab and Haryana are more willing to use it in comparison with the other Indian states.
- Farmers suggested that the thicker, more durable, easily available, stronger and impermeable jute fabric may give better performance.

7   Challenges and Limitations [38, 42]

- Nonwoven machines are costly. Nonwoven machines for natural fibre processing are not available and are mainly imported which is very sophisticated to process synthetic fibres. To reduce the cost of machines and nonwoven production, indigenous, less sophisticated and low-cost machines are to be developed. It should be suitable for natural fibre processing. It may encourage the small-scale sector also to produce nonwoven. Another part is on metallurgy for making indigenous machines. It is difficult to make tough and light needle boards as well as quality needles. In some decentralized and small-scale sectors, attempt was made to prepare needles. But those needles are not good enough for jute needling.
- As jute fabric is costlier than other mulches, the initial cost of cultivation increases. It demotivates the farmers to use jute mulch. But in the end, if every other natural fact is in favour, cost calculation shows that farmers earn more profit using waste mesta nonwoven soil cover which is 70% higher than without cover and 38% than plastic cover due to less requirement of pesticide, fertilizer, de-weeding cost, etc.; better yield and higher revenue from fruit. Cost calculation also shows that this is economically viable for high-valued crops only like strawberry, tomato during summer, medicinal plants or off-season vegetables.
- In a comparison of different agro-climatic zones, jute nonwoven is more effective in the arid and semi-arid zone where water scarcity is a great problem for farming.
- Farmers' awareness is very much required for using jute nonwoven in agrotextiles. Training sessions may be organized in collaboration with the Krishi Vigyan Kendras in different places, to educate the farmers about the use of jute nonwoven as mulch and other areas of agriculture.
- Jute nonwoven is mainly produced in organized big sectors by very few mills for some specific purpose. So, non-availability of that fabric discourages farmers to use nonwoven. Jute nonwoven should be easily available in all Krishi Vigyan Kendras so that farmers are not reluctant to use it due to difficulty in procurement. It should be easily available in the market also. Free trial samples may be given to the farmers and nurseries so that they can understand the benefits of practical use.
- Squirrels and rats are the main problems of using jute nonwoven in agriculture as they destroy it. This can be minimized if a diluted mixture of mud with water is poured over the nonwoven just after the laying. Termite attack is also a great problem in some fields.
- The effectiveness of jute nonwoven is not the same for different crops and agro-climatic zone. Therefore, extensive research is required on agrotextile and it should be crop specific and agro-climatic zone specific, and more effective in the arid and semi-arid zone where water scarcity is a great problem for farming. Standardization is required to get the best performance of jute agrotextiles.
- In some uses, air and water impermeability or less permeability is required. For that purpose, sawdust-like filler or natural rubber-like thin coating can be used.

- Jute nonwoven is for one-time use and is less durable than plastic. Durability can be improved by blending with coir or sisal.

8   Potentials and Economy [38, 42]

- Different studies show that properly designed jute needle-punched nonwoven may be used as agrotextiles.
- Nonwoven retains water (six times) and reduces moisture evaporation from the soil, resulting in needless water requirement (about 26%) during cultivation.
- The government may enforce laws/rules towards the usage of environment-friendly products like jute nonwoven for geo-environmental and agricultural purposes. This will ensure a bright future for jute nonwoven products.
- Since jute is biodegradable and eco-friendly, the government may promote the usage of jute nonwoven for geo-environmental and agricultural purposes by giving subsidies for its use.
- It is very much suitable for different agricultural applications. Hence, jute nonwoven may be a suitable or even better alternative to plastic material from a performance point of view and it will help to make a green environment.
- Jute nonwoven fabric almost degrades (loses strength) after about two months but makes a solid coating over the surface which performs well up to four to five months, [8, 19]
- Jute holds the loose soil particles in its place and resists the shifting of soil particles considerably (about 35%) during rain/thunderstorm/blowing of strong air/watering. Similarly, it reduces soil nutrient loss.
- Jute fabric acts as a thermal insulator to the soil. This is beneficial for the plants in extreme temperatures. In strawberry, it was found that the fruit was available on the land even at the end of April with jute nonwoven mulch only whereas for synthetic mulch or no mulch, it is finished by end of March. Farmers can earn more money from these off-season crops.
- On jute mulching, water is not stacked, but on plastic, due to rain or other reason water may be stacked and this is harmful to crop like strawberries 1. Jute absorbs water and it goes into the soil. Jute nonwoven has good permittivity and transmitivity resulting in even distribution of water in the covered area.
- Low-quality plastic material is generally used in mulching. It tears into pieces during use. Then, its removal is very difficult. Labour cost and much effort are involved to remove all the plastic pieces from the soil where jute nonwoven has no such problem because it degrades after use and becomes a soil nutrient.
- Jute nonwoven fabric laying for mulching is much easier than plastic. Mulches other than jute have a tendency to be shifted during strong air current.
- Jute nonwoven mulching reduces labour costs (21%).
- Jute nonwoven has a high potential to be used as agromulch. Based on four crops, i.e. tomato, brinjal, chilli and watermelon for which mulching practice is popular,

an estimated area of 367,590 ha can be mulched with jute nonwoven. To cover this area, 162,63,41,761 m^2 of jute nonwoven would be required.

Practical Significance/Usefulness

Previously, jute conventional products like hessian, sacking and carpet backing were secured by mandatory packaging act of Government of India and 40 lakhs families were earning their bread and butter directly or indirectly by jute sector. With the slow abolishment of mandatory packaging and capture of the jute market by synthetic one, if jute can be introduced in agrotextiles for eco- or green cultivation, a major share of jute will be utilized in this area saving the life of lakhs of families in future.

Jute needle-punched nonwoven is costlier than other mulching material as its material cost is higher. Trial shows that farmers can earn more profit using jute nonwoven agrotextiles. The cost calculation of nonwoven mulch on strawberries shows 70 and 38% higher profit than without mulch and plastic, respectively, due to less requirement of pesticide, fertilizer, de-weeding cost, etc.; better yield and higher revenue from fruit, etc. Therefore, it is economically viable for high-valued crops.

9 Conclusion

Jute needle-punched nonwoven is an effective and sustainable proposition for agro-textile due to its water-holding and retention capability, breathability and transmitivity, eco-friendliness, low carbon footprint and thermal insulation. It is possible only if scientists, manufacturers and all stakeholders promote jute needle-punched nonwoven joining hand to hand to make green earth in future days. The government also has a great role to promote jute in agrotextile. Indigenous, low-cost and less sophisticated (fully mechanical without electronics, sensors or automation) needle-punching machines are essential to manufacturing nonwoven in small/medium-scale sectors of jute growing areas. Support is also required from the organized sector for bulk supply and availability in the market.

References

1. Annual Report 2011–2012, National Institute of Research on Jute & Allied Fibre Technology, ICAR, Kolkata, 2012: 72–73& 56–59
2. Conjetta MB (1993) US Patent 5189833—A Turf-growing process
3. Debnath CR (1978) Indian Text J 89(3, 5, 6):137
4. Debnath CR (1986) Indian J Text Res 11(3):129
5. Debnath CR (2001) Manmade Text India 44(11):443
6. Debnath CR, Roy AN (2001) Manmade Text India 44(12):481
7. Debnath S, Ganguly PK, De SS, Nag D (2010) Control of soil moisture and temperature by light weight jute fabrics. J Inst Eng (India), Text Eng 90(2):16–19

8. Debnath S (2014) Jute-based sustainable agrotextiles: their properties and case studies. In Muthu SS (ed) Roadmap to sustainable textiles and clothing technology. Springer Science, Singapore, pp 327–355
9. Debnath S, Madhusoothanan M (2009) Compression properties of polyester needle punched fabric. J Eng Fibre Fabrics 4(4). https://doi.org/10.1177/155892500900400404
10. Debnath S, Madhusoothanan M (2012) Compression creep behaviour of jute-polypropylene blended needle-punched nonwoven 82(20):2116–2127
11. Debnath S (2017) Sustainable production and application of natural fibre-based nonwoven. In: Subramanian SM (ed) Sustainable fibres and textiles. Woodhead Publishing, pp 367–391
12. Ganguly PK, Sengupta S, Samajpati S (1997) Indian J Fibre Text Res 22(3):169
13. Hearle JWS, Purdy AT (1972) J Text Inst 63:475
14. Horrocks AR, Anand SC (2000) Handbook of technical textiles. Woodhead Publishing Limited in association with The Textile Institute, Cambridge England
15. Lunenschloss J, Albrecht W (1985) Nonwoven bonded fabrics. Wiley, New York
16. LupuIuliana G, Cramariuc O, Hogas Horatiu I, Hristian L (2013) Parameters optimization for the production of needle-punched nonwoven agrotextiles. J Text Inst 104(10)1125–1131
17. Majumdar AK, Bhattacharyya SK, Saha SC, Goswami K (1999) Use of jute nonwoven in protecting riverbanks and agro-horticultural practices, Paper presented at National Seminar on production and characterization of natural and manmade fibres. Central Institute of Research on Cotton Technology and Indian Fibre Society, Mumbai, India, 3 July
18. Manna K, Kundu MC, Saha B, Ghosh GK (2018) Effect of nonwoven jute agrotextile mulch on soil health and productivity of broccoli (Brassica olaracea L.) In lateritic soil. Environ Monit Assess 190(82):1–10
19. Nag D, Choudhury TK, Debnath S, Ganguly PK, Ghosh SK (2008) Efficient management of soil moisture with jute non-woven as mulch for cultivation of sweetlime and turmeric in red lateritic zone. J Agric Eng (India) 46(3):59–62
20. Purdy AT (1983) Developments in nonwoven fabrics. Text Progress 12(4):1–31
21. Ranganathan SR, Quayyum Z (1993) New Horizon for jute, National Information Centre for Textile and Allied Subjects, Ahamedabad
22. Rawal A, Majumdar A, Anand S, Shah T (2009) Predicting the properties of needle punched nonwovens using artificial neural network. J Appl Polym Sci 112(6):3575–3581
23. Rawal A, Shah T, Anand S (2010) Geotextiles: production, properties and performance. Text Prog 42(3):181–226
24. Roy AN, Ray P (2005) Manmade Text India 48(11):435
25. Roy AN, Debnath CR (1996) Indian Text J 107(1):38
26. Russell SJ (ed) (2007) Handbook of nonwovens. Woodhead Publishing Limited, England
27. Russell SJ, Smith PA (2016) Technical fabric structures—Nonwoven fabrics. In: Richard Horrocks A, Subhash C, Anand (eds) Handbook of technical textiles. Woodhead Publishing, pp 130–188
28. Samajpati S, Mitra BC, Ganguly PK, Sengupta S (1998) Water absorbency of jute needle punched nonwoven, Annual report of NIRJAFT, 10
29. Sarkar A, Barui S, Tarafdar PK, DeS K (2018) Jute Agro Textile as a Mulching Tool for Improving Yield of Green Gram. Int J Curr Microbiol App Sci 7(5):3604–3611
30. Sarkar A, Ghosh A, Tarafdar PK, DeS K (2019) Assessing the water and thermal use efficiencies of groundnut (Arachishypogaea) grown under different strength of Jute Agro Textile Mulch in India. Curr J Appl Sci Technol 37(4):1–10
31. Sengupta AK, Sinha AK, Debnath CR (1985) Indian J Text Res 10 (3,4):91, 147, 141
32. Sengupta S, Samajpati S, Ganguly PK (1999) Air permeability of jute based needle-punched nonwoven fabrics, Indian J Fibre Text Res, 24 (2), 103–110.
33. Sengupta S, Debnath S (2018) Organic soil cover to save environment and water. In: International conference on innovative and emerging technologies for farming—energy & environment-water, Vellore Institute of Technology, Vellore, Imdia. 12–14 Oct
34. Sengupta S, Debnath S (2018) Production and application of engineered waste jute entangled sheet for soil cover: a green system. J Sci Ind Res 77(1):240–245

35. Sengupta S (2005) Compressional behaviour of jute needle punched nonwoven textiles, Ph.D. thesis of University of Calcutta in Textile Technology
36. Sengupta S, Ray P, Majumder PK (2005) Asian Text J 14(6):69
37. Sengupta S, Ray P, Majumder PK (2005) Indian J Fibre Text Res 30(4):389
38. Sengupta S, Debnath S (2013) Development of low cost dense jute non-woven fabric, Final Project report, Jute Technology Mission, Mini-Mission (IV), Scheme 7.1, No 5
39. Sengupta S, Debnath S (2019) Study on needle punched jute nonwoven as an artificial medium for germination of seed: effect of bulk density. J Nat Fibers 16(4):494–502
40. Sengupta S, Debnath S (2020) Effect of processing parameters of mesta sheet for use as eco-friendly agrotextiles. J Sci Ind Res 79(3):256–260
41. Sengupta S (2018) Study on some functional properties of Mesta needle punched nonwoven fabrics using central composite rotatable design. J Nat Fibers 15(1):131–145
42. Sengupta S (2020) Potential of jute based needle-punched nonwoven: properties and applications, In: Nonwoven fabric: manufacturing and applications, Rembrandt Elise Ed., Nova Science Publishers, Inc., pp 37–100
43. Sengupta S, Bhattacharya GK (2008) Needle punched nonwoven fabric—a new dimension to jute diversification. In: ICAR-National Institute of Research on Jute and Allied Fibre Technology, Kolkata
44. Subramanian SM, Gardetti M (eds) (2016) Sustainable fibres for fashion industry. Springer, Singapore
45. Wilson A (2010) The formation of dry, wet, spunlaid and other types of nonwovens, In: Applications of Nonwovens in Technical Textiles, R.A. Chapman Ed., Woodhead Publishing, pp 3–1745

Popularization of Agrowaste Fibres—Banana and Areca Nut Fibre—A Sustainable Approach

Thomas Ruby Mariamma and Honey S. Nair

Abstract The most widely used plant fibres in the textile industry are cotton and linen and certainly they are the major fibres. There are a few minor fibres such as hemp, jute, bamboo, and ramie that have gained much commercial value because of their eco-friendly nature and excellent serviceability. Still there remain several more promising lignocellulosic fibres that can be exploited for textile benefits. Advances in research and technology have stated that annual crops or agricultural wastes can be used as favourable alternative sources as the raw material for the natural fibre. Biomass is generally used inefficiently, with very few higher value-added product markets. Banana plant provides one such biomass which can be utilized effectively. The fibres obtained from the variety nendran are filament fibres of adequate strength after degumming, which can be joined together end to end and used as a weft yarn with cotton or silk. The soft fibres found inside the hull of areca nut also have a wide scope in textile fibre industry. A waste bioresource with soft handle can be spun to give a yarn of adequate strength. Hence, bio-based renewable resources can provide raw materials for many new and growing industries besides stimulating rural development and job creation. A greater reliance on bio-based resources and biological processes is an inevitable part of an overall sustainability transition.

Keywords Fibre · Agrowaste fibres · Banana fibre · Areca nut fibre · Lignocellulosic fibre · Sustainable fibre

1 Introduction

Textile industry is varied in nature of its application. The Textile Industry is divided into three types i.e. by **Application** into Clothing, Industrial/Technical Application, and Household Application, by **Material** into Cotton, Jute, Silk, Synthetics, and Wool and by **Process** into Woven and Non-woven.

T. R. Mariamma (✉) · H. S. Nair
Department of Home Science Vimala College (Autonomous), Thrissur, Kerala, India

© The Author(s), under exclusive license to Springer Nature Singapore Pte Ltd. 2022
S. S. Muthu (ed.), *Sustainable Approaches in Textiles and Fashion*, Sustainable Textiles: Production, Processing, Manufacturing & Chemistry,
https://doi.org/10.1007/978-981-19-0878-1_4

Fibres are the basic unit of most fabrics. Fibres influence product appearance, durability, comfort, retention, care, environmental impact, sustainability, and cost. Successful textile fibres must be readily and continuously available and cost effective. They must have sufficient length, strength, pliability, and cohesiveness to be processed into yarns, fabrics, and products that satisfy consumer needs. Textile industry is shared between natural fibres and manmade fibres. Most common natural fibres are cotton, wool, silk, and linen.

Natural fibres are aesthetically attractive and are variable in performance (a common feature of all natural fibres). These fibres are trending to be higher in price. They have several advantages; for example, they have good tensile properties, reduced energy consumption, recyclable and biodegradable. It consists of two sub-categories, one includes fibres derived from animal sources; the most important are wool and silk, but there are also many minor fibres which include fur and hair fibres, such as alpaca, camel, cashmere, hog, llama, mohair, and vicuna. The other group consists of those obtained from vegetable sources. Here the most important are cotton, flax, and jute. The many others include those of the bast and leaf varieties (made from the inner tree bark) hemp, kapok, raffia, ramie, sisal.

Plant-based fibres contain cellulose and non- cellulosic materials such as hemicelluloses, pectin, and lignin; hence they are also known as lignocellulosic or cellulosic fibres. Cellulosic fibres are the most abundantly occurring natural fibre and it occupies an important position amongst the raw materials for the textile industry.

A good part of the manmade fibre is synthetic fibre. It is made of petrochemicals. These fibres are very commonly used. But the truth is that these fibres are harmful to human body and to the environment too. Nowadays people have become much aware of the pollution caused by synthetic materials on the environment system and it has led to the development of eco-friendly materials. The researchers carry on hardwork to find out some materials to replace all harmful synthetic products.

The globally used unsustainable fibre mix is comprised of synthetics majorly and the rest of cotton and other plant-based materials. Through pulping and extrusion of wood and bamboo, some natural polymer structures are obtained. This cellulose has functional advantage on natural fibres that it requires less processing requirements. But it has got two disadvantages. Firstly, it causes fossil fuel dependency secondly, it results in high deforestation. So researchers were prompted to find out alternatives such as reuse and recycling of textile products. But recycling has turned unbeneficial and it is slowing down.

The COVID-19 pandemic posed a great challenge in 2020 with trade restrictions and lockdowns in several parts of the world. But the textile industry sailed through this huge storm and proved to be an ever-growing market even though it experienced sudden drop in international demand for textile products. The loss had a greater impact in countries where the textile industry accounted for a larger share of the exports. According to the study by the International Labour Organization (ILO) the global textile trade collapsed during the first half of 2020. Also, exports to the major buying regions in the European Union, the United States, and Japan fell by around 70%. The industry also suffered several supply chain disruptions due to the shortages of cotton and other raw materials (mordor intelligence).

The textile industry is greatly responsible for huge amounts of waste water effluents, carbon emissions, and loads of waste. Fast fashion has led to vast amounts of textiles as in garments that are discarded which take up landfills or are incinerated. And majority of the discarded garments are plastic as they are made of synthetics like nylon and polyester. Thus this signifies the necessity to adopt environment-friendly fabrics and a circular economy in fashion.

Shifting the focus to newer alternatives such as bio-based fibres, novel fibres, and agrowaste fibres as raw materials in making commercial fabric is the need of the hour. These fibres are sustainable materials which are easily available in nature and have advantages like low-cost, lightweight, renewability, biodegradability, and high specific properties. Arguably more important than biodegradability is the concept of 'sustainability'. Sustainable living means not utilizing potentially renewable resources from the natural world that can be replenished naturally and not overloading the capacity of the environment to cleanse and renew itself by natural processes [7]. Resources are sustainable if they can be renewable without getting exhausted; for instance, oil resources are gradually decreasing whereas the wind can be harnessed to produce energy continuously [12].

In terms of textile fibres, a sustainable fibre is one that is manufactured using chemicals that are renewable along with energy used in its processing is obtained from non-fossil fuel. Hence sustainable development is maintained by providing various options of renewable sources of polymeric materials through technological advancements that are economical and ecologically safe.

The important factors that a sustainable material should possess are:

- The product must be manufactured from renewable resources.
- The raw materials should have low ecological footprint.
- Use of non-toxic chemicals that are safe to both humans and the environment.
- It should be equivalent to the product it replaces in all its function and could be utilized as well as or better than the existing product;
- Its availability and price should be more attractive
- The product should not impose negative impacts on food or water systems [44].

A green material means to be fully green, throughout the life cycle of the product. It involves

1. Preservation of fossil-based raw materials
2. Reduction in the volume of waste materials leading to conceptualizing of Zero waste
3. Protection of climate through decreasing the level of carbon di-oxide released
4. Reduction and elimination of health hazards
5. Elimination of environmentally toxic products released at any point in the life cycle of the product
6. Innovations in the development of materials from biopolymers/use of biotechnology in textile industry
7. Finding out renewable resources in developing new raw materials e.g. Use of agricultural residues.

2 Agro-Waste Fibres

Nowadays the fashion industry is highly dependent on fossil fuels. Hence a transition to a circular and regenerative system has become inevitable. Banana pseudostems, pineapple leaves, corn husks, and areca hulls are created enormously during harvest. This biomass causes landfills and when it is burnt down it leads to air pollution. Agricultural wastes can be used as the raw material for natural fibre. Major portion of this biomass is not efficiently used. Very few value-added products are manufactured. But these bio-based renewable resources can provide raw materials for many new and growing industries. It can speed up rural development by increasing employment opportunities and income generation. It enables reduction of greenhouse gases. Therefore, greater reliance on these agrowastes would help an overall sustainability transition of any agricultural country.

Agro-based countries would be generating a lot of agricultural and agro-industrial residues every year. Thus use of these in textile industry sets up a viable option as far as sustainability is concerned. For e.g. India generates over 500 million tons of agricultural and agro-industrial residues every year, according to official data of the Ministry of New and Renewable Energy (MNRE).

Though cotton is a major fibre used in Asia, it is a known fact that most aspects of conventional textile manufacturing of cotton immensely damage the environment. Hence there is a growing need that consumers and manufacturers both want to shift to better materials. But often the choice between environment and economics has been difficult.

This is where agro-waste fabrics come into the picture. When crop waste serves as a raw material for textile manufacturing, both sectors become environment-friendly. Agrowaste fibres are being used in the composite industry replacing synthetics and cotton as these meet the factors of a sustainable product. The final product shows equivalent and sometimes higher performance in terms of strength properties. This new textile innovation is offering hope to textile and fashion industry especially to countries that heavily depend on unsustainable fibres and apparel manufacturing practices.

Agricultural residue or agro-residue describes all organic material produced as by-products after harvesting and processing agricultural crops. Agro-residues are non-wood lignocellulosic and a rich source of cellulose with lignin. These may include stalk, cane, seed pod, and leaves. Agro-residues are annually renewable and a low-cost source for natural cellulosic fibres.

Agro-residues are of two types:

Field residues: These are materials left in an agricultural field after the crop has been harvested. Examples include pineapple leaves, banana pseudostems and leaves, cornhusk, cotton seed pods, kapok seed pods, areca hulls, and cotton stalks.

Process residues: They are materials left after the crop is processed into a usable resource. After isolation of the primary or main agricultural product, a huge volume of residue is generated simultaneously. Examples of process residues are bagasse, wheat, and rice straw.

The biochemical composition of agro-residues varies owing to the composition of the crop from which they are sourced and the crop-specific climatic conditions they are grown in. The pure cellulose component of these agro-residues has the potential to be used in textile industry either as natural cellulosic fibre or as regenerated cellulosic. Though it is a well-acknowledged fact, it was hardly exploited owing to lack of dedicated research and technology in the past. Bagasse and bamboo are two important stalk agro-residues, and cornhusk and banana leaves are important leaf agro-residues. The cellulose and lignin composition of the four residues testifies that bagasse holds the highest amount of processable cellulose (57%) amongst all other residues [9].

Altmat is a company based in India dedicated in producing fabrics from agro residues. It started as an idea which brought researchers together and led to a pilot plant. Today it has emerged into an industry, manufacturing commercial fabrics with agrowastes as raw materials, that would be launched soon in the domestic market.

3 Banana Pseudostem a Rich Source of Fibre

History narrates of the complex textile tradition of banana fibre weaving in the Ryukyu Archipelago of Japan. Ryukyus spinners and weavers were engaged in producing textiles of cotton, silk, mixed silk and cotton, ramie and banana fibre on a large scale. They produced dyed cloth either in a brown made from yarns or deep blue from indigo [11].

The stem of the banana plant is usually thrown away once the fruit is harvested. The pseudostem forms a major waste material in large-scale banana plantations and disposal of these pseudostems is a major problem for the farmers. It is estimated that 1.5 million tonnes of banana fibre could be potentially extracted from 30 million tonnes of pseudostem waste produced annually each year across the country [35].

The pseudostem is a clustered cylindrical aggregation of leaf stalk bases. Banana fibres obtained from the pseudostem of banana plant are leaf fibres that are complex in structure (Figs. 1 and 2). They are generally lignocellulosic, consisting of helically wound cellulose microfibrils in amorphous matrix of lignin and hemicellulose [25]. These lignocellulosic by-products could be principal source for fibres, chemicals, and other industrial products [32].

The banana fibre, which otherwise goes waste could be a very good substitute to cotton and other natural fibres. Banana fibre textiles could lessen the demand for cotton to a large extent although unable to completely replace cotton.

The usefulness of banana wastes in manufacturing of bio products have been studied. The production of many essential food products, fertilizers, bio chemicals, paper etc. from banana waste (pseudostem) has been tried out [20]. The banana fibre has also used in the development of handicrafts and textiles, paper, composites, and packaging industries [43].

Industries based on banana fibre will certainly be a boom in waste utilization, which will result in more income to the farmers and more employment opportunities to the coming generations. In rural areas, women self-help groups can involve in fibre extraction, which will add credits to women empowerment programmes. There is much scope in utilizing the biowaste and setting up handicraft, paper, and textile industries.

3.1 Methods of Extraction

There are three methods used for extracting the fibre. Banana fibres are mainly extracted by physical methods (manual and mechanical). Retting the leaf sheaths using chemical, water or biological methods is also practiced but these have limitations. Quality of fibres depend significantly on the method of extraction [18].

3.1.1 Manual Extraction

When the leaf sheaths are still fresh and succulent, the pseudostems are cut into standard length and width. Then these pieces are placed on a wooden plank, and using a scraper which has a flat and blunt blade it is scraped and the strips of the leaf sheaths are drawn out. This is a tedious and time-consuming process and the fibre recovery is low. But it produces high quality fibre without much pith adherent to it [38].

3.1.2 Mechanical Extraction

Usually, plant fibres are extracted by the Raspador machines. The same machine is used for extracting banana fibres mechanically. A drum is fitted to a carbon steel angle blade. In front of the drum there is a scraping plate. The shaft carrying the drum is connected with ball bearings mounted on the framework. An electric motor is connected to help the drum to rotate with 700–800 rpm. Between the adjustable rollers and the rotating drum the leaf sheath is inserted to an extent of 3/4 of the leaf sheath and then drawn back. Then the other end of the sheath is inserted and the fibre from the whole sheath is extracted. The power Raspador machine had a serious drawback, that was suitable for hard tissues and not for soft herbaceous tissues like banana leaf sheaths. After decorticating to get fine quality fibres, hand extraction was needed. So further experiments were carried and a more user-friendly machine was developed by the Krishi Vigyan Kendra, Central Tobacco Institute Rajamundry, Andhra Pradesh. The new machine was called as Banana Fibre Extractor.

The Banana fibre extractor consists of a rigid frame and a rotating roller drum. The roller drum consists of few horizontal bars with blunt edges. A belt pulley mechanism is connected to the roller with one HP motor. This machine proved more

Fig. 1 Banana fibre

Fig. 2 Banana Pseudostem sheaths

advantageous as it reduced drudgery, it was user-friendly and economic. It gave high fibre recovery and produced good quality long fibre [38].

3.2 Properties of Banana Fibre Extracted from Pseudostem

Banana fibre has high strength, good lustre, light weight, good moisture absorption and is biodegradable. It is a multicellular fibre consisting of xylem and phloem, schlerenchyma and parenchyma cells. The fibre is arranged in the form of a helix at an angle of 11–12° [17]. Transmission electron microscopy of banana fibres confirmed layers structure of cell walls comprising of primary cell wall and secondary (S1, S2 & S3) wall layers [14].

The chemical composition of banana fibre as determined by elemental analysis is cellulose 31–35%, hemicellulose 14–17%, and lignin 15–16% [6]. The bio extracted and physically extracted fibres have almost similar composition except for slight variation in hemicelluloses, lignin, and ash contents which were lower for bio extracted fibre [18]. The composition also varies on the basis of the variety of banana plant. The alpha cellulose and hemicellulose contents decreased from the outer fraction to the core. Inner sheaths and core contained higher ash and extractive contents. SEM analysis of the fractions showed a compact arrangement for the outer and inner sheaths. In the middle sheath fibres, the non-fibrous material covered the bundle [29].

The lignin content of the banana fibres was 18.6% according to [14]. Lignin content of different varieties did not show significant variation although it ranged from 18.55 to 22.46% [35]. Banana fibre is more lignified than other soft fibres such as flax, ramie, and hemp [15].

The properties of banana fibre, pineapple leaf fibre, coir, wheat straw, barley straw, and rice straw were compared and resulted that pineapple and banana fibres had higher cellulose content 70–82% and 60–65% respectively [32] Highest amount of alpha cellulose was seen in PALF followed by banana fibre [14].

Cellulose content differed significantly between the varied genomic groups. AAB group had mean cellulose content of 22% whilst it was 27.2% in AB, 26.5% in AAA, 28% in BB, and 26% in ABB groups. The varieties differed from each other with respect to cellulose content. The cellulose content ranged between 55 and 58% which was similar to other plant fibres like jute and hemp [41]. However, in commercial varieties there was significant variation from a low of 53.17% in Poovan to a high of 55.57% in Nendran [35]. Amongst Grand Naine, Poovan, Nendran, and Monthan highest cellulose content was seen in Nendran fibres [30].

The length of the banana fibre varies from 89 to 110 cm. Its fineness is about 130 denier, moisture regain almost 8.63%, and tensile strength 40.5 gm/tex [36]. The tenacity of fibres was higher (46.36 and 54.92gm/tex) than cotton fibre [19]. The diameter of the fibre ranged between 80 and 250 μm with elongation percentage of 1.0–3.5 [32].

Mechanically extracted fibres were conditioned at 65% RH and 21^0C for a day and tested for diameter and tensile strength at various stain rates to analyse the variability and fracture behaviour. Results indicated that majority of fibres followed a normal distribution in diameter measurements. 25% of the fibres fell in the range 0.17–0.19 mm and 66% of the fibres fell in the range of 0.14–0.23 mm which is characteristic of all natural fibres. However, the standard deviation decreased with an increase in diameter which indicated that coarser fibres were more regular in nature [22, 24].

Studies of fibre composition and morphology have found that cellulose content and microfibril angle tend to control the mechanical properties of cellulosic fibres. It has been reported that the quality of matrix composed mainly of lignin and hemicellulose exerts strong influence on tenacity [23]. Hence different cultivars showed a variation in tenacity. The tenacity of plantain fibre (385–583 Mpa) was higher than that of banana fibre (380–650 Mpa) [2].

Tensile strength of Red banana fibre was high (535 Mpa) followed by Nendran (486 Mpa), Rasthaly (346 Mpa), Morris (252 Mpa), and Poovan (175 Mpa) [16].

The strength characteristics of banana fibre extracted from Nendran showed a higher tenacity (60.8 gm/tex) for fibres anaerobically retted as compared to enzymatically retted fibres (35.19 gm/tex). Thus the strength values compared favourably with that of other natural fibres like pineapple, sisal, and jute [35]. The tenacity of fibres were almost equal and higher for physically extracted fibre than the bioextracted fibre [18]. Natural retted fibre showed higher tenacity than microbial retted fibre [2].

Irrespective of the method of extraction, there was a decrease in tenacity with increase in the diameter of the fibre [18, 28]. It was observed that the period of bio extraction affected the quality of the banana fibre. There was significant loss of strength as the duration of bio extraction process increased when tested for fibres extracted in 6 and 12 days [18].

Highest breaking load and breaking extension were observed in Nendran peduncle fibres followed by Nendran pseudostem fibres. Nendran peduncle fibres showed higher tenacity values followed by Monthan peduncle fibres [30].

3.3   Fibre Yield

Though fibre can be extracted from all banana and plantain cultivars, the fibre yield showed great variability according to genomic status (AA, AAA, AB, AAB, ABB, AAABB, AAAB, ABBB, BB), ploidy level (2x, 3x, 4x) and the variety. The relative combinations of the A and B genomes contributed greatly to plant biomass production. In general, the B genome produced strong, bulky plants with greater biomass. Therefore, plants with a greater proportion of the B genome, for instance ABB, produced more biomass than their other triploid counterparts, such as AAA or AAB. Although it had been stated that fibre is extracted from the pseudostem, only the top 8–10 layers of the leaf sheath yield fibre. The fibre-extractable pseudostem and fibre yield percentages were found to be 46.4% of the total plant weight and 0.53% of the extractable pseudostem in dessert cultivars and 55.2% and 0.78%, respectively, in culinary types [40].

Fibre yield varied significantly with all the cultivars. The lowest value (0.26%) was recorded for Karpuravalli and the highest 0.88% was from the cultivar Pachanandan [41].

Methods of extraction also showed a variation in fibre yield. It was found that the fibre recovery percentage was highest for cold retting using 2% Sodium hydroxide. Hot retting using 1% Sodium hydroxide was on par with machine extraction, whilst lowest fibre recovery was recorded for hand extraction [38]. A fibre yield of 0.25% to 0.55% was obtained for fibres extracted by chemical retting.

A comparative study on fibres extracted mechanically from the peduncle and pseudostem of four banana cultivars viz. Grand Naine, Poovan, Nendran, and Monthan was done. The highest fibre recovery percent was obtained from the pseudostem of Poovan (2.71%) followed by pseudostem of Nendran (2.3%) and the lowest was recorded for peduncle of Nendran (0.283%) [30].

3.4   Dyeability of Banana Fibre

Dyeing is an ancient process practiced by men. It is done to add value and enhance the appearance of the fabric, dyeing is done by using either natural or synthetic dyes. Synthetic dyes are manufactured from petro chemicals and many of them are banned due to the presence of carcinogenic compounds. Germany was the first country to take initiative to put ban on numerous specific azo dyes for their manufacturing and applications. Though Netherlands, India, and a few more countries followed the ban even then the large number of synthetic dyes are in use.

Recently dyes and colours of natural origin are gaining worldwide recognition as substitute for synthetic dyes. This renewed interest in the use of natural dyes is primarily due to increase in health concern of many of the synthetic dyes because of their toxic nature and adverse environmental impacts.

The choice of dyes and dyeing methods for textiles depends upon the type of fabrics, their utilization, fastness properties, and the type of preparatory and finishing processes.

Investigations on the dyeing behaviour of banana fibres with natural dye extract obtained from turmeric plants were done. The fibres were scoured in 3% NaOH and mordanted using three different mordants Copper sulphate, Alum, and Ferrous sulphate separately. The fibre samples were then dyed with dye extract from turmeric using ultrasonic and conventional methods. The results showed that the ultrasonic method gave better dye uptake than the conventional method. Amongst the 3 mordants used, the sample mordanted with ferrous sulphate gave good colour yield. The fastness properties showed low light fastness for all the mordanted samples in both the methods of dyeing whereas wash fastness was high for all mordants. They recommended to try different mordants and dyeing conditions in order to improve the dyeability as the colours obtained by the use of curcuma plants were slightly lower than the industrial requirements for these mordants used in the study [8].

When Banana fibres that were mechanically extracted were dyed with 4 different natural dyes viz. Marigold, Lac, Onion, and Madder at 3 depth of shades (10, 20, and 30%) using 10% Alum as a mordant. The fibres were tested for the colour strength value and fastness properties. The dyes showed good affinity for the fibres as indicated by an increase in K/s values with increase in percent shade. The fibres showed stability towards photo degradation as the overall light fastness ranged between 5 and 7 for all dyes. The wash fastness also proved to be good to very good. Hence the banana fibres can be successfully dyed using natural dyes [39].

The dyeing behaviour of banana fibres with 3 different reactive dyes. The banana fibres were initially scoured with NaOH, H_2O_2 for 30 min, neutralized with acetic acid washed and dried at 60 °C. The fibres were dyed with the dye at three different concentrations 1, 2 and 3% owf. The colour strength and fastness properties were recorded. Alkali treated fibres showed an increase in roughness due to fibrillations and exposing the inner fibrillar surface as compared to the SEM micrographs of raw banana fibres. The colour strength was found to be directly proportional to the concentration of the dye. Wash fastness was very good showing values 4–5 for all tested fibres. Alkali pretreated banana fibres showed excellent dyeability for all the dyes, owing to the dye structure and concentration [27].

The dyeability of Jute Banana blended fabric using sulphur and reactive dyes (15% owf) was studied. The colour strength was higher for Reactive green followed by Sulphur blue. The wash, perspiration and light fastness was excellent to good for both the dyes. The rub fastness was also good to excellent except for Sulphur blue which gave a grade value of 3–4 [3].

3.5 Weaving of Banana Fabric

Banana fibre can be effectively used in making utility items and handicrafts. It is used to prepare ropes, bags, mats, footwear, purses, and other household items.

From the thirteenth century onwards, in Japan, bananas have been cultivated for clothing and household uses. Soft fibres were used for making kimonos and kamishimos.

In the Philippines, it is woven into a thin, transparent fabric which is the sole material in some regions for men's shirts. The banana fibres are reported to be elegant and highly versatile. It is also used to make wedding gowns, barrons, and handkerchiefs.

In Sri Lanka, it is fashioned into soles for expensive shoes and used for floor coverings. In 2007, LIFEI developed banana fibre series yarns, banana fibre series fabric, and banana fibre clothing. These fibres are used in Europe for making socks [38].

A group of weavers in Anakaputhur have been weaving attractive pieces of cloth from banana fibre. The fibres were cleaned by a bleaching process and fat content removed. Single strands of the fibre were removed and woven into fabrics. The fibre had affinity for colours so attractive designs were woven.

At the Banana Research Station in Kerala, India weaving trials were done using untreated Banana fibre to make a sari using banana fibres in the weft (each strand tied using weaver's knot) and cotton yarns in the warp on a handloom [38].

Treated banana Nendran fibres were made into twisted filament yarn of adequate strength (Fig. 3). Weaving trials were done with treated banana Nendran filament fibres in the weft and silk warps. Owing to the length of the nendran fibres, they were used as single filament fibres with silk in warp producing union fabrics. The fabric had a good sheen and had good strength properties but was stiff (Fig. 4).

3.6 Applications in Composite Industry

Composites are materials that comprise strong load carrying material known as reinforcement imbedded in weaker materials known as matrix. Reinforcement provides strength and rigidity, helping to support structural load. The matrix or binder (organic

Fig. 3 Banana
twisted filament yarn

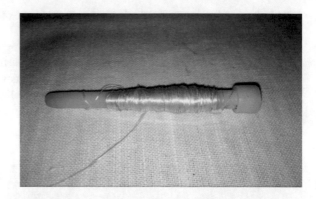

Fig. 4 Silk banana union
fabric

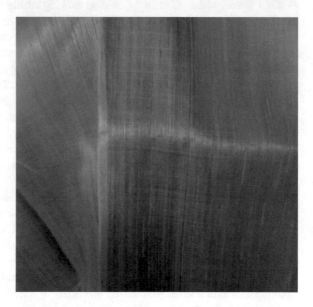

or inorganic) maintains the position orientation of the reinforcement and transfers the external load to the reinforcement.

Polymer matrix composites commercially produced composites in which resin is used as matrix with different reinforcing materials. Ecological concerns have evolved in a renewed interest in natural and biodegradable materials and therefore issues such as, sustainability and environmental safety are becoming important. Thus material components such as natural fibres, biodegradable polymers are considered as environmentally safe alternatives for the development of new biodegradable composites.

The interest in natural fibre reinforced polymer composites is rapidly growing both in terms of their industrial applications and fundamental research. Many lignocellulosic fibres like banana, sisal, bamboo, hemp are more often used as the reinforcement of composites. Considerable attention is being paid to banana fibre, taking into account the wide availability and the fact that it is a crop waste.

Banana fibre possesses good specific strength properties comparable to those of conventional materials like glass fibres. It has lower density than glass fibre. Banana and glass fibre reinforced nanochitosan composites were developed and studied their mechanical strength properties. These composites were prepared in two forms for both glass and banana fibre i.e. short fibres and long fibres. Results indicated that glass fibre reinforced composites had more crystallinity, showed better strength and higher dispersion than the banana fibre composites. Banana fibre composites had higher thermal stability due to high conductivity when compared with the glass fibre composite [21].

Alkali treatments have proven effective in removing impurities from the fibre, decreasing moisture absorption, and enabling mechanical bonding and thereby improving matrix reinforcement interaction [31].

The use of banana fibre as reinforcing agent in cement and polymer-based composites has been studied for their properties and structure. Due to low density, high tensile strength, high tensile modulus, and low elongation at break of banana fibres, composites based on these fibres have very good potential use in the various sectors like construction, automotive, machinery etc. [42].

The tensile strength of three varieties of banana cultivar on the influence of alcohol soaking treatment and epoxy resin were compared. Composite specimen was prepared using banana fibre and resin in volume fraction of 50:50 and alcohol in 70, 95 and 96% soaking treatment was carried out. The samples were tested for tenacity and elongation and compared amongst the three varieties used and also with the untreated sample. In the untreated sample variety Ambon showed the highest tensile stress of 2429 kgf/mm^2 and strain of 3%. Alcohol soaking treatment made the fibres brittle and fragile. Epoxy resin when used with the banana fibre in the ratio 50:50, increased the tensile strength and made the composite stronger [33].

4 Areca Nut Fibre

Areca nut is the fruit of betel nut palm or areca nut palm. The scientific name of Areca nut is Areca catechu. Areca is a variant of a Tamil word meaning a cluster of nuts. The original country of this palm is not known with certainty, it is cultivated along the coastal belt up to approximately 300 km from sea. The trunk is straight up to 30 m high and 15 cm thick. Leaves are similar to coconut tree 1–1 1/2 m long. The fruit bearing starts after 5 years of planting. Nuts are harvested when they are three quarters ripe. The number of harvests varies from three to five in one year depending upon the season and place of cultivation. The fruit is more or less spherical 3–5 cm across, smooth, reddish. The tree is cultivated for its nuts. The fruit is an important article of commerce in India. It is used either by itself or mixed with tobacco wrapped leaves of the betel wine for chewing all over the country. The wood of the tree is valuable as it is hard. Leaves are used for thatching [34].

Areca Catechu husk fibres are agro-waste fibres. These husk fibres from the nut shell have limited applications. Karnataka is the India's largest Areca nut producing state which has a share of around 50% Areca production in the country. The Areca nut is covered with a husk which constitutes 60–80% of the total volume and weight of the fruits (fresh weight basis). Not much applications are available for these Areca husk fibres. However, a part of these fibres are used as fuel and most of them are left as a waste in landfills and are difficult to manage. Several processes have been developed for utilization of Areca husk for making hard boards, plastic, and brown wrapping paper. Areca husk is used as a substrate for mushroom cultivation too. Amongst all the natural fibres Areca nut fibres, a type of nut shell fibres, are more promising because it is inexpensive, abundantly available, and very high potential perennial crop but have limited applications.

Areca nut fibres are hypoallergenic and have inherent antimicrobial characteristics. The average filament length is around 4 cms. which is too short compared to other bio-fibres. Mainly 2 types of fibres are present—one very coarse and the other very fine. The coarser ones are ten times coarser than jute. The portions of the middle layer below the outermost layer are soft fibres, which are very similar to the jute fibres. Areca husk fibres are predominantly composed of hemicelluloses and not of cellulose [13]. Areca fibres contain 13–24.6% of lignin, 35–64.8% of hemicelluloses, 4.4% of ash content and remaining 8–25% of water content. The Areca nut fibres can be processed to develop new applications for these fibres [5].

4.1   Properties of Areca Nut Fibre

The properties of the Areca Fibres include: Fibre length seems to decrease with an increase in concentration of NaOH. According to the classification of cotton fibres by their staple length defined by USDA, the fibres >34.9 mm can be designated as extra-long. Large diameter of Areca fibres are crisp, rough and stiff and the fibre will be finer if the diameter is small. Fine Areca fibres are soft and pliable and fabric made of them drape more easily.

The linear density of the treated Areca nut fibre seems to be decreased as the concentration of the alkali increased. This can be attributed to the fact that alkali treatment helps in the removal of hemicelluloses and lignin thereby losing their cementing capacity and making them finer.

With the partial removal of lignin from the natural fibre surface, more cellulose components of natural fibres are exposed. This was evident in the increase of cellulose in areca nut fibre after alkali treatment.

Fine areca fibre has good physical properties in terms of length, water absorption, maturity which make it suitable for spinning. However, in case of strength and fineness it showed ordinary value compared to cotton fibre. Areca/cotton 50:50 blend 48 tex ring and rotor yarn has been produced and yarns showed tenacity 7.1cN/tex and 6.88 cN/tex accompanied by 6.81% and 6.35% elongation, respectively. The water absorption of areca/cotton blend yarn showed 4.8 times higher value than 100% cotton yarn. [4].

The blended cotton and areca nut fibre yarn gives high strength and elongation compared with coconut fibre. These types of yarn can be used for technical textile applications and home textiles in the future []. New applications of agro waste fibres, like Areca nut husk fibres, will contribute to better management of agro-waste and facilitate sustainability.

4.2   Method of Extraction

The fibres are packed strongly with hemicelluloses, lignin and with slight deposition of wax and inorganic elements. In early days to extract fibre, areca husk was soaked in water about 2–3 days, this wet areca husk is exposed to sunlight for drying. The fibres were soaked in NaOH and these fibres were further washed with water containing few drops of acetic acid. Finally, the fibres were washed again with water and dried. By this process slight degumming takes place which makes areca husk soft and loosens fibre. These loosened fibres are extracted manually by hand. This way of extracting fibre from areca husk is time consuming and cannot be adopted in producing fibre in large scale. Also this method requires more labour and hard work.

In traditional method areca fibre is extracted by retting process. Retting process is employing the action of moisture and bacteria on areca husk to dissolve or rot away much of cellular tissues and other substances, thus facilitating separation of the fibre from areca husk. Basic methods include dew retting and water retting [26].

4.3   Dyeability of Areca Fibre

Absorbency of soft areca fibre is more compared to hard areca fibres and Untreated fibre also absorbs equally or more than treated fibres. It was found that adsorption is highly dependent on the contact time, adsorbent dose, and dye concentration. Being a plant based-fibre, the dyeability would be similar to other cellulosic fibres. More dyeability studies need to be carried out to understand its dyeing behaviour.

4.4   Mechanical Properties

Areca nut fibres have been treated with *P. chrysosporium* and *Phanerochaete* sp. and observed tensile strengths of fibres were 125.2 Nm m^{-2} and 116.5 Nm m^{-2} respectively and untreated fibre had a tensile strength of 92.7 Nm m^{-2}, this shows the versatility of biosoftened fibre procured after the treatment with both the living beings was practically identical. The fibre treated with *P. chrysosporium* demonstrated 35.1% and *Phanerochaete* sp. indicated 25.7% expansion in quality when contrasted with the untreated fibre. Lignin is the part, which grants weakness to the fibre. Incomplete evacuation of lignin causes alternate parts of the filaments like cellulose to wind up noticeably more conservative and along these lines expands the quality and adaptability of the areca nut fibre.

Some of the researchers had utilized KOH (Potassium Permanganate) treated filaments in the composites, these composites have shown more ductility than the untreated filament composites and also alkali treated strands have shown good elasticity properties [1].

The mechanical properties of tensile, flexural and impact test of treated areca nut fibre composites are more compared to untreated areca nut fibre composites. In compression and hardness test the untreated areca nut fibre composites have shown better results than compare to treated areca nut fibre composites [37].

The mechanical properties like strength and elongation of chemically treated Areca nut fibres showed better results when compared to natural untreated fibres. The lignin percentages dropped following alkali treatment, yielding a higher proportion of cellulose in alkali—treated Areca nut fibres. The fibre length tends to decrease with increase in concentration of sodium hydroxide. A negative increase in fibre diameter with alkali treatment was observed. Fibre treated with 0.5% NaOH showed superior physical and mechanical properties than other treated fibre samples. Cotton rich alkali treated yarn samples show superior mechanical properties than untreated Areca nut fibre rich yarn samples. Cotton rich untreated yarn samples have greater absorbency as compared to treated yarn samples. Treated yarn sample with 30% cotton and 70% Areca nut fibre proved to be the best amongst all the yarn samples.

4.5   Applications of Areca Fibre

Areca fibre has been used in composite industry lately. Areca fibre and maize powder reinforced PF composite materials can be used in packing industries, low-cost housing, and domestic purposes and can be used as a commutative material for plywood. A structural and non-structural application such as suitcases, post-boxes, grain storage, automobile interiors, partition boards, and indoor applications untreated chopped natural areca sheath fibre reinforced polymer matrix bio-composites are best suited. [1].

Spinning and weaving trials have also been carried out though at an experimental level. Spinning trials were carried out using open end spinning with fine inner areca fibres and cotton fibres in three ratios (50:50, 30:70, and 70:30). Weaving trials are at a latent stage. However greater research can help in developing fine breathable fabrics with excellent properties.

5 Conclusion

Streamlining new and novel fibres that are sustainable along with the major fibres will ensure high acceptance amongst the consumers as they are more environment conscious. Value addition of these fibres will also help in replacing the harmful fibres. India being one of the largest agro-based industries and second largest Textile industry, should search for viable options for raw materials for textiles that are eco-friendly. Agrowaste which causes landfills or high carbon emissions through burning them down has a negative impact on the environment. These waste being rich in cellulose must be utilized in all application types.

These agrowaste fibres extend a promising future ensuring sustainability, circularity, and building economy of the nation.

Popularizing it will help rural development, generation of revenue and a circular management system which reduces the impact on the environment.

References

1. Ashok RB, Srinivasa CV, Basavaraju B (2018) A review on the mechanical properties of areca fiber reinforced composites. Sci Technol Mater 30(2):120–130
2. Akubueze EU, Ezeanyanaso CS, Orekoya EO, Ajani SA, Obasa AA, Akinboade DA, Oni F, Muniru OS, Affo G, Igwe CC (2015) Extraction and production of agro sack from banana (musa sapientum) and plantain (musa paradisiaca) fibres for packing agricultural produce. Int J Agric Crop Sci 9(1):9–14
3. Balan V (2011) Development of preparatory processes for making of banana fibre blended fabrics and their evaluation. Doctoral Dissertation, S.N.D.T Womens' University, Mumbai
4. Begum HA, Saha SK, Siddique AB, Stegmaier T (2019) Investigation on the spinability of fine areca fiber, 2019
5. BhanuRekha V, Ramachandralu K, Vishak S (2015) Areca Catechu Husk fibers and polypropylene blended nonwovens for medical textiles. Int J PharmTech Res 8(4):521–522
6. Bilba K, Arsene MA, Ouensanga A (2007) Study of banana and coconut fibres: botanical composition, thermal degradation and textural observations. Biores Technol 98(1):58–68
7. Blackburn RS (2005) Biodegradable and sustainable fibers, xx, xxi
8. Canbolat S, Merdan N, Dayioglu H, Kocak D (2015) Investigation of the dyeability behavior of Banana fibres with Natural dye extract obtained from Turmeric plants. Marmara J Pure Appl Sci 1:40–44
9. Channa B, Parmar MS, Sachdeva PK (2016)Agro-residues: beyond waste, potential fibres for textile industry-FibretoFashion, Oct 2016
10. Dhinakaran M, Govidan R (2005) Areca nut fibre and cotton blend well for technical textile application. Textile Asia 46(4):29–32
11. Hendrickx K (2007) The origin of banana—fibre cloth in the Ryukus, Japan, Leuven University Press, Leuven, Belgium, p 336
12. http://ecology.org/biod/library/glos_index.html
13. Keerthi A, Shaik I, Mendonca MR, Keerthan KS, Pavana Kumara B (2015) Processing and characterization of epoxy composite with arecanut and casuarina fibers. Am J Mater Sci 5(3):97
14. Khalil HPSA, Siti Alwani M, Omar AKM (2006) Chemical composition, anatomy, lignin distribution, and cell wall structure of Malaysian plant waste fibres. Bioresource 1(2):220–232
15. Kirby RH (1963) Vegetable fibres: World Crop Series. London Hill, London, and Interscience, Newyork, U.S.A.

16. Kiruthika AV, Veluraja K (2009) Experimental studies on the Physico chemical properties of Banana fibre from various varieties. Fibres Polym 10(2):193–199
17. Kulkarni AG, Sathyanarayana KG, Rohatgi PK, Vijayan K (1983) Mechanical properties of Banana fibres (Musa sapientum). J Mater Sci 18:2290–2296
18. Manilal VB, Sony J (2011) Banana pseudostem characteristion and its fibre property evaluation on physical and bio-extraction. J Nat Fibres 8:1–12
19. Mishra V, Goel A (1999) Softening Banana fibres. Indian Text J 110(11):34–35
20. Mohiuddin AKM, Saha MK, Hussain MS, Ferdoushi A (2014) Usefulness of Banana (Musa paradisiaca) wastes in manufacturing of bio products: a review. Agriculturist 12(1):148–158
21. Mubashirunnisa A, Vijayalakshmi K, Gomathi T, Sudha PN (2012) Development of banana/glass short hybrid fibre reinforced polymer composites. Pharm Lett 4(4):1162–1168
22. Mukherjee PS, Sathyanarayana KG (1984) Structure and properties of some Vegetable fibres. J Mater Sci 19:3925–3934
23. Mukherjee PS, Sathyanarayana KG (1986) An empirical evaluation of structure—property relationships in natural fibres and their fracture behavior. J Mater Sci 21:4162–4168
24. Mukhopadhyay D, Fangueiro R, Arpac Y, Sentrik U (2008) Banana fibres—variability and fracture behaviour. J Eng Fibres Fabr 3(2):139–145
25. Mukhopadhyay S, Vijay G, Talwade R, Dhake JD, Pegoretti A (2006) Some studies on banana fibres. In: International Conference on advances in fibrous materials, Nonwovens and Technical Textiles.
26. Naik K, Swamy RP, Naik P (2014) Design and fabrication of areca fiber extraction machine. Int J Emerg Technol Adv Eng 4 (7):860–861
27. Oliveria L, Cordeiro N, Evtuguin DV, Torres IC, Silvester AJD (2007) Chemically composition of different morphological parts from "Dwarf cavendish Banana plant and their potential as a nonwood renewable sources of natural products." Ind Crops Prod 28:163–172
28. Osorio JCM, Baracaldo R, Elorez JJO (2012) The influence of alkali treatment on banana fibre's mechanical properties. Ing E Invest 32(1):83–87
29. Pereria ALS, Nascimento DM, Souza MDM, Cassales AR, Morais JPS, Paula RCM, Rosa MF, Feitosa JPA (2014) Banana (Musa sp.Cv. Pacovan) Pseudostem fibres are composed of varying lignocellulosic composition throughout the diameter. Bioresource 9(4):7749–7763
30. Preethi P, Murthy BG (2015) Physical and chemical properties of banana fibre extracted from commercial banana cultivars grown in Tamilnadu State. Agrotechnology SII:008. https://doi.org/10.4172/2168-9881.811-008
31. Raghavendra S, Lingaraju, Shetty PB, Mukunda PG (2015) Mechanical properties of short banana fibre reinforced natural rubber composites. Int J Innov Res Sci Eng Technol 2(5):1652–1655
32. Reddy N, Yang Y (2005) Biofibres from agricultural byproducts for Industrial applications. Trends Biotechnol 23(1):22–27
33. Sahari GNA, Buku A (2015) Textile strength of fibre for some type bananas (Ambon, Kepok, Susu). Int J Res Eng Technol 4(8):183–196
34. Santapan H (1966) Common trees (India—the land and people: National Book Trust, India,). New Delhi, pp106–107
35. Shivashankar S, Nachane RP, Kalpana S (2006) Composition and properties of fibre extracted from pseudostem of banana (Musa sp.). J Horticult Sci 1(2):95–98
36. Shroff A, Karolia A, Shah J (2015) Biosoftening of banana fibre for nonwoven application. Int J Sci Res 4(4):1524–1527
37. Somashekhara J, Ramesh BT, Belagavi V, Madhu HT (2018) Investigation and study of mechanical properties of areca shell fiber and palm powder natural composites. J Mech Civ Eng, 62–73
38. Suma A (2009) Banana fibre and utilization. Kerala Agricultural University, Thrissur, p 27p
39. Teli MD, Sanket PV, Mahajan JS (2015) Dyeing of banana fibre with Marigold, Lac, Madder and Onion dyes. J Text Assoc 75(5):341–344
40. Uma S, Kalpana S, Sathiamoorthy S (2003) Banana fibre. National Research Centre for Banana, Tiruchirapalli, India

41. Uma S, Kalpana S, Sathyamoorthy S, Kumar V (2005) Evaluation of commercial cultivars of Banana (Musa spp.) for their suitability for fibre industry. Plant Genetic Res Newslett 142:29–35
42. Venkateshwaran N, Elayaperumal A (2010) Banana fibre reinforced polymer composites—a review. J Reinf Plast Compos 29(15):2387–2396
43. Vigneswaran C, Pavithra V, Gayathri V, Mythili K (2015) Banana fibre: scope and value added product development. J Text Apparel Technol Manage 9(2):1–7
44. Vink ETH, Rabago KR, Glassner DA, Gruber PR (2003) Polym Degradation Stab 80(3):403

Development of Union Fabrics from Lotus Petiole Waste

Madhu Sharan and Sumi Haldar

Abstract Lotus (Nelumbo Nucifera Gaertn.) is a potential aquatic cash crop grown and consumed across the globe. The flower has a huge economic importance across the India and used as offerings for god and goddess. Researcher started the study right from the base that is finding out the wetlands near and around Vadodara region of Gujarat, contacting the people associated with the marketing and cultivation of Lotus. It was observed that lotus cultivators generally cut the flowers with little length of the stem (petiole) leaving the rest entire as a "waste" and abandoned for a natural decay. In the present study, the effective utilization of this aquatic biomass was studied and explored in the field of textiles. 300 g of fibers were extracted manually to prepare yarns using the traditional spinning system. Union fabrics were prepared on handloom using cotton and silk as a warp and lotus as a weft. The properties of these fabrics were compared with 100% cotton and silk fabrics. Thickness, GSM, Tensile Strength, Fabric Count, Crease Recovery, Bending length, and Shrinkage test was done as per ASTM and ISO test standards. Results revealed lotus fiber has a great potential to be used in developing soft and light weight woven fabrics. Product like stole was prepared using lotus yarn by extra-weft weaving technique.

Keywords Lotus · Petiole · Biomass · Textile fiber · Spinning · Yarn · Union fabric · Weaving · Extra-weft

1 Introduction

Textile fiber is an integral part of human civilization. These textile fibers and their use predate the recorded history of textiles. Civilization has taught us to make clothes using common natural fibers such as cotton, wool, silk, and linen [1]. From the very beginning of human history, they have served to protect people from the adverse elements of the environment in perfect comfort [2]. Natural fibers are a renewable

M. Sharan (✉) · S. Haldar
Department of Clothing and Textiles, Faculty of Family and Community Sciences, The Maharaja Sayajirao University of Baroda, Vadodara, Gujarat, India
e-mail: madhusharan@yahoo.co.in

© The Author(s), under exclusive license to Springer Nature Singapore Pte Ltd. 2022   101
S. S. Muthu (ed.), *Sustainable Approaches in Textiles and Fashion*, Sustainable Textiles:
Production, Processing, Manufacturing & Chemistry,
https://doi.org/10.1007/978-981-19-0878-1_5

resource, thus providing a better sustainable sourcing solution, as they are low cost, low density, lower processing costs, no health risks, and better mechanical properties. The most important properties of natural fibers are their biodegradability and non-carcinogenic properties, making it fashionable again, with the advantage of being economical [1].

There are global concerns about the use of non-renewable materials in manufacturing and increasing environmental legislation. There is pressure, both from consumers and regulators, for products that are more environmentally friendly and to reduce dependence on fossil fuels. As a result, the use of natural fibers in products will continue to grow as major contributions to the bio-based economy. The transition to a biased economy and sustainable development as a result of the Kyoto protocols on greenhouse gas reduction and CO_2 neutral production offers good prospects for the natural fiber market. A fundamental change in attitude is needed for the development of a sustainable world economy that enables improved purchasing power and living standards without depleting resources for future generations. For ecological reasons, products based on photosynthetic CO_2 fixation should be prioritized. The advantage of these sustainable resources is that they can grow into a foreseeable future without adversely affecting global biodiversity. Therefore, we need to develop competitive products based on renewable raw materials. It is of high quality, has excellent technical performance, and has less environmental impact than current products based on petrochemical materials [3].

Cellulose is one of the major resources available on earth. Minor cellulosic fibers are the fibers which are the agricultural by-product (biomass), less explored, seasonal, and suitably grown in some region with the particular soil and climate requirement and need some additional finishing to be used successfully in textiles. Some of the examples of cellulosic minor fibers are Banana, Sisal, Ramie, Pineapple Abaca, Hemp, Coir, Okra, Nettle, Kenaf, and Roselle which are grown in the terrestrial regions. Apart from the terrestrial plants a significant number of wetland plants are considered as bio-resources. There are major and minor plant resources harvested from the wetlands of rural India [4]. One such aquatic plant is Lotus (Nelumbo nucifera) comes under the family Nelumbonaceac, beautiful flowering aquatic plant has been honored in the history of three countries—China, India, and Egypt [5]. In India, it occurs from Kashmir in north to Kanyakumari in South showing huge phenotypic diversity with different shapes, sizes, shades and it is a national flower of India and Vietnam [6]. Lotus plants require a humid tropical climate, wetland habitats, and floodplains, ponds, lakes, lagoons, swamps, and backwaters of reservoirs. The plant grows 1.5 m high and up to 3 m horizontally. Its roots stay firmly on the muddy bottom, and leaves up to 60 cm in diameter slide on the surface of the water. Flowers can reach a diameter of up to 8 inches. The seeds are about 1 cm in diameter and are placed in a wood container like a shower head. Lotus plants grow by spreading rhizomes that creep to the bottom of the body of water over anaerobic sediments. Rhizome carries the node that produces the leaves [7].

The entire lotus plant endowed with numerous medicinal properties. Pharma-cological studies of the plant revealed that the whole plant possesses antidia-betic, antipyretic, anti-inflammatory, anticancerous, antimicrobial, antiviral, and anti-obesity properties [8]. The cultivated lotus is generally divided into three cate-gories, namely rhizome, seed, and flower, according to their usage in reality. The rhizomes and seeds are consumed as a vegetable and lotus flower is mainly applied in ornamentation and environmental improvement [9].

In India, lotus flower has a huge economic importance in the religious functions like saraswati puja, bishwakarma puja, durga puja as well as in daily worshipping. Cultivators of lotus generally cut the flower with little length of petiole (a part between flower and rhizomes) leaving the rest entire as a "waste" and abandoned for a natural decay. Such waste may cause environmental hazards shown in Fig. 3 [10]. Petiole can grow up to the height of 6 m, depending upon the depth of the water. The color of the petiole varies from light green to dark green depending upon the water quality and maturation. The stem of a lotus plant has two types of xylem cells, the tracheids and vessels. The tracheids are an elongated tubular cell with a hard, thick, woody wall, and a large cell cavity. The vessels have a cylindrical tubular structure with thin lignified wall.

These cells help in transportation of water and some nutrients from the roots to the leaves and also provide mechanical support to the plant. Lotus fibers are arranged in these cells in the form of "Helix" shown in Fig. 2 [11].

Lotus plant has symbol of religion and spiritual power in history of Myanmar. Ancient Myanmar Kings used lotus flower, bud, stems, and petals in wall painting and statue. All parts of lotus plant are useful, stem for producing lotus fiber, flower for offering to Buddha, leaves for packing food, meat etc. Lotus leaves are also used as plate to decorate food. Lotus root and seed are also edible. Lotus fiber has been extracted from lotus stalks since 1910. Fibers from the lotus flower of the lake in Myanmar are hand-spun and woven into silky fabrics within 24 h. The whole process is tedious and time consuming. In the late 1990s, Japan set up a workshop to create an overseas market for handmade textiles, but the fabric was extraordinary [12].

In India lotus farming is practiced nearly in all states. Lotus cultivators used to sell flowers, seeds, fruits, rhizomes. But they are less aware about the petioles which are generally thrown away while cutting and bundling the flower and sometimes also left in the pond during picking the flower. As per the review the manual extraction process followed in Myanmar is—first the fibers are initially spun by palm twisting and again it is taken on the hand charkha. In the present study, the researcher has tried to extract the fiber from the lotus petiole and directly winded on the pern without the palm twisting and later subjected for final spinning on *Peti charkha*.

Some wetlands in central Gujarat, such as Ahmedabad, Kheda, Anand, and Vado-dara, and southern Gujarat (Valsad and Navsari) are reported to show remarkable growth of lotus plants [13]. In recent years the fabrics of different blends and union fabrics are great demand in market for various end uses [14]. Union fabrics are where the fiber content of the warp is different from that of weft [15]. The main aim of the study was to prepare union fabrics from 100% lotus hand spun yarn as a weft and cotton as a warp and other was 100% lotus hand spun yarn as a weft and silk as

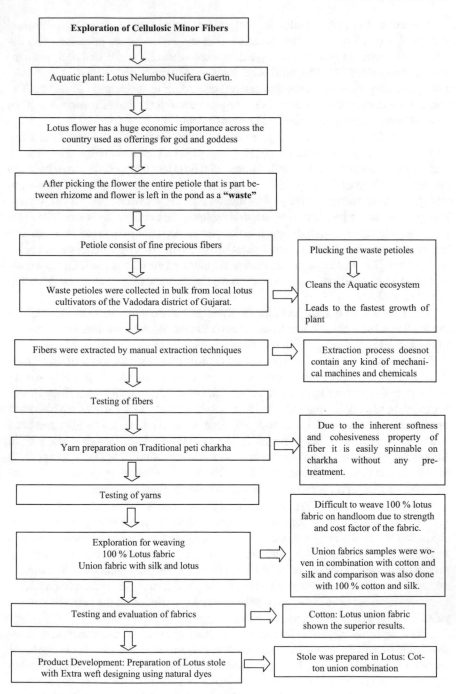

Fig. 1 Flowchart of the work from pond—fiber, fiber-yarn, yarn-fabric and fabric to fashion

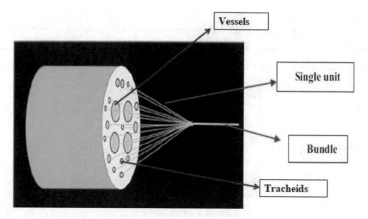

Fig. 2 Arrangement of fibers in Petioles

a warp. To compare the properties 100% cotton and silk warp and weft was also prepared. Testing results of all the samples were evaluated from which the Lotus: Cotton have shown a superior result. Later the lotus stole was woven with 100% lotus as a weft and 100% cotton as a warp with extra-weft weaving techniques.

1.1 Collection of Lotus Petioles

Petioles were collected from the local lotus cultivators—Mr. Poonam bhai Parmar and Mr. Isabbhai Rathod from the Vadodara district, Gujarat.

1.2 Extraction of Fiber

300 g of fibers were extracted manually by researcher. For the Extraction, the petioles were washed and wiped. It was not possible to extract fibers from single/two petioles so, a bunch of three petioles was taken, slit at 5–7 sections as per its length with the help of sharp knife. After one slit, the bunch of three petioles were slowly stretched and fibers were laid on the wooden slab. End to end points were joined by palm twisting and therefore, the yarns were wounded on the bobbin. The physical properties of the extracted fiber are shown in Table 1.

Table 1 Physical properties of lotus fiber

Sr. No	Test	Results	Standards
1	Fiber length	60–105 cm	–
2	Fiber diameter	2–6 μm	–
3	Length to breadth ratio	269,000:1	–
4	Fineness	Denier—32 Tex—4 Cotton count—166 s	ASTM-D 7025-09 [24]
5	Fiber strength	Maximum load (gf)—161.52 Extension at maximum (mm)—1.5589 Stress—5.0476 Strain (%)—1.5589	ASTM-D 3822-07 [25]
6	Moisture regain	11.8%	ASTM D 2495-07 [26]

Table 2 Physical properties of yarn used for the construction of union fabrics

Sr. No	Physical properties	100% cotton	100% silk	100% lotus
1	Yarn count (ASTM-D 885) [27]	2/80'S	102.21'S	50'S
2	Average twist (TPI) (ASTM-D 885) [27]	45	0.91	1.66
3	Twist direction	S	S	S
4	Maximum load (gf) (ASTM-D 885) [27]	233	192	288
5	Extension at max (mm)	3.0	7.8	1.7
6	Stress (gm/den)	0.96	3.6	2.66
7	Strain (%)	3.0	7.8	1.7

1.3 Preparation of Yarn

100% Lotus hand spun yarn was prepared by using traditional spinning system on Peti Charkha. The spinning of lotus yarn was done by Mr. Bakul Shah, member, Sardar Bhavan Trust, Vadodara. Another yarns used for making union fabrics were cotton and silk which were procured from Mr. Pachan Premji, Artisian Designer, and weaver Bhujodi Kutch, Gujarat. The physical properties of all the yarns i.e. Lotus, Cotton, and silk are given in Table 2.

1.4 Constructional Details of Union Fabrics

In the present study, union fabrics were woven on handloom by Mr. Pachan Premji, Artisian designer, and weaver from Bhujodi Kutch, Gujarat. The specification of handloom for the union fabric is mentioned in Table 3. Constructional details of the union fabric are given in Table 4.

Table 3 Specification of Hand loom for the construction of union fabrics

Type of loom	Treadle loom
Number of shaft	4
Loom width	34 inches
Reed number (for cotton warp)	24
Reed number (for silk warp)	18

Table 4 Constructional details of union fabrics

Union fabrics	Fabric code	Weave type	Fiber content	
			Warp	Weft
100% cotton	C_1	Plain	Cotton	Cotton
Lotus: cotton	LC_2	Plain	Cotton	Lotus
100% silk	S_3	Plain	Silk	Silk
Lotus: silk	LS_3	Plain	Silk	Lotus

The union fabrics were tested for important mechanical properties like fabric count, fabric thickness, GSM, tensile strength, stiffness test, crease recovery, shrinkage, and Cover factor (K). Methods with standards for each test are as follows:

1.4.1 Fabric Count

Fabric count in woven textile material is the number of ends and picks per unit area. Following ASTM D 3775–98 [16], the number of warp and weft yarns in one square inch of the fabrics is counted with the help of pick glass at five randomly selected places across the width and along the length of the test specimens. The region near the selvedge should be avoided because the spacing of the thread is often a little different than the body of the cloth [17].

1.4.2 Cover Factor

Cover factor is a numerical value indicating the extent to which area of the fabric is covered by the component yarn [22]. Cover factor was calculated with the following formula:

$$\text{Fabric Cover factor}\,(K) = K_1 + K_2 - \frac{K_1 K_2}{28}$$

where,

$$K1 = \frac{EPI}{\sqrt{Warpcount}}$$

$$K2 = \frac{PPI}{\sqrt{Weftcount}}$$

1.4.3 Fabric Thickness

Thickness is the distance between the upper and lower surface of the material measured under the specified pressure, expressed in mm. The specimen chosen were free from folds, crushing or distortion, and wrinkles [17]. Specimen was tested on Universal thickness tester. The specimens were tested as directed in ASTM D 1777-96 [18].

1.4.4 GSM

Cloth weight is expressed as mass per unit area in g/sq.mt. A sample of 5 × 5 cm was cut and weighted on an electronic weighing balance to determine the weight per sq.m [17]. GSM was calculated using the formula

$$GSM = \frac{\text{Weight in grams of sample} \times 100 \times 100}{5 \times 5}$$

1.4.5 Tensile Strength

Tensile strength of the fabrics was tested on Universal tensile tester (UTM) by ravelled strip test method using ASTM D 5035-95 [19]. Specimen size of 150 mm × 25 mm (15 cm × 2.5 cm) was cut. Gauge length was kept 75 ± 1 mm (7.5 cm) with the speed of 300 mm/min.

1.4.6 Stiffness Test/Bending Length

Fabric stiffness is the resistance of the fabric to bending. The samples were tested as per ASTM D 1388-18 [20] using Shirley's stiffness tester. Rectangular strip of fabric, 6 inch × 1 inch was mounted on a horizontal platform in such a way that it hangs as a cantilever and bends downwards. The strip of the fabric was started to droop over the edge of the platform and the movement of the templates and the fabric was continued until the tip of the specimen viewed in the mirror cuts both index lines. Bending length was read off from the scale mark opposite to a zero line engraved on the side of the platform [21].

1.4.7   Crease Recovery

Crease recovery is the allowance of the fabric to recover from crease. Samples were tested as per IS method: 4681-1968 [23] by using Shirley's crease recovery tester. Samples were cut both in warp and weft. The sample size was 2 inch long and 1 inch wide. It was creased by folding into half and placed under the weight of 2 kg for 5 min. The weight was removed and the specimen was transferred to the fabric clamp on the instrument using forceps and allowed to recover from the crease for 5 min. As it recovered the dial of the instrument was rotated to keep the free edges of the specimen in line with the knife edge. At the end of the time period it was allowed for recovery, usually 1 min the recovery angle in degrees was read on the engraved scale. Readings were recorded for both warp and weft separately [17].

1.4.8   Shrinkage

Dimensional stability of the fabric is measured in terms of shrinkage percentage. The fabric sample of 25 × 25 cm was taken and initial length of 20 cm was marked both in warp and weft direction. The test samples were soaked in the soap solution of 2 gpl at room temperature for one hour rinsed thoroughly in cold water and dried under the shade. The dried samples were pressed gently without stretching [17]. Further final lengths were measured and shrinkage percentage was calculated using the formula:

$$\text{Shrinkage } \% = \frac{\text{Shrunken Length} - \text{Original length}}{\text{Original length}} \times 100$$

1.5   *Preparation of the Product*

As per the test result and evaluation of the union fabrics, lotus: Cotton stole was prepared in handloom using extra-weft weaving techniques. Specification of Lotus: Cotton stole is given in Table 5.

Table 5 Specification of the product—Lotus: Cotton Stole

Size	200 × 56 cm
Fiber Content	Warp : 2/80'S Cotton Weft : 2/50'S Lotus
Weave	Plain weave with extra—weft figuring motifs
Dye used	For the base fabric, both cotton (warp) and lotus (weft) yarns were dyed with natural dye extracted from onion peels For extra-weft designs: Lotus yarns were dyed with direct dye

2   Results and Discussions

2.1   Construction of Fabrics

Initially the 100% lotus warp and weft fabric sample was prepared on bead loom shown in Fig. 3a, b. Later 100% lotus (warp and weft) fabric weaving was also tried on handloom. But due to the low elongation property of yarn there were more breakages faced in warp yarns on handloom. So the union fabrics were prepared on handloom using cotton and silk as a warp and lotus as a weft however to compare the properties 100% Cotton and silk fabric was also prepared shown in Fig. 4. As per the visual appearance and feel, lotus yarn has a silk like luster and smoothness so due to this reason silk was selected to use in the warp.

 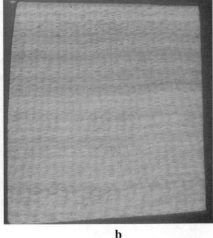

a   b

Fig. 3   a Construction of 100% lotus fabric on bead loom, **b** 100% lotus fabric

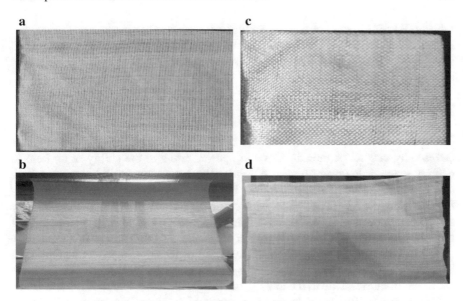

Fig. 4 Images of constructed union fabrics: **a** 100% Cotton Fabric (C1); **b** Lotus: Cotton (LC2); **c** 100% Silk (S3); **d** Lotus: Silk (LS3)

2.2 Evaluation of the Constructed Fabrics

Fabric properties like fabric count, thickness, cover factor, weight per unit area (GSM), tensile strength, bending length, crease recovery and shrinkage were evaluated for all the constructed fabrics.

2.2.1 Fabric Count

Fabric count is one of the most important considerations. It has impact on comfort, quality, and value. Table 6 shows the fabric count of the union fabrics. It was observed that there was not much difference between the fabric count of Cotton (C1) and Lotus: Cotton (LS2). By adding lotus in the weft the count increases in Lotus: Cotton (LC2) fabric, the structure of the fabric was more compact as compared to 100% cotton

Table 6 Fabric count of constructed union fabrics

Sr. No	Fabric sample	Fabric code	EPI (ends per inch)	PPI (picks per inch)	Fabric count
1	100% cotton	C_1	50	42	50×42
2	Lotus: cotton	LC_2	50	44	50×44
3	100% silk	S_3	47	45	47×45
4	Lotus: silk	LS_3	45	37	45×37

fabric (C1). But in case of silk, by adding lotus in weft the fabric count decreases. Lotus: Silk (LS4) fabric was found less compact as compared to 100% silk fabric. It was observed that lotus silk fabric has a sheer appearance.

2.2.2 Cover Factor

Cover factor is a number that indicates the extent to which the area of a fabric is covered by one set of threads. Covering power refers to the ability of an item to occupy space or to cover an area. A fabric with better cover will be warmer, look and feel more substantial, and durable. For any woven fabric, there are two cover factors: a warp cover factor and a weft cover factor. It depends on **thread density and thread count** [28]. As the cover factor increases, the fabric has a denser and closer structure and when the cover factor decreases, the fabric has an open structure.

Table 7 and Fig. 5 shows the cover factor (K) of the constructed union fabrics. It was observed that 100% Cotton (C1) has a less cover factor from rest of the three set of union fabrics. Due to the lotus yarn in weft, the cover factor of Cotton: Lotus (LC2) increases. This could be due to fineness of the yarn; lotus yarn used in weft was coarser than 100% Cotton yarn (C1). In case of silk, Lotus: silk has less cover factor than 100% silk fabric (S3). Due to this reason the Lotus: Silk (LS4) Fabric has an open and sheer appearance like net fabric.

2.2.3 Thickness

Figure 6 shows the thickness of all the constructed union fabrics. It was observed that there was slight difference in thickness between 100% Cotton (C1) and Lotus: Cotton (LS2). In case of silk, there was a much difference in thickness observed in Lotus: Silk (LS4) as compared to 100% silk fabric (S3). Thickness increased because lotus yarn was coarser as compared to silk yarn.

2.2.4 Weight Per Unit Area (GSM)

Table 8 shows the GSM of the union fabrics. It was observed that by using lotus yarn in the weft, the GSM increases in Lotus: Cotton (LS2) and Lotus: Silk (LS4) fabric as compared to 100% cotton (C1) and 100% Silk (S3) fabrics. GSM between 1 and

Sr. No	Fabric sample	Cover factor (K)
1	100% cotton (C1)	0.23
2	Lotus: cotton (LC2)	10.56
3	100% silk (S3)	11.72
4	Lotus: silk (LS4)	8.85

Table 7 Cover factor of the constructed union fabrics

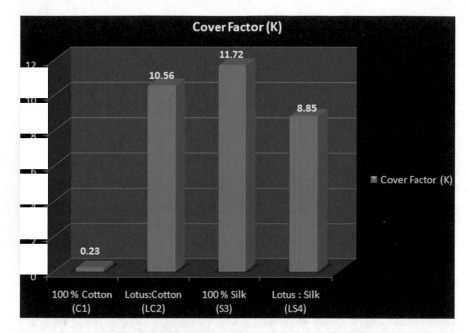

Fig. 5 Cover factor of the constructed Union fabrics

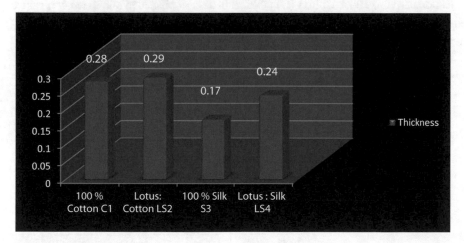

Fig. 6 Thickness of constructed union fabrics

Table 8 GSM of constructed union fabrics

Sr. No	Fabric samples	Fabric code	GSM (in grams)	GSM (in ounces)
1	100% cotton	C_1	70.8	2.48
2	Lotus: cotton	LC_2	80.8	2.85
3	100% silk	S_3	68.8	2.42
4	Lotus: silk	LS_3	75.6	2.66

Table 9 Classification of different weights of the fabric and its appropriate end uses

Sr. No	Category	Ounces (Oz)	Appropriate end uses
1	Extremely light weight	0–1	Chiffon, tulle, sheers, some laces
2	Light weight	1–4	Blouses, shirting, and light summer dresses
3	Medium weight	4–7	Heavier tops, lighter bottoms, and slacks
4	Medium to heavy weight	7–9	Bottom weight slacks suiting and light weight summer dresses
5	Heavy weight	>9	Blankets, heavier coats, and denims

4 oz, fall under the category of light weight fabrics. However, GSM of Lotus: Cotton (LS2) and Lotus: Silk (LS4) fall under the category of light weight fabrics and its suggested applications are shown in Table 9 [29].

2.2.5 Tensile Strength

Table 10 and Fig. 7 shows the tensile strength of the constructed union fabrics. It was observed that Load and extension value increased in Cotton: Lotus (LC2) fabric as compared to 100% Cotton (C1) fabric. In case of silk, load and extension values of silk: lotus fabric in weft direction was considerably lower than warp direction indicating lower strength of lotus yarn as compared to silk.

Table 10 Tensile strength of constructed union fabrics

Fabric samples	Maximum load (kgf)	Max extension (mm)	% strain
100% cotton C1 warp	6.45	6.41	8.52
100% cotton C1 warp	3.69	3.21	4.28
Lotus: cotton LC2 warp	7.75	6.79	9.05
Lotus: cotton LC2 weft	7.49	4.6	6.13
100% silk S3 warp	31.33	13.61	18.15
100% silk S3 weft	23.58	12.83	17.1
Lotus: silk LS4 warp	24.24	14.14	18.85
Lotus: silk LS4 weft	2.49	3.1	4.13

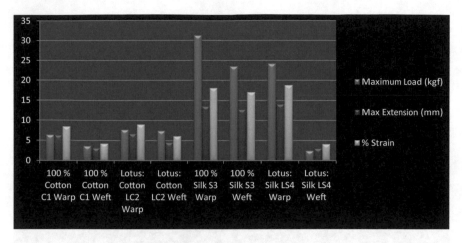

Fig. 7 Tensile strength of constructed union fabrics

Table 11 Bending length of the constructed union fabrics

Fabric samples	Bending length			
	Warp direction		Weft direction	
	Face to face	Back to back	Face to face	Back to back
100% cotton C1	2.7	2.8	2.7	2.9
Lotus: cotton LC2	2.5	2.6	2.8	2.8
100% silk S3	1.2	1.3	1.2	1.5
Lotus: silk LS4	2.5	2.3	3	3

2.2.6 Bending Length

Stiffness or flexural rigidity is one of the important physical properties of textile materials and widely used to judge bending rigidity and fabric handling. The bending behavior of woven fabric is very important for fabric producers, designers of clothing, and also for apparel production [32]. From Table 11 and Fig. 8, it was observed that bending length of weft yarn increased as compared to the warp yarns in Lotus: Cotton (LC2) and Lotus: Silk (LS2). This could be due to the presence of lotus yarn in weft. Maximum difference was observed in Lotus: Silk (LS2) in weft direction. But there was a minor difference in bending length of Lotus: Cotton (LC2) as compared to 100% Cotton (C1) fabric.

2.2.7 Crease Recovery

Crease recovery is the property of a fabric that enables it to recover from folding deformations. It is one of the most important properties expected out of any apparel

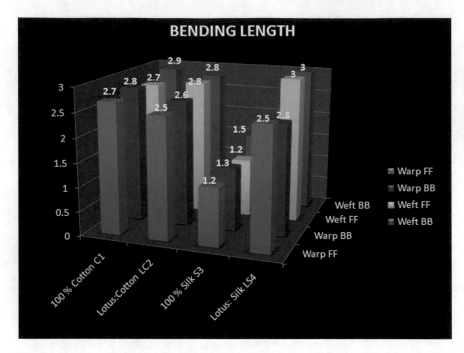

Fig. 8 Bending length of the constructed union fabrics

Table 12 Crease recovery of the constructed union fabrics

Fabric samples	Crease recovery (angle)			
	Warp way		Weft way	
	Immediate	After 5 min	Immediate	After 5 min
100% cotton C1	105	110	105	112
Lotus: cotton LC2	115	125	115	130
100% silk S3	130	140	135	137
Lotus: silk LS4	139	147	138	150

textile to make it easy to care [30]. Table 12 and Fig. 9 depicted crease recovery angles of all the constructed union fabrics.

2.2.8 Shrinkage

The dimensional stability of fabric refers to the change of fabric size when it is used or reprocessed due to the properties of a material and the potential thermal contraction force in the process of processing. The fabric with good dimensional stability is worn and washed for many times, the original pleating and shape are unchanged, and the

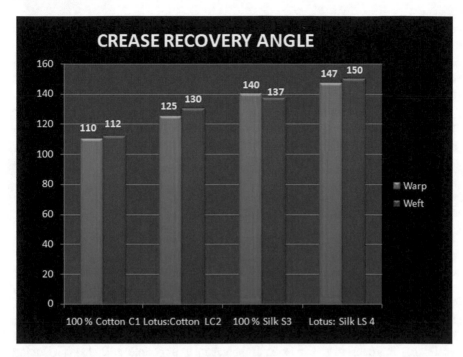

Fig. 9 Crease recovery angles of constructed union fabrics

dimensions don't shrink or elongated, which don't affect the user experience. The fabric with poor dimensional stability is usually called as shrinkage, such as shrinkage in sewing, ironing, washing, and so on. Among them, the wash shrink is the problem that consumer pays close attention to very much [31]. The relaxation shrinkage of all the union fabrics is shown in Table 13. Different factors affecting to the fabric shrinkage are—moisture absorption of the fibers, tightness of yarn, fabric structure, and processing tension. Natural fibers like cotton, wool and majorly regenerated fibers have a maximum shrinkage rate. It was observed that Lotus: Cotton (LC2), 100% Silk (S3) and Lotus: Silk (LS4) has a negligible shrinkage.

Table 13 Shrinkage of the constructed union fabrics

Fabric samples	Original length	Shrunken length		% Shrinkage	% Shrinkage
	20 cm	Warp	Weft	Warp	Weft
100% cotton C1		17.5	17.8	12.5	11
Lotus: cotton LC2		25	25	0	0
100% silk S3		25	25	0	0
Lotus: silk LS4		25	25	0	0

2.3 *Preparation of Product*

As per the evaluation from the test results and visual appearance of all four union fabrics. Lotus: Cotton (LC2) has given an encouraging result like good strength, 0% shrinkage, compact appearance, stiffness was also maintained. So the stole was prepared on handloom using lotus yarns as a weft and cotton as a warp shown in Fig. 10. To make it completely sustainable the fabric was dyed with natural dye— onion peels. Extra-weft figuring was also done on the stole.

3 Conclusion

From the study it has been concluded that Lotus: Cotton (LC2) fabric has a compact appearance, high tensile strength, and elongation both in warp and weft direction as compared to 100% Cotton (C1) fabric. Lotus: silk fabric (LS4) has a sheer appearance and has a highest crease recovery angle. In the present scenario, where there is a demand of environment friendly process and products along with the sustainability, the use of lotus plant waste for fibers is the perfect solution. The quest of novel fabric is always there in the fashion world. Exploration of minor fibers for making fabric will satisfy this quest to certain extent. Fibers from lotus petiole will help in producing novel fabrics. Cost is another factor which always hits the fashion and textile world. Lotus fiber fabric can be substitute for costly fabrics like silk which have similar in properties but less expensive is the solution. Silk: lotus union fabric can substitute silk to some extent for specific purpose like its luster and smooth appearance and thus less expensive fabric can be used with same effect. Textile industry has been classified as the second largest polluting industry. The process from extraction to the development of fabric does not add pollution to the environment, while converting the waste into useful textile. Lotus union fabric with cotton and silk will be a new fabric for the designers and design students to explore for newer collection. Lotus union fabrics can cater the kids (infant) wear segment due to its extreme softness. India is a developing country training regarding lotus fiber extraction to rural peoples will generate income as well it will add skill development. Clusters of rural women can learn this process of extraction along with their house hold work which will be a way to women empowerment. Fibers have a potential to be used in craft sector, handloom sector, newer developments in research and development institutes and industries.

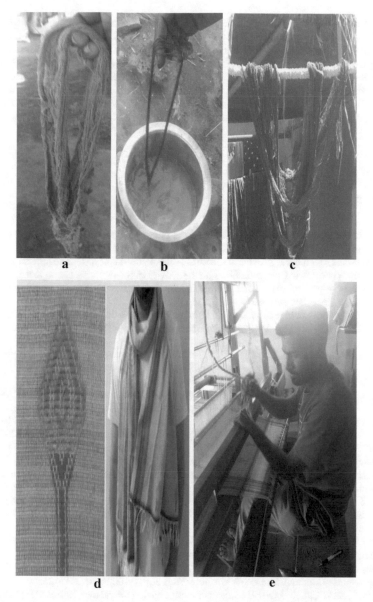

Fig. 10 Preparation of product: lotus: cotton stole—**a, b, c** Dyed yarns; **d** Weaving of stole on handloom; **e** Lotus stole

Acknowledgements The authors are thankful to Mr. Poonambhai Parmar, Mr. Sanjaybhai Mali, Mr. Isabbhai Rathod (Cultivators of Lotus), Mrs. Shantaben Mali (Vendor of lotus flower), and Mr. Bakul Shah (Member of Sardar Bhavan Trust) of Vadodara district. Mr. Himanshu Agarwal (Deputy General Manager), Mr. Nishith K. Sheth (Asst. Manager Tyrecord and CSY plant) of Century Rayon (Under the management and operation of GRASIM Industries Limited) Shahad, Mumbai and Bombay Textile Research Association (BTRA) Ghatkopar, Mumbai, Mr. Jagadish Haldar (Manager-Marketing and operation, KCS group of Companies, Gujarat) and Mr. Shiv Dayal Sharma (Branch manager, Inland world logistics Gujarat) for helping me in various stages of research.

References

1. Konwar M, Boruah RR (2018) Natural fibers as sustainable textiles: a review. Int J Pure Appl Sci 6(6):504–507. https://doi.org/10.18782/2320-7051.7069
2. Zimniewska M, Huber J, Krucinska I, Torlinska T, Kozlowski R (2002) The Influence of clothes made from natural and synthetic fibers on the activity of the motor units in selected muscles in the forearm—preliminary studies. Fibres and Textiles in Eastern Europe 55–59 https://www.researchgate.net/publication/258158332
3. https://www.prnewswire.com/news-releases/global-natural-fibers-market-report-2021-revenues-for-2020-2030-by-fiber-types-industries-and-region-301211682.html Last accessed 8 May, 2021
4. Mishra MK, Panda A, Sahu D (2012) Survey of useful wetland plants of South Odisha, India. Indian J Traditional Knowl 11(4):658–666. http://nopr.niscair.res.in/bitstream/123456789/14963/1/IJTK%2011%284%29%20658-666.pdf
5. Paudel KR, Panth N (2015) Phytochemical profile and biological activity of Nelumbo Nucifera. Evid-Based Contemp Altern Med. 1–16https://doi.org/10.1155/2015/789124
6. Sheikh SA (2014) Ethno-medicinal uses and pharmacological activities of lotus (Nelumbo Nucifera). JMed Plants Stud 2(6):42–46. https://www.researchgate.net/publication/293183331
7. Zaidi A, Srivastava AK (2019) Nutritional and therapeutic importance of Nelumbo Nucifera (Sacred Lotus). Era's Jo Med Res 6(2):1–5
8. Ventakatesh B, Dorai A (2011) Antibacterial and antioxidant potential of white and pink Nelumbo Nucifera Gaertn Flowers. Int Conf Biosci Biochem Bioinform (5):213–217
9. Lin Z, Zhang C, Cao D, Damaris RN, Yang P (2019) The latest studies on lotus (nelumbo nucifera)—an emerging horticultural model plant. Int J Mol Sci 22(3680). https://doi.org/10.3390/ijms20153680
10. Chen Y, Wu Q, Huang B, Huang M, CAi X (2015) Isolation and characteristics of cellulose and nanocellulose from lotus leaf stalk agro-wastes. Bio Resour 10(1):684–696
11. Mengxi W, Hua S, Qunfeng C, Lei J (2014) Bioinspired green composite lotus fibers. Angewandte Commun 53:3358–3361. https://doi.org/10.1002/anie.201310656
12. Myint T, San D, Phyo U (2019) Lotus fiber a value chain in Myanmar. Helvatas:1–40. Retrieved from themimu.info>node
13. https://www.dnaindia.com/ahmedabad/column-flowers-of-some-wetland-plants-elegance-personified-2748658. Accessed 16 Dec 2020
14. Taga G, Kalita B (2013) Study on Eri Union fabrics for various end uses. J Acad Indus Res 2(5):252–256
15. http://mytextilenotes.blogspot.com/2011/10/difference-between-blended-fabrics-and.html union fabric. Accessed 15 May 2021
16. ASTM D 3775-98 (1998) Standard Test method for fabric count of woven fabric ASTM International, West Conshohocken, PA

17. Sharma N, Jain M, Kashyap R (2016) Assessment of physical properties of cotton with ahimsa and conventional silk union fabrics. Int J Sci Res, 458–462.https://doi.org/10.21275/3041807
18. ASTM D 1777-96 (2019) Standard test method for thickness of textile materials, ASTM International, West Conshohocken, PA
19. ASTM D 5035-95 (1995) standard test method for breaking force and elongation of textile fabrics (strip method), ASTM International, West Conshohocken, PA
20. ASTM D1388-18 (2018) Standard Test Method for Stiffness of Fabrics, ASTM International, West Conshohocken, PA
21. Booth JE (1996) Principles of textile testing. CBS Publishers and distributors Pvt. Ltd. New Delhi, India, pp 282–287
22. https://www.onlinetextileacademy.com/what-is-cloth-cover-factor-derivation-of-cover-factor-formula/. Accessed 15 May 2021
23. IS 4681-1968 method for determination of wrinkle recovery of fabrics (By measurement of Crease Recovery Angle)
24. ASTM D 7025-09 e1 (2015) Standard test method for assessing clean flax fiber fineness, ASTM International, West Conshohocken, PA
25. ASTM D3822-07 (2007) Standard test method for tensile properties of single textile fibers, ASTM International, West Conshohocken, PA
26. ASTM D2495-07(2019) Standard test method for moisture in cotton by oven-drying, ASTM International, West Conshohocken, PA
27. ASTM-885/D885-10 (2014) A e1 Standard test methods for tire cords, tire cord fabric, and Industrial filament yarns made from Manufactured Organic base fabrics: ASTM International West Conshohocken, PA
28. https://textilelearner.net/cover-factor-of-woven-fabric. Accessed 16 May 2021
29. Young D (2014) Swatch reference guide for fashion fabrics. Bloomsburg Publications New York
30. https://www.fibre2fashion.com/industry-article/7729/effect-of-chemical-treatments-to-improve-crease-recovery-on-cotton-fabric Accessed 17 May 2021
31. https://www.testextextile.com/fabric-dimensional-stability-shrinkage-test/ Accessed 17 May 2021
32. Eryuruk SH, Bahadar SK, Kalaoglu F (2015) The evaluation of stiffness and drape behaviour of wool fabrics. J Text Eng 22(98):24–32.https://doi.org/10.7216/130075992015229803

Evaluating the Potential of Pineapple Leaf Fibre Fabrics and Its Blends for Sustainable Home Textile Applications

R. Surjit, P. Kandhavadivu, and S. Ashwin

Abstract This chapter excogitates the potential of pineapple leaf fibres (PALF) for usage in home textile applications. PALF is extracted from the leaves of a pineapple plant and is biodegradable in nature. It is environment-friendly and has better strength thereby leading to development of textile products. In this work, the pineapple leaf fibre is blended with cotton fibres for making fabrics. Home furnishing products are made using these fabrics. Table mats and curtains are made and are looked as a potential environment-friendly alternative in the current scenario. The PALF and cotton fibres were spun using rotor spinning in three blend proportions. The yarn properties were tested and it is found that the 50/50 pineapple/cotton fibre had better elongation of 7.81% and had less U% and hairiness compared to the other two blends of 70/30 pineapple cotton and 60/40 pineapple cotton yarn. The strength of 50/50 pineapple cotton yarn was found to be 21.22 g/tex. Hence, the blend proportion of 50% pineapple fibre and 50% cotton fibre was selected for making woven fabrics. The woven fabric was tested for its thickness, areal density, tensile strength, tearing strength, abrasion resistance, pilling, dimensional stability and flammability. The results were compared with commercially available fabrics used for table mat and curtain. It was found that the developed fabric had better properties than the commercially available fabrics as it had a tensile strength of 22.95 kgf, tearing strength of 61.72 lbs and flammability of 23 s. The table mats and blinds were produced using the developed fabric. From the results, it is found that the developed fabric has a good potential for usage in home furnishings and hence this work has provided a viable alternative for synthetic products and an effective use for PALF which is otherwise thrown as waste.

Keywords Pineapple leaf fibre · Sustainable fabric · Table mats · Blinds · Home textiles · Cotton

R. Surjit (✉) · P. Kandhavadivu · S. Ashwin
Department of Fashion Technology, PSG College of Technology, Coimbatore, Tamil Nadu 641004, India
e-mail: rst.fashion@psgtech.ac.in

© The Author(s), under exclusive license to Springer Nature Singapore Pte Ltd. 2022   123
S. S. Muthu (ed.), *Sustainable Approaches in Textiles and Fashion*, Sustainable Textiles:
Production, Processing, Manufacturing & Chemistry,
https://doi.org/10.1007/978-981-19-0878-1_6

1   Introduction

Sustainability is more than a buzz word today as it harnesses the present and future needs intransigently. Sustainable practices are required at each and every stage of a human's life to save the planet and get it inherited better by the future generations. Sustainability spans across every activity that is carried out in the world. In fact sustainable development goals have been advocated for a sustainable planet by the United Nations Organization in which 17 goals have been advocated [5]. The sustainability is involved in every walk of life. The sustainability involves sustainable practices, sustainable processes, sustainable procedures, sustainable materials, sustainable products, sustainable governance, sustainable environment, sustainable production and the list goes on. At this present juncture, there are millions of people working towards building a better sustainable future in making changes in the above-mentioned areas. One such area mentioned is the sustainable materials. It is imperative to use sustainable materials especially for manufacturing which will avoid any impact to the environment and avoid pollution. The ecological balance restoration is required at the present stage due to more unsustainable materials present in the world. The sustainable materials involve any material used for production of a product. Also sustainable materials are involved in basic necessities of a human which include food, shelter and clothing. Our emphasis here is on the clothing as there is a huge scope in this area to use sustainable materials for manufacturing. Textile products are produced using natural and synthetic material. Any effort in reducing synthetic material and increase in natural material usage is very much appreciated as it is the need of the hour. In today's context, using sustainable fibres for manufacturing is a growing need.

2   Sustainable Fibres

The fibres that are used can be broadly classified as the natural and synthetic fibres. The synthetic fibres have gained a huge market share as detailed in the preferred fibre and materials market report 2019 [81]. According to the report, in 2018, 107 million metric tons of fibres have been produced globally and in that the synthetic fibres occupy a share of 62% which amounts to 66.6 million metric tons of fibres. In the synthetic fibre category, polyester occupies a share of 51.5% amounting to 55.1 million metric ton. The synthetic fibres overtook natural fibre market in 1990s and it was due to their superlative properties or because of possessing properties that are not in natural fibres. New synthetic fibres are being produced even at this point in time. However synthetic fibres lead to lot of risks especially with respect to their disposal. Even they pollute the water when a synthetic fibre made fabric is washed every time [65]. Microplastic pollution is in rise and during washing of synthetic fibres, micro fibres come out and cause serious pollution issues [64]. In a study done by Noren et al. [54], it was observed that 90% of the microplastics found in the Swedish coasts are textile fibres. The majority of the microplastics ends in oceans thereby causing serious damages to the aquatic life and disturbs the ecological balance. Another study reported that polyester, acrylic and nylon fibres were the major contributors to the microplastics in ocean [52]. The microfibres have reached our food chain as they have been found in honey, beer, table salt, mussels, seafood and chickens [64]. Apart from the presence of microplastics in ocean, it is also amounting to huge landfills as it takes a very long time for it to decompose. A study by Loppi et al. [43] reported that 68% polyethylene terephthalate, 26% polyethylene and 5% polystyrene was present in a landfill and numerous research works have been published with respect to this. Although 13% of the polyester used in the world is recycled polyester as per 2018 fibre report, the avoidance of microplastics is very difficult and various mitigation strategies are being tried out.

In the above context, it becomes very essential to find fibres which are eco-friendly and sustainable in nature. Although, it is very difficult to get all the properties of synthetic fibres in natural materials, efforts should be made to utilise the natural fibres to the fullest possible extent and wherever possible. The need for sustainable fibres grows day by day manifold times. There are many new natural fibres that are being found everyday with varied usage, however there is long way to go to make these fibres used effectively and efficiently. The natural fibres other than cotton and wool are utilised to a very less extent and only 6 million metric ton of other plant fibres are produced in a year [81]. The bast fibres do find some amount of production whereas leaf fibres are produced at a very low level. The reason for low usage of natural fibres is the lack of awareness of the benefits or properties of those natural fibres. It is also because the processing of fibres involves lot of time and enough machinery is not available for producing them. In an attempt to increase the natural fibre usage, many fibres are modified and made suitable for various end uses by mechanical and chemical means. Apart from those, there are various fibres available with properties similar to synthetic fibres which have to be explored more and suitable usages can be obtained from those fibres. One such fibre which has unlimited potential and excellent properties is a leaf fibre obtained from pineapple. It is called as pineapple leaf fibre (PALF). This fibre rivals synthetic fibres because of its excellent properties.

3 Pineapple Leaf Fibres (PALF)

The PALF is obtained from pineapple (Ananas comosus) plant. The plant is known for producing pineapple fruit and the leaves of the plant are arranged in a spiral order over the fruit. The pineapple plant grows to a maximum height of six feet. The pineapple plant is mainly cultivated across the world for the fruit and the leaves are discarded during harvesting of the fruit [57]. In fact, the leaves are dumped as waste material. The leaves can be made into useful textile material. The leaves are also used for making papers and composites. The leaves are by-product of the pineapple plant. Costa Rica has the highest pineapple production in the world. The pineapple production amounts to 23.33 million metric ton in a year with more plantations in

Asia than other continents [13]. Many throw the leaves as a waste instead of utilising them to extract the fibre [46]. Only by proper education of the benefits of using this fibre, the production and development can be increased. The lucrative benefits associated with the fibre production should be explained. The fibre extraction process involves lot of time and it requires lot of patience.

3.1 Extraction of PALF

The pineapple leaves are thrown as agro waste. From the waste, the PALF is extracted. The PALF is extracted by water retting and then manually fibres are scrapped using a tool called 'ketam' [79]. In another method, fresh leaves of Spanish or native pineapple variety are hand scrapped using a broken porcelain plate followed by washing with tap water and air drying [57]. The fibres obtained by this method resemble sisal fibre and have a length of 55–75 mm. The yield will be less than 2.5% in this method. The decorticators are also used for extraction where the leaves are fed into a decorticator machine which has blades in it and it does the work of scrapping. After removal of fibres by a decorticator, the fibres are washed and air dried. Liniwan and bastos variety of fibres are produced by decortications method [88]. The fibres produced using decorticating machine were found to be of white creamy colour instead of brown and fibres were also soft. There is lot of wax available on the leaf which is removed by scrapping or retting or by chemical treatments. Once the fibres are extracted, then fibrous stands are split up by using retting or degumming process.

In the process of retting, the fibre bundles are separated from the cortex and the material that binds the fibre is dissolved during retting process. Retting is done using water and chemicals. In the case of water, the extracted foliages are tied together and immersed in water. The fibres are soaked in water for close to four weeks. In some cases, for quick retting urea and ammonium phosphate is added [6]. The process of retting involves development of microorganisms that dissolve the pectins which are cement like substances that bind the extracted leaves together thereby leading to easy separation of the fibres. Initially the fibres swell and later microorganisms grow and remove the pectins. The fibres after retting are washed in water and then air dried. In the degumming process, the extracted fibre bundles are immersed in sulphuric acid, and then they are washed and treated with Sodium hydroxide in boiling condition followed by washing, bleaching and drying. Using this process dissolves the hemicelluloses. Finest fibres are obtained by using NaOH of 5% quantity and immersing the fibres in it for 12 h. The chemical degumming process reduces the time of extraction.

In all these efforts of extracting PALF, the yield is very less and it is around 2 to 3% [2, 30]. More technical advancements are being made to increase the yield percentage. There are various advancements that have come in respect to extraction using machines where the old decorticator machine is modified with feed roller, scratching roller and serrated roller. The scratching roller removes waxy material and serrated roller produces breaks in the leaf by grinding for easy retting process [6].

In another method, the PALF is extracted using an indigenously developed machine within 3 days of harvesting to preserve its quality and then scoured using another indigenously developed scouring machine [87]. In another method employed by Biagiotti et al. [7], the fibres are separated physically by scutching and swingling where the effectiveness of the technique was measured based on length and diameter of the fibre obtained [57].

3.2 Properties of PALF

The PALF fibres extracted have been analysed for their physical, chemical, morphological, thermal, analytical properties by many researchers. Many studies have also been performed to modify the fibre properties by chemical treatments. The composition of the fibre is similar to other leaf fibres and vegetable fibres and it has cellulose, hemicelluloses, pectins, lignin, fat, waxes and colouring matters [19, 57]. The chemical composition of PALF has been studied and reported by Singha et al. [71]. The PALF has 73.13% carbon, 24.17% oxygen, 2.7% nitrogen and in some cases, traces of calcium, phosphorus, iron, potassium, magnesium and copper are found. The FTIR spectra of PALF has been analysed and it is found that –OH class was around 3349.9 cm^{-1} and it was the highest peak similar to the kenaf fibres [82]. The C–H stretching was in the region of 2903.8 cm^{-1} and C=O stretch was in the region of 1737.4 cm^{-1} and this is due to the carbonyl stretching of ester and hemicelluloses [38, 42, 56]. These hemicelluloses can be removed by NaOH treatment or with any alkali as they get dissolved in alkali as mentioned above [45]. There are two other peaks witnessed around 1608.3 cm^{-1} and 1374.2 cm^{-1} which correspond to C=C stretching and C–H stretching respectively similar to other bast fibres. Another study has been done to compare the FTIR spectra of PALF and pineapple leaf microfibres (PALMF) [74] in which the main difference is the peaks that occur at 1737 cm-1 and 1246 cm^{-1} are noticed in PALF and not in PALMF.

The morphological structure of the fibre has been assessed and it is found that the fibre bundles of fibrovasculars have a crescent shape when the cross-section is observed and another bundle of mesophylls are found which are little round in structure. Both the bundles consist of numerous microfibres with an average diameter of 4–5 mm [74]. The PALF cross-section indicates 44% fibrovascular bundles and 56% mesophyll bundles. The microfibre wall consists of one primary wall and 3 secondary walls as sublayers [37]. The longitudinal view of PALF indicates tubular structure with irregularities in it. It is not completely regular as noticed in a synthetic fibre.

The TGA studies have been performed on PALF and it is found that the curves are very similar to what is observed for flax fibre [34, 74]. The weight loss occurs in two steps at 70 degree centigrade and immediately after 250 degree centigrade. The loss at 70° c can be attributed to loss due to moisture and at 250 °C can be attributed to the decomposition of the fibres. These fibres decompose quickly than animal fibres which are stable up to 450 °C [62]. The fibres have been treated with NaOH and then subjected to TGA and it was found that the thermal decomposition started after 270 °C thereby increasing its thermal properties.

The WAXS studies have been conducted for crystal structure understanding. The PALF had very high orientation similar to jute, ramie fibre. It was observed that the type I of cellulose is found in PALF based on diffractions assigned at (004) (100) (200) [74]. Another recent study by Gaba et al. [20] also identified the type of cellulose as type I and the crystallinity of the NaOH treated PALF sample was 76%. The PALF fibres had a very high crystallinity compared to animal fibres which have very low crystallinity. This crystallinity has a significant impact on the fibre properties like strength and elongation as higher crystalline structure results in higher strength and lower elongation.

The mechanical properties of the fibres have been studied in detail by many researchers and it was found that the tensile strength of PALF fibres ranged between 148 and 174 MPa for selected varieties. The young's modulus ranged between 7.45 and 18.94 GPa and elongation at break was between 0.52 and 1.41% [79]. Another study by Jose et al. [32] reported 31.32 gf/tex tenacity for retted fibre and 25.29 gf/tex for degummed fibre with elongation at break of 6.1% for degummed and 4.5% for retted fibre. The degumming process removes the hemicelluloses and the matrix holding the fibre bundles thereby reducing the strength. The fineness of fibres has been reported as 4.1 and 3.5 tex for retted and degummed fibre. The effect of extraction on the physical and mechanical properties of PALF has been reported where the fineness varied between 3.4 and 5 tex based on the type of mechanical extraction technique. The PALF obtained from mechanical extraction technique had a tensile strength of 613.75 MPa which was higher than the tensile strength of 393.7 MPa for the PALF obtained from hand scrapping technique [60]. The fibres obtained from PALF had coarser diameter ranging from 10 to 100 μm [79]. The finer fibres were difficult to obtain whereas coarser fibres were easily available. The PALF fibres had a coarser diameter compared to other leaf fibres [30].

The density of PALF is 1.543. The pineapple fibres lose strength and elongation in wet condition. Although the fibres are coarser in nature, they have good lustrous properties. They have high flexural and torsional rigidity. Their length ranges between 10 and 90 mm. PALF has low resistance to strong acids as they dissolve in 60% sulphuric acid in 5 min. They have good resistance to mild alkalis similar to other bast fibres. PALF are dyed with direct, reactive and vat dyes. It has better dye absorption capacities than cotton. They have good strength and are softer than hemp. PALF fibres can be blended with other fibres like silk, cotton to aid in improving the yarn and fabric properties.

3.3 Spinning of PALF

There is no dedicated spinning system available for PALF spinning. Different systems have been used for spinning PALF fibres separately and also in blends with other fibres. PALF fibres have been spun in jute spinning system after blending with jute fibres and also with flax spinning system. It was found that the U% was very high at 29.5 and 27.5 when spun with jute and flax systems respectively. However, there was improvement in weight irregularity when flax system was used. The PALF has been blended with wool and yarns were produced using semi worsted spinning system with a very small proportion of PALF in the blend [87]. Apart from this cotton spinning system has been used for spinning to a large extent as both ring and rotor spinning is used to produce the yarns. However, the 100% PALF spinning is very difficult due to fibre losses and only after chemical treatments to PALF and modifications in the machine, it is possible. Normally blended yarns are produced using cotton spinning system using binary blending technique [58]. Another indigenous system has been developed for yarn spinning known as Jantra in Indonesia. It was used for spinning silk initially and later it accommodated short fibre spinning and PALF spinning.

3.4 Weaving of PALF

The PALF fabrics have been woven using automatic looms and handlooms. Normally, PALF blended yarns are woven than 100% PALF yarns weaving. The woven fabrics are used in composites as laminates to a very great extent. The produced fabrics are also used in home textiles. The PALF woven fabrics in blends with silk, kenaf, flax, jute, cotton have been woven. The plain and twill weave fabrics are the most common weaves that have been produced using PALF and its blends [14]. Apart from weaving, there is not much literature reported about knitting of PALF fabric.

3.5 Applications of PALF

The PALF has extensive applications and it can be a replacement for many synthetic materials. Hazarika et al. [24] created various garments using PALF and silk waste blends. The PALF fibres have been extensively used in composites. There are various literatures available which explain the role played by PALF in a composite based on its higher strength and other properties. In fact, many new avenues of usage of PALF are explored due to chemical modifications done in PALF to improve its properties significantly. The PALF is used as a green acoustic observer. It has been made into non-wovens and used as a thermal and sound insulator [78]. The cellulose nanostructures obtained from PALF finds extensive use and has excellent thermal properties. The PALF is used in automotive applications. It is also used in pulp and paper production. It can be used extensively in biotechnology and biomedical applications like drug delivery, tissue engineering, vascular grafts, etc. due to its cellulose nanostructure [31]. The PALF and few other natural fibres have been compared with aramid, glass and carbon and their properties have been studied after delignification. It is observed that the results are promising and provide more scope of PALF in replacing synthetic fibres for high performance applications [9].

Although significant applications of PALF have been witnessed, its usage in clothing sector is very little. It has not made a significant impact with respect to apparel. In a clothing store, a PALF-based fabric can be very rarely found. Although many studies have been done to use the PALF in apparel as mentioned above by Hazarika et al. [24] in mixing PALF with silk waste and in green fashion [15], still it is a long way to go due to the properties of PALF not suiting clothing like how cotton or polyester suits. In this stage, it is essential to find more usage for PALF due to its superlative properties than many other fibres which may fit some other application. Hence an application has to be found for PALF where there will be more demand for it and will lead to increased income for the fibre extractors. It will create more interest for the PALF extractors and producers to come out with fibres with improved properties to suit the requirements if there is more demand. One sector which looks promising and ideal for usage of PALF is home textiles sector where there is a huge demand for natural fibres which can replace synthetic fibres and still provide desired properties. There are quite a few studies done on using PALF in home textiles as table mats, furnishings, etc. [35, 75]. However, its complete potential is not explored. The reasons for PALF suiting home textiles sector are its higher strength, abrasion resistance, improved flammability properties compared to other bast and leaf fibres, lustre and durability. Hence, this study becomes quintessential in today's context where home textile products are going to be developed using PALF in a basis where it can lead to significant reproduction in future and have manufacturers utilise PALF in home textiles. It will lead to sustainable fibre usage and also replace synthetic fibres to a great extent.

In this current work, it is tried to use PALF in home textile products. The PALF is blended with cotton in three different proportions and yarns are produced using rotor spinning. The yarn properties are studied in detail to identify the blend with better yarn properties. The selected yarn blend is to be further made into woven fabrics using handloom and then their properties are going to be compared with the commercially available table mat fabric and curtain fabric. Based on the properties, PALF blended fabric will be used to create home textile products which will help in increased usage of the PALF fibre.

4 Materials and Methods

Materials

The pineapple leaf fibres were sourced from an agricultural farm in Guntur, Andhra Pradesh, India. The diameter of the sourced fibre was 22 μm as per the supplier. The cotton fibres with a staple length of 32 mm and diameter of 20 μm were sourced from M/s. NSC Spinning mill, Tirupur, India. Sodium hydroxide (anhydrous, ACS, 99.5% min, Sigma-Aldrich, India) was procured for fibre softening. The commercially available table mat and curtain made of 100% cotton with warp and weft count

Fig. 1 **a** Fibre cut to a length of 32 mm **b, c** softening process with NaOH

of 14 Ne were sourced from M/s. Sri Durga decors, Tirupur for comparison with the
produced pineapple leaf fabric.

4.1   Fibre Cutting and Softening

The PALF were cut manually to a length of 32 mm approximately and due to their
roughness and coarseness, the fibres were treated with sodium hydroxide for softening [20, 72]. The fibres were softened by treating them with 5% NaOH at 70
degree centigrade for 2.5 h with a material to liquor ratio of 1:15. Every time, 500 g
of fibres were softened and then the fibres were shade dried for 24 h. The cut fibres
and softening with NaOH are shown in Fig. 1.

4.2   Fibre Blending

The PALF was blended with cotton to increase the spinnable characteristics of
the fibre [27]. The PALF was blended with cotton in three different combinations as shown in Table 1. The cotton fibres were chosen for their better spinnable
characteristics.

Table 1 Blend proportions of PALF and cotton fibres

Blend proportion	Blend code
50% PALF50% cotton	50/50P/C
60% PALF40% cotton	60/40 P/C
70% PALF30% cotton	70/30P/C

4.3 Fibre Morphological Structure

The longitudinal and cross-sectional view of the PALF was determined at defect analysis laboratory, SITRA Coimbatore, India and the views were compared with the fibre compendium in SITRA to confirm the PALF.

4.4 Yarn Production

The yarn production was done on a spinning line which is a lab model. Cotton spinning was used to spin the fibres as the length of the fibres was maintained at 32 mm. The process flow is provided in Fig. 2. Trytex spinning line was used for yarn preparation. The various machines used and the process of spinning are shown in Fig. 3. The slivers were prepared on a carding machine of miniature model and draw frame machine. The slivers were passed through draw frame for two times to increase parallelisation of fibres. The yarn was then produced using rotor spinning

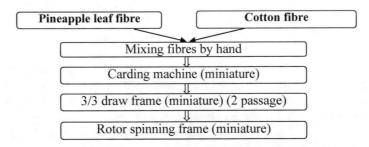

Fig. 2 Flow process chart of spinning yarn using rotor spinning

Fig. 3 Yarn production process **a** fibres fed into carding machine, **b** web formation in carding machine, **c** rotor spinning machine

Table 2 Process parameters for yarn spinning using cotton system

Process parameters	Details
Carding (Trytex)	
Licker-in speed	140 rpm
Cylinder speed	370 rpm
Doffer speed	5 rpm
Delivery hank	4.5 Ktex (0.13 Ne)
Draw frame (Trytex)	
Delivery rate	70 m/min
Fluted roller setting (front/back)	36/40 mm
Total draft	4.5
Break draft	1.3
No. of doublings	6
Delivery hank	4.5 Ktex (0.13 Ne)
Rotor spinning frame (Trytex)	
Opening roller speed	10,000 rpm
Rotor speed	39,000 rpm
TM	4.5
TPI	18
Wire type/wire angle	OS 21/18°
Rotor diameter	43 mm
Count	37 Tex (16 Ne)

machine. The process parameters of spinning the yarn are provided in Table 2. The count of yarns spun were 16 Ne with a TPI of 18. The opening roller of saw tooth type OS21 having a conventional wire angle of 18° (OK40) used for cotton fibres was used in rotor spinning. The yarns were spun in three different blend proportions.

The yarns produced in three blend proportions are shown in Fig. 4.

4.5 Yarn Properties Testing

The yarns produced were tested for their characteristics. The yarn properties that were tested are given below:

4.5.1 Yarn Count

The yarn count was measured with the help of a wrap reel for winding the lea and weighing machine with an accuracy of 0.01 mg to weigh the lea. The electronic balance used for weighing was AUW 220D Shimadzu semi microbalance. The yarn

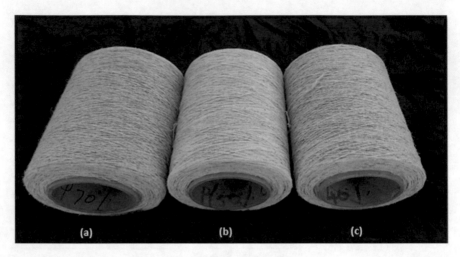

Fig. 4 Yarns produced **a** 70/30 P/C, **b** 50/50 P/C, **c** 60/40 P/C

count was determined in Ne. The standard used for measurement of yarn count was ASTM D1907-12. Then the yarn count in Ne was determined. Thirty tests were done with 5 tests in each cheese package.

4.5.2 Yarn Twist

Mec Twist single yarn twist tester was used for measuring the yarn twist using ASTM D 1422-13 standard test method where the yarn is untwisted-retwisted with a tension of 1.5gf/tex. For each blend proportion, 30 tests were done.

4.5.3 Single Yarn Strength and Elongation

Eureka single yarn strength tester was used for testing the yarn strength and elongation based on ASTM D2256-15 standard test method. Single yarn strength testing was done in the instrument where the specimen for testing was held between two jaws arranged vertically. The test specimen is tensile stressed continuously until it broke. Along with the strength, elongation of yarn was measured simultaneously based on specimen extension between the two jaws during the yarn tensile test and the elongation percentage was determined. The number of tests was limited to 30 for each blend proportion.

4.5.4 Yarn Evenness, Imperfections and Hairiness

The yarn evenness, hairiness and imperfections consisting of neps, thick places and thin places were measured using Premier Tester 7000 of Premier Evolvics based on ASTM D1425-14 standard test method. The speed of testing was 100 m/min. The yarn irregularity U% was also determined along with the above-mentioned parameters. The number of tests was limited to 5 for each blend proportion.

4.5.5 Yarn Quality Index (YQI)

The descriptive number Yarn quality index (YQI) is a useful parameter to infer the quality of yarn in place of many separate parameters. It is a single estimate of yarn quality based on the yarn tenacity, elongation % and U%. The standard of yarn will be higher if YQI is higher [68]. The YQI is calculated as follows:

$$YQI = \frac{(Yarntenacity\left(\frac{g}{tex}\right)XElongation(\%)}{U\%}$$

The yarn testing was carried out after conditioning the test specimens in standard atmospheric conditions of $27.0 \pm 2\,°C$ temperature and $65 \pm 2\%$ relative humidity. The yarn properties were further checked for statistical significance between them, if any, by using one-way ANOVA.

4.6 Woven Fabric Production

Based on the yarn properties, the 50/50 P/C fabric was further woven into a fabric. Handloom weaving was used to produce the fabrics to promote sustainable business. The handloom weaving machine used for fabric production and the 50/50 P/C fabric produced are shown in Fig. 5. The ends per inch and picks were inch was maintained at 36 and the cloth cover factor obtained is 15.11. The weave structure produced was 1×1 plain weave. The width of the fabric produced was 24 inches.

4.7 Sourced Commercial Fabrics

The sourced commercial fabrics used for table mat and curtain with the specifications as mentioned in materials section are shown in Fig. 6.

Fig. 5 **a**, **b** Handloom weaving machine and **c** produced 50/50 P/C fabric

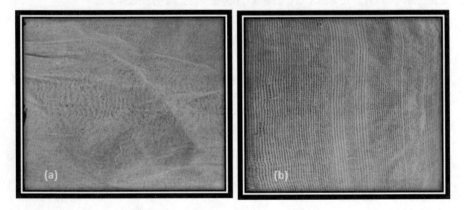

Fig. 6 Commercially sourced fabric used for **a** table mat, **b** curtain

4.8 Woven Fabric Properties Testing

The properties tested for the produced and commercially sourced woven fabrics are provided below. The tests were done in standard atmospheric conditions similar to the yarn properties testing.

4.8.1 Areal Density

The GSM cutter is used to cut the fabric and then the fabric is weighed in a balance to determine the areal density. The areal density is given in grams per square metre for the fabric and is measured as per ASTM D3776-09a.

4.8.2 Fabric Thickness

The fabric thickness was measured as per ASTM D1777-96 standard using an Essdiel thickness tester. $20gf/cm^2$ with an accuracy of 0.01 mm was maintained for thickness testing in the instrument.

4.8.3 Fabric Tensile Strength and Elongation

The fabric tensile strength and elongation was tested as per ASTM D5035 standard. The raveled strip sample was used for testing in an Instron tensile testing machine.

4.8.4 Fabric Tearing Strength

The tearing strength of the fabric was measured using Elmendorf tearing strength tester. The standard followed for testing was ISO 13937-1. The sample size used for the test is 100 mm × 75 mm.

4.8.5 Fabric Stiffness Measurement

The fabric stiffness was measured using MAG stiffness tester working on cantilever principle. The flexural rigidity of the fabrics was determined as per ASTM D1388-18 with a sample size of 1 inch × 8 inches.

4.8.6 Fabric Crease Recovery

The fabric crease recovery angle was measured using MAG crease recovery tester. The crease recovery angle of the fabrics was determined as per ISO 2313 with a sample size of 15 mm × 40 mm.

4.8.7 Fabric Abrasion Resistance

The fabric abrasion resistance was measured using Martindale abrasion tester manufactured by MAG. ASTM D4966 is the standard test method used for measuring abrasion resistance.

4.8.8 Fabric Pilling Resistance

The fabric pilling resistance was measured using ICI pill box tester manufactured by MAG. ISO 12945 is the standard test method used for measuring fabric pilling resistance.

4.8.9 Fabric Dimensional Stability

The dimensional stability of the fabrics was also assessed to find the shrinkage and growth in the fabrics. ISO 6330 standard test method was used to assess the dimensional stability of the fabrics. The specimen size is 50 cm × 50 cm. The fabric specimen is washed in a washing machine as per the standard procedure and then dried, conditioned. The conditioned specimen is then measured to calculate the shrinkage or growth percentage.

4.8.10 Fabric Flammability

The vertical flammability tester was used to measure the flammability characteristics of the fabrics. ASTM D6413 is the standard test method for finding the flammability properties of the fabrics. The specimen size used is 1 inches × 3 inches for flammability testing.

The fabric properties were further checked for statistical significance between them, if any, by using one-way ANOVA. Using the above production methods, the fabric produced is made into home textile products. Table mats and curtains were stitched using single needle lockstitch sewing machine.

5 Results and Discussions

The PALF fibres which are obtained from a natural source are utilised in this study to make textile products so that an effective usage can be ensured for these unused fibres and to also increase the growth of sustainable products in the globe. The PALF obtained was checked for confirmation as many fibres obtained from plant sources have similar appearance and identifying PALF by visual mode is a complex process. The morphological structure of the sample was studied to confirm the PALF fibre and then the fibres were used to make yarns in different combinations with cotton as indicated in the methods.

5.1 Morphological Structure of PALF

The morphological structure of PALF was determined and Fig. 7 shows the longitudinal view of PALF sourced for the study and the photographs of longitudinal view of PALF preserved in SITRA, Coimbatore for fibre confirmation. Similarly, Fig. 8 shows the cross-sectional view of sourced PALF and the photograph used for confirmation.

The longitudinal view and cross-sectional view of fibres are similar to the earlier researches reported [27, 74]. The cross sectional view shows a crescent like shape with a lot of irregularities and longitudinal view shows long narrow cylindrical tubes with slight convolutions. The PALF has many bundles of microfibres and as previous studies report, these bundles do break up during the extraction process [36, 53]. These fibre bundles are joined together by a matrix material and the cross-sectional view shows the same. It looks like some matrix material is surrounding the fibre bundles [74]. This matrix like material dissolves in NaOH solution [67, 85] and since the morphological structure was analysed before the NaOH treatment, the matrix can be seen in the cross-sectional view.

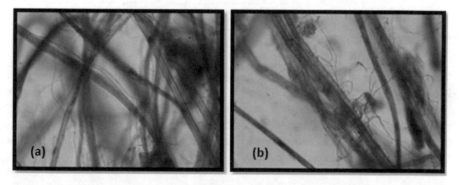

Fig. 7 Longitudinal view of **a** sourced PALF, **b** PALF photograph with SITRA

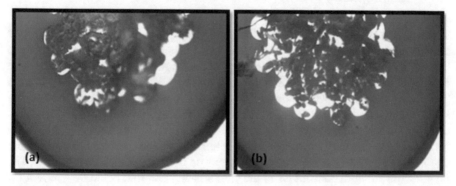

Fig. 8 Cross-sectional view of **a** sourced PALF, **b** PALF photograph with SITRA

Table 3 Yarn characteristics of PALF and cotton blends

Yarn code Yarn properties	PALF and cotton		
	50/50 P/C	60/40P/C	70/30 P/C
Count (Ne)	16.12	15.89	15.76
Count CV%	5.82	6.34	6.55
Twist (TPI)	18.5	18.3	18.12
Twist CV%	6.12	6.34	7.12
Tenacity (g/tex)	21.22	17.89	15.32
Tenacity CV%	11.42	8.43	8.83
Elongation (%)	7.81	6.7	6.3
Elongation CV%	8.43	8.38	7.72
U%	15.08	16.42	16.58
Thin (−50%)	75	82	88
Thick (+50%)	319	344	414
Neps (+280%)	377	392	481
Imperfections/Km	771	818	973
Hairiness	7.81	9.18	11.08

5.2 Yarn Count and Twist

The yarns were spun with an average count of 16 Ne in three different blend proportions. The yarn counts spun are shown in Table 3. The fibres obtained were coarser in nature and hence the coarser counts were spun. The twist was kept at around 18 per inch with a high twist multiplier of 4.5 as it leads to higher strength in the case of rotor yarns [49]. The reason for producing coarser yarn can be attributed to the fibre diameter as higher diameter leads to spinning coarser yarns [30]

5.3 Yarn Tenacity

The yarn tenacity of all the three yarns produced was measured and it was found that the yarn with lesser PALF content had more strength than the other two blends. The 50/50 P/C blend had a higher tenacity of 21.22 g/tex as shown in Table 3. The 60/40 P/C had a tenacity of 17.89 g/tex followed by 15.32 g/tex for 70/30 P/C yarn. The tenacity increases with increase in cotton fibre content in the yarn. The tenacity is found to be higher than the results reported by Hazarika et al. [24] and Ismoilov et al. [27] whereas it was similar to the results reported by Jalil et al. [30]. The reason for higher yarn tenacity can be due to the better cohesiveness between the PALF and cotton fibres as yarn tenacity majorly depends on fibre blend composition [18]. The higher fibre strength contributes to the yarn tenacity and that could be a reason for higher tenacity of 50/50 P/C blend yarn [51]. Another reason that can be attributed to higher strength is the higher crystallinity percentage of the blended fibres which lead to higher strength [63]. The crystallinity of PALF has been reported as 76% thereby leading to higher strength [20, 40].

5.4 Yarn Elongation

The yarn elongation values ranged between 6.3 and 7.81% for the blended yarns produced. The elongation was higher for the 50/50 P/C blend proportion at 7.81% followed by 60/40 P/C at 6.7% as shown in Table 3. The elongation was lower for 70/30 P/C blend proportion at 7.3%. The elongation reduced with increased content of PALF fibre and this could be due to the higher crystalline nature of the PALF fibre as higher crystallinity leads to lower elongation [50]. The elongation % is more than 80% lesser than the elongation properties of animal fibres like wool and chiengora (obtained from dog hair) due to the higher crystallinity percentage [61]. The elongation values are similar to the studies reported earlier [24, 27, 30]. The reason for more elongation when cotton fibres are more could be the more number of fibres per unit cross-section of yarn [30].

5.5 Yarn Evenness

The U% was measured for all the three yarns and it was observed that the increase in the PALF content increased the unevenness in the yarn. The U% was lower for 50/50 P/C blend at 15.08% and higher for 70/30 P/C blend at 16.58% as shown in Table 3. The 60/40 P/C blend had a U% of 16.42%. The yarn unevenness increases with increase in fibre diameter and coarseness of fibre and these can be the reason for having higher yarn unevenness. The yarn evenness is due to the irregularity in the raw material [55] as observed in the cross-section of PALF where the shapes have been irregular. However, it is found that the unevenness is lesser than the study reported by Ismoilov et al. [27]. It could be due to proper mixing done at the blending stage that has reduced the unevenness. The U% has been lesser than the U% reported for PALF yarns spun using jute and flax spinning system where the U% was 29.5% and 27.5% respectively [87].

5.6 Yarn Imperfections

The total yarn imperfections that include the thick, thin places and neps are found for the blended yarns. The total yarn imperfections were higher for the 70/30 P/C blend at 973 followed by 60/40 P/C blend at 818 and 50/50 P/C blend at 771. The average of yarn imperfections is rounded off to a whole number and is mentioned in Table

3. Similar to the yarn unevenness, the imperfections reduced with lesser content of PALF in the blend. The yarn imperfections are normally higher when rotor spinning is used as coarser fibres with higher diameter are spun on it [49]. However, the total imperfections are much lesser than the previous study where it is mentioned that the higher imperfections can be due to improper mixing as like increase in yarn evenness [27] P/C yarn and comparatively higher thin places of 88 for 70/30 P/C yarn. The 60/40 P/C yarn had 82 as the count for thin places. Similarly, the thick places were 319, 344, 414 for 50/50 P/C, 60/40 P/C and 70/30 P/C yarns respectively. The nep count was slightly higher for all the yarns as the higher diameter and coarser nature of fibres cause neps. The nep count was 377, 392, 481 for 50/50 P/C, 60/40 P/C and 70/30 P/C yarns respectively. The increase in cotton content reduces the nep count, thick and thin places due to their better strength and spinnability. The entangled nature of natural fibres compared to synthetic fibres may be a reason for more imperfections [39].

5.7 Yarn Hairiness

The yarn hairiness results were also similar to the unevenness and imperfections reported. The same trend was observed where the increase in PALF content increased the yarn hairiness leading to lesser quality of yarn. The hairiness was lesser for 50/50 P/C amongst the blends at 7.81 and higher for 70/30 P/C at 11.08 as shown in Table 3. The 60/40 P/C blend had a hairiness of 9.18. The higher linear density of the fibres leads to coarseness in fibres thereby contributing to hairiness during mechanical processing. The reason for more hairiness in higher constituents of PALF can be due to the lower number of fibres in a unit cross-section due to larger fibre diameter [1]. However, the hairiness values reported are lesser than the hairiness reported for 80% cotton 20% PALF fibre [30] and that could be due to larger differences in fibre diameter leading to move the coarser fibres to periphery thereby causing yarn hairiness. In fact, hairiness is a desirable factor when going for fabrics with high thermal insulation as hairiness leads to better thermal insulation in fabrics [4].

5.8 Yarn Quality Index (YQI)

The yarn quality index is a useful measure to find and assess the overall quality of yarn. It helps in comparing the quality of different yarns and thereby finding the better ones amongst the available ones. YQI is a number obtained using the measured yarn properties. The YQI is lesser for 70/30 P/C blend at 7.13 as shown in Fig. 9. The 60/40 P/C had a slightly better YQI of 9.13. The 50/50 P/C had the highest YQI of 11.86. The YQI increases with increase in cotton content. The YQI is based on the linear density of the yarn as lower counts lead to lower YQI [23]. The YQI shows

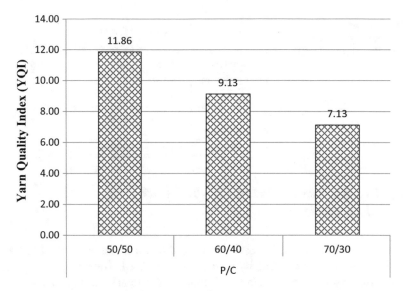

Fig. 9 Yarn quality index (YQI) of PALF cotton blends

that 50/50 P/C has better properties than other yarn blends thereby making it a strong contender for further usage.

5.9 Statistical Significance of Yarn Properties Between PALF Cotton Blends

It was required to further proceed with taking one blend out of the three with better yarn characteristics to make a fabric. It was essential to find if there was a significant difference between the yarn properties statistically. One-way ANOVA was utilised to find the significant difference in yarn properties between blends if any. The significant difference was measured between the blend proportions. It was found that there was a significant difference between all the yarn properties including strength, elongation, U%, total imperfections and hairiness thereby indicating that blend proportions are contributing to a difference in yarn properties. The P value and F values of the three blend proportions for all the yarn properties at 95% level of significance are shown in Table 4.

Table 4 Statistical significance of yarn properties between PALF cotton blends

Yarn property	F values	P values (significance)	Significant difference
Yarn strength	113.238	0.002	Yes
Yarn elongation	46.461	0.003	Yes
Yarn evenness	8.318	0.001	Yes
Yarn total imperfections	32.573	0.000	Yes
Yarn hairiness	24.936	0.000	Yes

5.10 Overall Assessment of the Produced Yarn Characteristics

The yarn properties results indicate that the reduction in the quantity of PALF in a blend improves the overall quality of the yarn produced. The strength, elongation, evenness reduces with increase in PALF content. The total imperfections in yarn and yarn hairiness increase with increase in PALF content thereby getting the quality of yarn down. The YQI is also higher for 50/50 P/C blend yarn. The coarser nature, higher diameter, surface roughness, irregularities in its cross-sectional shape and higher crystallinity of PALF can be some of the reasons for the reduction in yarn quality when compared with the quality of yarn with higher proportions of cotton in a blend. The one-way ANOVA also shows that there is statistical significance in yarn properties between the blends. Based on the above findings, it can be assessed that 50/50 P/C yarns have comparatively better quality amongst the three blends and hence it is further utilised for making woven fabrics. These woven fabrics can further be made into various textile products apart from apparel like home textile products. With the above results, it shows that 50/50 P/C yarns have good strength, lesser hairiness and U% thereby leading to use them for creating home textile products like table mats and curtains. The surface roughness and coarser nature of the fabric has limited the use of PALF in apparel and that is why not many products are seen in the mainstream market. The usage of PALF is very less in the apparel front when compared with composites. Hence, PALF can be used for alternate applications where the coarser nature will not play a major role on the end use and hence, an attempt is made to create home textile products.

5.11 Woven Fabric Properties

The 50/50 P/C yarn was taken and converted into a fabric using handloom weaving. The fabric was woven with a 1×1 plain weave structure. It is easy to produce fabrics through handloom weaving as it is possible to have tight control over the production process [83]. Handloom weaving leads to sustainable way of fabric production and it protects the artisans qualified in this trait with enough work to sustain their business. There is no to very less work available on the standards and properties required for home textile products like table mats, curtains [3, 12, 22]. There are no definite standard values required with respect to strength, stiffness, dimensional stability, etc. of the fabrics. Hence, an attempt is made here to compare the produced 50/50 P/C fabric with commercially available table mats and curtains to see its usability. This attempt will help in knowing the values of each property for the commercially available fabrics as that will be helpful in future researches. The 50/50 P/C fabric is hereafter referred to as PALF fabric for easy comparison.

5.11.1 Fabric Areal Density

The areal density of the PALF fabric was found to be 190 g which was similar to the sourced fabrics which had an areal density of 208 g and 201 g for table mat fabric and curtain fabric respectively. The coarser count of yarn used in the fabrics is the reason for higher areal density of commercially sourced fabrics. The areal density plays a major role in the fabric properties like tensile strength, flexural rigidity, etc. The EPI and PPI of the fabrics are also given in Table 5. The table mat fabric had a higher EPI and PPI.

Table 5 Characteristics of PALF, table mat and curtain fabrics

Fabric type Fabric properties	PALF, table mat and curtain fabric		
	PALF fabric	Table mat fabric	Curtain fabric
Ends per inch (EPI)	36	44	32
Picks per inch (PPI)	36	46	34
Areal density (g)	190	208	201
Thickness (mm)	0.72	0.88	0.76
Tensile strength warp way (kgf)	24.4	16.8	13.6
Tensile strength weft way (kgf)	21.5	14.2	11.2
Elongation warp way (%)	5	4.1	4.3
Elongation weft way (%)	3.2	2.8	2.7
Tearing strength warp way (lbs)	66.13	44.09	48.5
Tearing strength weft way (lbs)	57.30	35.2	36.22
Flexural rigidity warp way (mg cm)	36.7	35.65	26.77
Flexural rigidity weft way (mg cm)	38.2	35.12	20.11
Crease recovery angle warp way (°)	120	130	125
Crease recovery angle weft way (°)	125	120	135
Pilling resistance (grade)	4	4	4
Weight loss due to abrasion (%)	1.23	1.75	3.21
Dimensional stability (%)	−4	+2	−5
Flammability (sec)	23	17	12

5.11.2   Fabric Thickness

The fabric thickness was found to be lesser for the PALF fabric at 0.72 mm followed by curtain fabric at 0.76 mm and it was 0.88 mm for the table mat fabric. The thickness plays a major role in the stiffness and crease recovery angle of the fabrics. The abrasion resistance of the fabric is also affected by the fabric thickness as resistance increases with increase in fabric thickness.

5.11.3   Fabric Tensile Strength

The warp way and weft way tensile strength of the fabrics have been measured and is given in the Table 5. It is found that the tensile strength of the PALF fabric was higher than the table mat and curtain fabrics. The warp way tensile strength was 24.4 kgf and weft way was 21.5 kgf for the PALF fabric. The tensile strength of the fabric is a very important property as higher strength leads to greater fabric durability [16, 44]. The PALF fabrics are stiffer in nature [32] and that contributes to higher strength. The fibre crystallinity is another reason that can be attributed to the higher tensile strength of the PALF fabric. The count of the yarns used in PALF fabric is slightly higher than the commercially sourced fabrics and that could also be a reason for the higher tensile strength of the produced PALF fabric. The weave structure plays a role in fabric strength as 1×1 plain weave has lot of tensile strength [28] although coarser counts are used. The tensile strength of the produced PALF fabric is much lesser than the PALF silk blended fabrics. The produced PALF fabric is having 25% of the strength of the PALF silk blended fabric [24]. The presence of silk may be a reason for the higher strength of the PALF silk blended fabric. The loom tension also plays a role in the tensile strength of the resultant plain weave fabric. Lesser the loom tension, higher is the tensile strength of the plain weave fabric [48]. The sourced fabrics woven in automatic loom may have been subjected to higher loom tension than the PALF fabric woven in handloom and it can also be a reason for the higher strength of PALF fabric compared to table mat fabric and curtain fabric.

5.11.4   Fabric Elongation

The woven fabric elongation details are provided in Table 5. The average elongation percentage of all the three fabrics was less than 5%. The PALF fabric had a higher elongation of 5% in warp way and 3.2% in weft way comparatively. The curtain fabric had 4.3% and 2.8% warp and weft way elongation respectively. The table mat fabric had the lowest elongation amongst the three fabrics at 4.1% and 2.7% warp and weft way respectively. The elongation % of the produced fabrics is lesser than the elongation reported for PALF silk blended fabrics similar to tensile strength [24]. The elongation percentage had been found to be very less in these fabrics and this can be due to the higher crystalline nature of the PALF and cotton fibres as elongation gets reduced with increased crystallinity percentage in the fibre. The fibre properties play a major role in the fabric elongation as yarn elongation is also based on the fibre properties. Also, the woven fabrics do not elongate much unlike knitted fabrics [47].

5.11.5   Fabric Tearing Strength

The fabric tearing strength decides the woven fabric end usage as fabrics with lesser tearing strength are not used in home textile products [12]. The tearing strength of the three fabrics has been measured and the values are given in Table 5. It is observed that the tearing strength results are similar to the tensile strength results obtained where the tearing strength was higher in PALF fabric than the table mat and curtain fabrics. The warp and weft way tearing strength of PALF fabric was 66.13 lbs and 57.30 lbs respectively. The curtain fabric had the next best tearing strength of 48.5 lbs and 36.22 lbs in warp and weft way respectively followed by table mat fabric with 44.09 lbs and 35.2 lbs tearing strength in warp and weft way respectively. The tearing strength is higher due to the coarser count as coarser yarns used lead to higher tearing strength [70]. The tearing strength is normally lesser for plain weave fabric due to lesser floats [76, 77] in fabric unlike satin weave and in this case, the coarser count may be off setting the effect of weave on the tearing strength. The higher loom tension may be the reason for the lower tearing strength of the table mat fabric and curtain fabric produced in automatic loom [48] compared with the PALF fabric. It is found that the tearing strength of curtain fabric is higher than the table mat fabric and the reason that can be attributed to this phenomenon is the pick density. If the pick density is higher, it leads to reduction in tearing strength [17] and since the pick density is higher for table mat fabric at 46, the tearing strength is lower for them.

5.11.6   Fabric Stiffness

The flexural rigidity of the three fabrics is calculated and is shown in Table 5. The flexural rigidity was higher for the PALF fabric compared to the other two fabrics. The warp way and weft way flexural rigidity of the three fabrics were 36.7 mg cm, 38.2 mg cm for PALF fabric, 35.65 mg cm, 35.12 mg cm for table mat fabric and 26.77 mg cm, 20.11 mg cm respectively. The overall flexural rigidity of the fabrics was 37.44 mg cm, 35.38 mg cm, 23.20 mg cm for PALF fabric, table mat fabric and curtain fabric respectively. The coarseness of the PALF contributes to increased stiffness and that could be the reason for higher flexural rigidity of PALF fabric. The stiffness of the PALF fabric is higher similar to the result obtained with respect to PALF silk blends thereby indicating that the PALF leads to increased stiffness in fabrics. The cotton fabrics have less stiffness compared to other natural fibres especially the leaf-based ones and hence the stiffness is lesser in table mat fabric and curtain fabric. The plain weave fabrics with less thickness have low stiffness [11, 69] and that could be the reason for lower stiffness of the sourced fabrics.

5.11.7 Fabric Crease Recovery Property

The crease recovery angle of the fabrics is provided in Table 5. It is found that the crease recovery angle ranges between 120° and 135° for all the three fabrics. The crease recovery angle is more or less same for all the three fabrics with PALF fabric having warp and weft way crease recovery angle of 120° and 125° respectively. The table mat fabric had a crease recovery of 130° in warp way and 120° in weft way. The curtain fabric had a crease recovery of 125° and 135° in warp and weft way respectively. The bending length and fabric stiffness also play a major role in the crease recovery property of a fabric [84]. The crease recovery angle is slightly higher than the one reported for silk PALF blend fabric [24] and that could be due to the presence of cotton in all the fabric blends. The thickness of the fabric also plays a role in crease recovery angle and since the fabrics had lesser thickness, it could be a reason for higher crease recovery angle than the earlier reported PALF fabric blends results. The yarn twist is also an important factor contributing to the crease recovery properties of the fabric [73].

5.11.8 Fabric Pilling Resistance

The pilling is an undesirable property in a fabric. A pilled fabric may not be suitable for any application other than being used as a waste cloth. In this regard, it is essential to understand the pilling properties of PALF fabric as home textile products should be without pills as it affects the aesthetic property of the fabric thereby avoiding the intended usage of a fabric [3]. Pilling resistance has to be checked in home textile products and it is an important property to be evaluated similar to abrasion resistance of the fabric. The results of the pilling test are favourable for all the three fabrics where they are graded as 4. The grade 4 indicates that there is very slight surface fuzzing and no pills formed. The pilling happens due to inter fibre migration [26]. The reason for pilling grade to be 4 could be the yarn count as finer counts lead to compact structure thereby reducing the pilling [66]. Another reason can be the blending as when two fibres are blended, the fibres which are coarser in nature tend to move to the periphery causing surface fuzzing. In all the three fabrics, only very slight surface fuzzing is noticed thereby making them suitable for home textile products without affecting the aesthetic property of the final products.

5.11.9 Fabric Abrasion Resistance

Along with pilling, abrasion resistance is another important property for the home textile products. The home textile products are expected to have more abrasion resistance as they are subjected to lot of wear and tear during their usage in furnishings and kitchen fabrics [8]. The abrasion resistance of the three fabrics has been determined and it is found that the PALF fabric had an abrasion resistance of 1.23%. The table mat fabric and curtain fabric had an abrasion resistance of 1.75 and 3.21%. The percentage here refers to the weight loss of the fabrics after abrasion. It indicates that PALF fabric had better abrasion resistance and this could be due to the higher fibre strength. The abrasion resistance of fabrics improves with surface treatments [86]. However, in the case of these three fabrics, no surface treatments have been given and the abrasion resistance is found to be better for all the fabrics. The curtain fabric had lesser abrasion resistance and that could be due to the combination of lower fabric thickness and constituent fibre collectively contributing to it.

5.11.10   Fabric Dimensional Stability

The dimensional stability of all the three fabrics has been studied and it is found that there was a shrinkage of 4 and 5% in PALF fabric and curtain fabric whereas the table mat fabric had a growth of 2%. Normally, the woven fabrics are more dimensionally stable than the knit fabrics [25]. The dimensional stability is better for all the three fabrics and this could be due to the weave structure as fabrics with more interlacing have better dimensional stability [80]. The dimensional stability of the woven fabrics can be increased for making them more suitable for home textile furnishings with the help of special finishes [29], however, in this study, no finish has been imparted on to the fabrics.

5.11.11   Fabric Flammability Characteristic

The flammability property of the fabrics is estimated with the help of a vertical flammability tester. The home textile products are regulated under flammability regulations [59] and hence it becomes imperative to check the flammability characteristics of the three fabrics. It is found that the PALF fabric had better flammability characteristics as it had 23 s flame spread time which was higher than the other two fabrics. The table mat fabric had a flame spread time of 17 s and curtain fabric had a flame spread time of 12 s. The constituent fibre plays a major role in flammability property [10]. Cotton has a very low flammability property [21]. The fabrics with higher thickness have good resistance to flame and in this case, all the fabrics have low thickness. The flame retardant finishes are very much essential when cotton fabrics are used due to their poor flammability resistance characteristics [33]. However, in this case, the PALF fabric has comparatively higher flame spread time of 23 s due to presence of PALF thereby making them suitable for home textiles application. In the future, the flame resistance of these fabrics can be increased by applying flame retardant finishes as suggested in the study by Lee et al. [41].

5.11.12   Overall Comparison of the Fabric Properties

The produced PALF fabric has indicated that it has similar properties when compared with the commercially available home textile products. In fact, with respect to properties like strength and abrasion resistance, the PALF fabric is better than the other two fabrics. The pilling resistance of the PALF fabric is similar to the commercially purchased fabrics. The flexural rigidity of the PALF is slightly higher thereby leading to higher stiffness of the fabric. However, it can be seen that the flammability characteristics of the PALF fabric are better and hence, by considering the overall properties, it can be said that PALF fabrics can be used effectively as table mats and curtains. Many other uses can also be identified for PALF in home textile sector. Based on the above results, table mat and curtain were stitched and its usage was checked.

5.12   PALF Table Mat and Curtain

Figures 10 and 11 show the table mat and curtain prepared using the PALF fabric respectively. It is found that the curtains and table mats fit the purpose and it shows that PALF fabrics can be used in home textile applications in the future. The PALF which is otherwise thrown as waste in bulk can be converted into useful home textile products using handloom weaving as this gives scope for sustainable production of a sustainable fibre for a sustainable usage to safeguard the ecology and economy of the world in the future.

Fig. 10 Produced PALF table mat

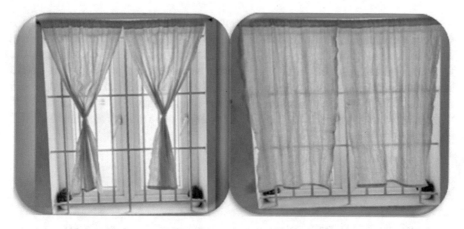

Fig. 11 Produced PALF curtain

6 Conclusions

The PALF is available abundantly and its use in areas other than composites has been very less. The chances of using the PALF for various applications like home textile products are there as it provides an effective use for PALF as it is not the main end use of a pineapple plant. A sustainable approach will lead to finding usage of this by-product from pineapple plant. An attempt has made in this work to spin PALF yarns in three blend compositions along with cotton. The morphological structure of PALF was also studied. The yarns were spun with a count of 16 Ne and TPI of 18. The yarn properties were studied and it was found that the yarn strength reduced with increase in PALF content in the blend proportion with 50/50 P/C having a higher tenacity of 21.22 g/tex. The U% also increased with increase in PALF content and it was 15.08% when the blend proportion was 50/50 P/C. The total imperfections in yarn and hairiness increased with reduction in cotton content in the blend. The YQI was higher for 50/50 P/C blend at 11.86. There was also a significant difference between the yarn properties of three blends.

Based on the yarn properties, it was found that 50/50 P/C had better yarn properties and those yarns were woven into handloom fabrics. The woven fabric was compared with commercially available table mat and curtain fabric made of cotton. The areal density of the produced fabric was 190 g. The PALF fabric thickness was similar to the purchased fabrics and hence it formed a base for easy comparison of the other fabric properties. The fabric properties like tensile strength, tearing strength were higher for the produced fabric at 22.95 kgf and 61.72 lbs. The flexural rigidity of the fabrics increased with increase in PALF content thereby indicating that the PALF fabrics had higher stiffness than the table mat and curtain fabrics available commercially. There was shrinkage of 4% in the fabric after washing when checked for dimensional stability. The abrasion resistance and pilling resistance of the PALF fabric was similar to the purchased fabrics and hence it meets the requirements for usage in home textiles. Another significant factor is the flammability of the PALF fabric which was higher at 23 s thereby making it a suitable contender for usage in home textile products. The table mat and curtain was constructed using single needle lockstitch sewing machine and it was found to be confirming for use. Hence, useful home textile products can be created using PALF thereby indicating that PALF can be produced more as the demand will increase when versatile usage is found for it. This attempt will form a base and help in creating eco-friendly sustainable PALF products in the future.

References

1. Altas S, Kadoglu H (2006) Determining fibre properties and linear density effect on cotton yarn hairiness in ring spinning. Fibres Text Eastern Euro 14:48–51
2. Arora RK, Gupta NP, Patni PC (1985) Characteristics and processing performance of pineapple leaf fibre. Indian J Fibre Text Res 10:125–126
3. Babu VR, Sundaresan S (2018) Home furnishing. WPI Publishing
4. Balasubramanian N (2007) Hairiness of yarns: relative merits of various systems. Retrieved from http://www.indiantextilejournal.com/articles/FAdetails.asp?id=435
5. Ban KM (2016) Sustainable development goals
6. Banik S, Nag D, Debnath S (2011) Utilization of pineapple leaf agro-waste for extraction of fibre and the residual biomass for vermicomposting. Indian J Fibre Text Res 36:172–177
7. Biagiotti JMKJ, Puglia D, Kenny JM (2004) A review on natural fibre-based composites-part I: structure, processing and properties of vegetable fibres. J Nat Fibers 1(2):37–68
8. Bilisik K, Yolacan G (2009) Abrasion properties of upholstery flocked fabrics. Text Res J 79(17):1625–1632
9. Buitrago B, Jaramillo F, Gómez M (2015) Some properties of natural fibers (sisal, pineapple, and banana) in comparison to man-made technical fibers (aramid, glass, carbon). J Nat Fibers 12(4):357–367
10. Collier BJ, Epps HH (1998) Textile testing and analysis. Prentice Hall
11. Cooper DNE (1960) 24—The stiffness of woven textiles. J Text Inst Trans 51(8):T317–T335
12. Das S (2010) Performance of home textiles. Woodhead Publishing India Pvt, Limited
13. De La Cruz Medina J, García HS(2021) Post-harvest operations. 2015. Retrieved from, Medina http://www.fao.org/fileadmin/user_upload/inpho/docs/Post_Harvest_Compen dium_-_Pineapple.pdf. Accessed on 12 Octo 2021
14. Debnath S (2016a) Pineapple leaf fibre—a sustainable luxury and industrial textiles. In: Handbook of sustainable luxury textiles and fashion. Springer, Singapore, pp 35–49
15. Debnath S (2016b) Unexplored vegetable fibre in green fashion. In: Green fashion. Springer, Singapore, pp 1–19
16. Devarakonda VK, Pope CJ (1970) Relationship of tensile and tear strengths of fabrics to component yarn properties. No. C/PLSEL-TS-175. Army Natick Labs Ma Clothing and Personal Life Support Equipment Lab.7825014722
17. Dhamija S, Chopra M (2007) Tearing strength of cotton fabrics in relation to certain process and loom parameters, vol 32, pp 439–445
18. Fang HH, Lee CSP, Cassidy T (1997) Woolen ring-frame drafting. Text Res J 67:163–173
19. Franck RR (ed) (2005) Bast and other plant fibres, vol 39. CRC Press
20. Gaba EW, Asimeng BO, Kaufmann EE, Katu SK, Foster EJ, Tiburu EK (2021) Mechanical and structural characterization of pineapple leaf fiber. Fibers 9(8):51
21. Gandhi S, Spivak SM (1994) A survey of upholstered furniture fabrics and implications for furniture flammability. J Fire Sci 12(3):284–312

22. Gopalakrishnan M, Vijayasekar R, Ashok Kumar A, Saravanan D (2021) Value addition in handloom textile products for sustainability. In: Handloom sustainability and culture. Springer, Singapore, pp 119–141
23. Halimi MT, Hassen MB, Azzouz B, Sakli F (2007) Effect of cotton waste and spinning parameters on rotor yarn quality. J Text Inst 98:437–442
24. Hazarika P, Hazarika D, Kalita B, Gogoi N, Jose S, Basu G (2018) Development of apparels from silk waste and pineapple leaf fiber. J Nat Fibers 15(3):416–424
25. Hsu LH, Cheek L (1989) Dimensional stability of ramie, cotton and rayon knit fabrics. Cloth Text Res J 7(2):32–36
26. Hunter L (2009) Pilling of fabrics and garments. In: Engineering apparel fabrics and garments, pp 71–86
27. Ismoilov K, Chauhan S, Yang M, Heng Q (2019) Spinning system for pineapple leaf fiber via cotton spinning system by solo and binary blending and identifying yarn properties. J Text Sci Technol 5(4):86–91
28. Jahan I (2017) Effect of fabric structure on the mechanical properties of woven fabrics. Adv Res Text Eng 2(2):1018–1022
29. Jain H (2010) Effect of specialty finishes on handloom woven furnishing fabrics with different fiber contents. Man-made Text India 53(1)
30. Jalil MA, Moniruzzaman M, Parvez MS, Siddika A, Gafur MA, Repon MR, Hossain MT (2021) A novel approach for pineapple leaf fiber processing as an ultimate fiber using existing machines. Heliyon 7(8):e07861
31. Jawaid M, Asim M, Tahir PM, Nasir M (eds) (2020) Pineapple leaf fibers: processing, properties and applications. Springer Nature
32. Jose S, Salim R, Ammayappan L (2016) An overview on production, properties, and value addition of pineapple leaf fibers (PALF). J Nat Fibers 13(3):362–373
33. Kadolph SJ (1998) Quality assurance for textiles and apparel
34. Kannan TG, Wu CM, Cheng KB, Wang CY (2013) Effect of reinforcement on the mechanical and thermal properties of flax/polypropylene interwoven fabric composites. J Ind Text 42(4):417–433
35. Kannojiya R, Gaurav K, Ranjan R, Tiyer NK, Pandey KM (2013) Extraction of pineapple fibres for making commercial products. J Environ Res Dev 7(4):1385
36. Kengkhetkit N, Amornsakchai T (2012) Utilisation of pineapple leaf waste for plastic reinforcement: 1. A novel extraction method for short pineapple leaf fiber. Ind Crops Prod 40:55–61
37. Khalil HSA, Alwani MS, Omar AKM (2006) Chemical composition, anatomy, lignin distribution, and cell wall structure of Malaysian plant waste fibers. BioResources 1(2):220–232
38. Labbe N, Rials TG, Kelley SS, Cheng ZM, Kim JY, Li Y (2005) FT-IR imaging and pyrolysis-molecular beam mass spectrometry: new tools to investigate wood tissues. Wood Sci Technol 39(1):61–76
39. Lawrence CA (2003) Fundamentals of spun yarn technology. Crc Press, Florida, FL
40. Lee CH, Khalina A, Lee SH, Padzil FNM, Ainun ZMA (2020) Physical, morphological, structural, thermal and mechanical properties of pineapple leaf fibers. In: Pineapple leaf fibers. Springer, Singapore, pp 91–121
41. Lee SH, Lee CH, Ainun ZMA, Padzil FNM, Lum WC, Ahmad Z (2020) Improving flame retardancy of pineapple leaf fibers. Pineapple leaf fibers. Springer, Singapore, pp 123–141
42. Lopattananon N, Payae Y, Seadan M (2008) Influence of fiber modification on interfacial adhesion and mechanical properties of pineapple leaf fiber-epoxy composites. J Appl Polym Sci 110(1):433–443
43. Loppi S, Roblin B, Paoli L, Aherne J (2021) Accumulation of airborne microplastics in lichens from a landfill dumping site (Italy). Sci Rep 11(1):1–5
44. Malik ZA, Malik MH, Hussain T, Arain FA (2011) Development of models to predict tensile strength of cotton woven fabrics. J Eng Fibers Fabr 6(4):155892501100600420
45. Marchessault RH (1962) Application of infra-red spectroscopy to cellulose and wood polysaccharides. Pure Appl Chem 5(1–2):107–130

46. Mathew S (2017) Economic analysis of production and disposal of pineapple (Ananas comosus L.)in Dodamarg Tahsil of Sindhudurg district (MS) (Doctoral dissertation, DBSKKV., Dapoli)
47. Matković VP, Skenderi Z (2013) Mechanical properties of polyurethane coated knitted fabrics. Fibres Text Eastern Euro 4(100):86–91
48. Mebrate M, Gessesse N, Zinabu N (2020) Effect of loom tension on mechanical properties of plain woven cotton fabric. J Nat Fibers,1–6
49. Miao M (2009) High-performance wool blends. In: Johnson NAG, Russell IM (eds) Advances in wool technology, pp 284–307. Woodhead, Florida
50. Mishra SP (2016) Fibre structure. Woodhead, New Delhi
51. Mwasiagi JI, Huang XB, Wang XH (2008) Performance of neural network algorithms during prediction of yarn breaking elongation. Fibers Polym 9:80–86
52. Napper IE, Thompson RC (2016) Release of synthetic microplastic plastic fibers from domestic washing machines: effects of fabric type and washing conditions. Marine Pollut Bull 112(1–2):39–45
53. Neto ARS, Araujo MAM, Barboza RMP, Fonseca AS, Tonoli GHD, Souza FVD, Mattoso LHC, Marconcini JM (2015) Comparative study of 12 pineapple leaf fiber varieties for use as mechanical reinforcement in polymer composites. Ind Crops Prod 64:68–78
54. Noren F, Noren K, Magnusson K (2014) Marine Microscopic Debris, Survey along Swedish West Coast 2013 and 2014, IVL, Swedish Environmental Institute report no 2014: 52. County Administrative Board, West Gotaland, available at: https://sverigesradio.se/diverse/appdata/isi dor/files/406/14638.pdf
55. Nptel (2011) Causes of irregularity. Retrieved from http://nptel.ac.in/courses/116102029/32
56. Pandit P, Nadathu GT (2018) Characterization of green and sustainable advanced materials. Green Sustain Adv Mater Process Charact 1:35–66
57. Pandit P, Pandey R, Singha K, Shrivastava S, Gupta V, Jose S (2020) Pineapple leaf fibre: cultivation and production. In: Pineapple leaf fibers. Springer, Singapore, pp 1–20
58. Pardeshi S, Mirji MJ, Goud V (2012) Extraction of pineapple leaf fiber and its spinning: a review. In: Textile review. Saket Projects Limited, Ahmedabad
59. Patel DB (2015) Flammability and Its Influencing Factors. Int J Res Hum Soc Sci 3(4):33–36
60. Rafiqah A, Abdan K, Nasir M, Asim M (2020) Effect of extraction on the mechanical, physical and biological properties of pineapple leaf fibres. In: Pineapple leaf fibers. Springer, Singapore, pp 41–54
61. Ramamoorthy S, Ramadoss M, Kandhavadivu P, Kalapatti Jagannathan VV (2021b) Evaluation of yarn and fabric properties of environmentally friendly and sustainable chiengora fibers from Pomeranian dog breed for textile applications. J Nat Fibers, 1–14
62. Ramamoorthy S, Ramadoss M, Ramasamy R, Thangavel K (2020) Analysis of physical and thermal properties of chiengora fibers. J Nat Fibers 17(2):246–257
63. Ramamoorthy S, Ramadoss M, Thangavelu K, Kalapatti Jagannathan VV (2021a) Analysis on the yarn and fabric characteristics of Chiengora fibers from hairs of Lhasa Apso breed and its blends in comparison with wool and its blends. Part I—yarn characteristics. J Nat Fibers,1–14
64. Rathinamoorthy R, Balasaraswathi SR (2020) A review of the current status of microfiber pollution research in textiles. Int J Cloth Sci Technol
65. Rathinamoorthy R, Raja Balasaraswathi S (2021) Domestic laundry and microfiber shedding of synthetic textiles. Microplastic Pollut, 127–155
66. Ruhul AM, Rana MRI (2015) Analysis of pilling performance of different fabric structures with respect to yarn count and pick density. Ann Univ Oradea. Fascicle of Text 16(1):9
67. Saha SC, Das BK, Ray PK, Pandey SN, Goswami K (1993) Some physical properties of pineapple leaf fiber (PALF) influencing its textile behavior. J Appl Polym Sci 50(3):555–556
68. Sakthivel JC, Mukhopadhyay S, Palanisamy NK (2005) Some studies on mudar fibers. J Ind Text 35:63–76
69. Sebastian SARD, Bailey AI, Briscoe BJ, Tabor D (1986) Effect of a softening agent on yarn pull-out force of a plain weave fabric. Text Res J 56(10):604–611
70. Sharma IC, Malu S, Bhowan P, Chandna S (1983) Influence of yarn and fabric properties on tearing strength of woven fabrics 8:105–111

71. Singha K, Pandit P, Shrivastava S (2020) Anatomical structure of pineapple leaf fiber. In: Pineapple leaf fibers. Springer, Singapore, pp 21–39
72. Sinha MK (1982) A review of processing technology for the utilisation of agro-waste fibres. Agric Wastes 4(6):461–475
73. Steele R (1956) The effect of yarn twist on fabric crease recovery. Text Res J 26(10):739–744
74. Surajarusarn B, Traiperm P, Amornsakchai T (2019) Revisiting the morphology, microstructure, and properties of cellulose fibre from pineapple leaf so as to expand its utilization. Sains Malaysiana 48(1):145–154
75. Tamta M, Mahajan S (2020) Innovative applications of pineapple leaf fibre in textiles and other fields. Recent trends in textiles—a paradigm for innovative and sustainable fibers, yarns, fabrics and garments production, processing & designing aspects. Conference proceedings
76. Taylor HM (1959) 9—Tensile and tearing strength of cotton cloths. J Text Inst Trans 50(1):T161–T188
77. Teixeira NA, Platt MM, Hamburger WJ (1955) Mechanics of elastic performance of textile materials: part XII: relation of certain geometric factors to the tear strength of woven fabrics. Text Res J 25(10):838–861
78. Thilagavathi G, Muthukumar N, Neela Krishnanan S, Senthilram T (2020) Development and characterization of pineapple fibre nonwovens for thermal and sound insulation applications. J Nat Fibers 17(10):1391–1400
79. Todkar SS, Patil SA (2019) Review on mechanical properties evaluation of pineapple leaf fibre (PALF) reinforced polymer composites. Compos Part B Eng 174:106927
80. Topalbekiroğlu M, Kaynak HK (2008) The effect of weave type on dimensional stability of woven fabrics. Int J Cloth Sci Technol 20:281–288
81. Truscott L (2018) Preferred fiber & materials market report 2019. TextileExchange, https://textileexchange.org/wp-content/uploads/2019/11/Textile-Exchange_Preferred-Fiber-Material-Market-Report_2019.pdf
82. Uma Devi L, Joseph K, Manikandan Nair KC, Thomas S (2004) Ageing studies of pineapple leaf fiber–reinforced polyester composites. J Appl Polym Sci 94(2):503–510
83. Wan Ahmad WY, Salleh J, Ahmad MR, Ab Kadir MI, Misnon MI, Mohd Nor MA, Omar K, Tumin SM (2010) Kajian penghasilan polipina: the conversion of pineapple leaves to white fabric
84. Wang L, Liu J, Pan R, Gao W (2015) Exploring the relationship between bending property and crease recovery of woven fabrics. J Text Inst 106(11):1173–1179
85. Xiao B, Sun X, Sun R (2001) Chemical, structural, and thermal characterizations of alkali-soluble lignins and hemicelluloses, and cellulose from maize stems, rye straw, and rice straw. Polym Degrad Stab 74(2):307–319
86. Yang CQ, Qian L, Lickfield GC (2001) Mechanical strength of durable press finished cotton fabric. Part IV: Abrasion resistance. Text Res J 71(6):543–548
87. Yusof Y, Yahya SA, Adam A (2014) A new approach for PALF productions and spinning system: the role of surface treatments. J Adv Agric Technol 1(2):161–164
88. Yusof Y, Yahya SA, Adam A (2015) Novel technology for sustainable pineapple leaf fibers productions. Proc CIRP 26:756–760

Production of Sustainable Banana Fibers from Agricultural Wastes and Their Properties

Feristah Unal, Ozan Avinc, and Arzu Yavas

Abstract Agricultural waste of banana plant cultivated for only fruit production generally increases in each year due to the increment of banana cultivation to fulfill the need of increased world population. Evaluation of agricultural waste of banana plant as fiber increases the area of use of banana fiber. Because of its easy accessibility, renewability, biodegradability, and sustainability, the areas of use of banana fibers, which are especially preferred in composite materials, are increasing day by day. Banana fiber, ligno-cellulosic fiber, which has superior physical and mechanical properties, are now regarded as an important textile material. Banana fibers play a role in several different applications in many different fields such as textile, composite, paper, construction, thermal insulation, automotive and transportation, etc. Moreover, banana fibers can also be used in the production of clothing such as t-shirts, shirts, kaftans, kimonos, etc., and carpets, bags, handbags, wallets, purses, belts, shoes, ornaments, souvenirs, handicraft products, sanitary pads, etc. Banana fiber, whose popularity has increased recently, also appears in different textile products such as dresses, sarees, and trousers in textile fashion. In this chapter, the cultivation of the banana plant and the extraction methods of the banana fibers, the physical–chemical structure and characteristics of the banana fibers, and their usage areas were examined in detail.

Keywords Banana · Banana fiber · Banana fabric · Biodegradable · Sustainable · Renewable · Composite · Agricultural waste

1 Introduction

Today, sustainability means to consciously protecting the future and to meet our needs to live without disturbing the natural balance [1, 2]. Sustainability aims to protect the ecological balance and manage the resources correctly in order to leave a

F. Unal · O. Avinc (✉) · A. Yavas
Department of Textile Engineering, Faculty of Engineering, Pamukkale University, Denizli 20160, Turkey
e-mail: oavinc@pau.edu.tr

livable planet to future generations [1, 2]. The transfer of this consciousness to future generations is very important [1, 2]. Continuous increase in textile consumption with changing fashion and wastes left to the environment as a result of fast and intensive production disrupt the balance of ecology [2]. The use of natural resources alone is not enough to be produced completely harmless and natural products [1]. Every step to be taken in terms of sustainability is of huge value for the world; and therefore, sustainability issue has also become very important in textile production [3–12]. In order to ensure sustainability in textiles, questions of how to be more sustainable are asked at every step of textile production, and for this purpose, sustainable fibers (such as many different types of natural fibers, soybean, casein, zein, ardil, poly (lactic acid) fibers, etc.), sustainable materials, sustainable production techniques, sustainable textile wet-processing technologies and methods (such as ozone, UV/Ozone, UV, $ScCO_2$, ultrasound, etc.) are sought and are increasingly being implemented day by day [13–56]. It is necessary to minimize the wastes that cause harm to the environment and to recover by recycling and to use natural resources, environmentally friendly fibers, and materials as much as possible [1, 2, 57, 58]. Therefore, another important factor in sustainability is recycling [1, 2, 59–61]. By recycling, it is possible to convert products made from natural raw materials into products that can be reused like fabric and yarn [1, 58, 62]. For this reason, besides the production of widely used plant fibers such as cotton, flax, jute fibers, etc., the diversity, and usage areas of other sustainable, renewable biodegradable fibers such as naturally colored cotton, hemp, nettle, pineapple leaf, abaca, raffia, and banana fibers that find less use should be increased [1, 58, 63–81]. Natural fibers, which offer a wide range of consumers for use in daily needs, play an important role in human life [82]. Although synthetic fibers have been preferred over natural fibers in recent years because of their low costs, the fact that most of the synthetic fibers cannot be biodegradable causes serious pollution in nature [65, 82–92]. Therefore, various projects and works are underway worldwide to popularize the use of natural fibers known as recyclable, environmentally friendly fibers [57, 63, 65, 84, 86, 87, 89–94].

The pseudostem is a part of the banana plant which appears to be a trunk, which arises from a soft central core and tightly wrapped up to 25 leaf sheaths, and banana fiber is extracted from the banana tree which possesses circular leaf structure often named to be as bast [95, 96]. Therefore, banana pseudostem fibers could be characterized as leaf fibers and as bast fibers. For this reason, banana fibers are called leaf fibers, pseudostem fibers, or bast fibers in different sources [82–84, 88–90, 92, 95, 97–117]. Ligno-cellulosic banana fibers are among the fibers that are likely to find more use in future with their high potential and environmentally friendly structure obtained from the agricultural wastes of the banana plant [83, 86, 89, 90, 93, 111, 118]. It is also among the oldest and most important fruits grown in tropical regions of the world [57, 82, 83, 100, 101, 118–120]. It is known that growing banana plant cultivation day by day creates agricultural wastes which are growing at the same rate due to the lack of suitable technology for commercial use [57, 84, 86, 88, 90, 93, 111, 121–123]. Generally, these wastes are used as animal feed and fuel [82, 83]. Although alcohol production, starch extraction and numerous other uses are available, these wastes are known to be a good source of textile fiber [97, 98]. It is the right place to mention that in addition to the fibers acquired from the pseudostem of the banana plant, fibers could also be obtained from banana inflorescence bracts [87] and from the peel of the banana fruit [124].

India is the biggest banana producer country, accounting for 27% of our planet's banana production [120, 123]. The percentage of pseudostem waste generated after the harvest of banana plants is 51.18 million tons per year in India [82, 99, 121]. The amount of fiber obtained from this rate of waste was calculated as 3.87 million tons [82, 99, 125]. Universally 1,2 billion tons million tons of banana waste, banana peel, leaves, and stems are produced each year and are not used commercially [90, 126]. The banana fiber industry in Jamaica has been active since the mid-nineteenth century [82, 99]. Banana fiber, which is used day by day for different purposes in different countries, is woven as a transparent thin fabric known as agna in the Philippines and generally used for men's shirts [82]. In Sri Lanka, banana fiber is used on the insoles of expensive shoes [82].

Thanks to the developed processes, paper, money paper, rope, bags, hats, and handcrafted products can be produced from banana fiber, and thus many different products suitable for daily use can be produced from banana fibers [82, 87, 88, 90, 104, 109–111, 113, 120, 126–128]. Banana fibers also could be used as reinforcement in the polymer matrix to create composite materials [82, 87, 91, 129, 130]. In addition, cellulose nanoparticles obtained from banana peels can also be utilized as reinforcement of polymer matrices [57, 124, 130]. Here, it is very important to obtain cellulose nanoparticles from cheap, biodegradable, and renewable sources [57, 124]. Moreover, cellulose nanofibers obtained from banana peel by chemical and mechanical methods are used in polymer matrix composites [57, 82, 124, 129, 130]. Thanks to these fibers used in polymer matrices, a composite material which is better strength and toughness than reinforced plastics is obtained [63, 83, 84, 88, 90, 91, 94, 100, 129–133]. Furthermore, these fibers are more preferred as they are in the group of lighter, cheaper, renewable, and biodegradable fibers (natural fibers) than non-renewable and expensive chemical fibers such as nylon, glass, and carbon [63, 83, 84, 88, 90, 91, 94, 100, 129, 131–134].

Banana plant is considered as an important source of fiber [82, 101]. Pseudostems and leaves of the banana plant are left on the roadside at the time of harvest, and this causes environmental pollution [86, 90, 101, 110, 126–128]. After harvesting banana plants, it is expected to be a solution to environmental problems in the sense that by evaluating the stems and leaves of the plant left on the roadside as fiber [90, 101, 110, 126, 128].

For instance, the banana plant is grown in large quantities in also Turkey, and it is known that only the fruit of the plant is utilized [100]. However, recently, there are different projects initiated to make daily use possible by obtaining banana fiber from banana plant in Turkey [135]. Such steps and projects are recently taking place in many different countries in the world where the banana tree grows. In this chapter, the cultivation of the banana plant and the extraction methods of the banana fibers, the physical–chemical structure and characteristics of the banana fibers, and their usage areas were investigated in detail. In this chapter, information about cultivation of banana plant, banana fibers and their production, their yarn formation, chemical structure, physical and chemical properties, application areas is given, respectively.

2 Banana Plant and Its Cultivation

Bananas are one of the world's most popular and our planet's most consumed fruits (Fig. 1). Banana fruits comprise essential nutrients which could have a protective impact on human health [136]. Banana fruits grow on the banana plant. The appearance of a complete banana plant (from Turkey) with flowers, fruits, leaves, and stem is shown in Fig. 2. The banana plant belongs to the group perennial herbaceous plants belonging to the Musaceae family [83, 99, 102, 103, 111, 133, 137]. Musaceae is known as a flowering plant family consisting of the three genera [138]. This plant family is found in tropical regions such as Asia and Africa [128, 134, 138]. The banana plant belonging to the Musaceae family is also grown in tropical and subtropical regions [57, 58, 86, 103, 120, 125–127, 137, 139]. The most intense regions of banana cultivation in the world are Asia, Africa, and Australia [103, 139]. For instance, the production of the banana plant in Turkey is carried out mainly in a part of the Mediterranean region [100].

Fig. 1 Banana fruits and other exotic fruits in a fruit seller in Cuba

Fig. 2 The appearance of a complete banana plant (from Turkey) with flower, bracts, fruits, leaves, and stem (pseudostem)

Although the average length of these plants belonging to the Musaceae family is 2–5 m, there are species that grow up to 10 m [102, 111]. The leaves are 2.7–3 m in length and show a spirally aligned [111, 139–143]. A banana plant can have about 26–32 leaves [87]. Leaf width can be up to 60 cm [111, 139–143]. The flowers are covered with greenish and purplish fleshy bracts [87]. These appear in a spiral along the inflorescence axis in groups of 10–20. As the flowering development progresses, greenish and purplish fleshy bracts are shed. Spilled bracts are left to rot on agricultural land. However, fiber can also be obtained from the bracts, and the remaining mass can be utilized instead of soil in compost making or in plant pots [87]. The banana plant, known to have more than 40 species, is actually divided into two genera, Ensete and Musa [100]. Ensete and Musa are known as a genus of monocarpic flowering plants found in tropical regions [83, 144]. The most significant difference between these two types of bananas; Ensete-type bananas are not consumed as fruit, whereas Musa-type bananas are consumed as fruit [100]. Almost all modern edible bananas are derived from banana plants of the *Musa acuminata* and *Musa balbisiana* type [137]. *Musa paradisiaca* is the name of the hybrid *Musa acuminate* x *Musa balbisiana* type banana. The fruits obtained from the banana plant are usually long and curved [137]. However, they can differentness in size, color, and hardness.

It is anticipated that the annual total banana production in the world is 116 million tons [123]. India is the world leader in banana production with an annual production of approximately 30 million tons. India, which produced 14 million tons of bananas in the 2000s, produced approximately 30.8 million tons of bananas in 2018. In 2018, the country producing the highest banana production after India with 11.5 million tons in the world is China [123]. According to the data of FAO (Food and Agriculture Organization), only bananas production in Europe in 2013 was 615 thousand tons [119, 145, 146]. For instance, the same year, according to the TSI (Turkey Statistical Institute) data, banana production in Turkey was 215 thousand tons, 305,926 tons in 2016, that was recorded in 2019 at 548,323 tons [119, 123, 145, 146]. Banana plant is among the most agriculturally produced plants in the world [86, 119, 134, 137] and is the most traded fruit worldwide [86, 109, 127, 137]. According to the data of the FAO (Food and Agriculture Organization) in 2018, it shows that there are more than 5 million hectares of banana plantation in more than 130 countries in the world [58, 119]. In addition, bananas rank fourth among the most important global food products [111]. Most production of banana takes place in Costa Rica, Colombia, Philippines, India, Brazil, Ecuador, China, Indonesia, Mexico, and Thailand [82, 83, 100, 102, 111, 118, 120, 127, 128, 147]. According to the data of the United Nations Food and Agriculture Organization, a large part of this banana export has been realized by Latin America and Caribbean countries [119]. In addition, South Africa, Spain, Israel, and Turkey are conducted banana production in many countries [82, 83, 100, 102]. For instance, Anamur, Bozyazı, Alanya, and Gazipaşa are the districts where the highest banana production is realized in Turkey [148].

According to Food and Agriculture Organization (FAO) and Turkish Statistical Institute data, annual increase in banana production means an increase in annual agricultural wastes [82, 83, 119, 145]. The actually visible stem part of the banana plant is a pseudostem formed by the intertwining of the leaves [82, 83]. The pseudostem part of the harvested banana plant is cut, and the new banana plant grows. A 10–15 month cycle is required for the banana plant to bear fruit [87, 127]. 336 metric tons of bananas pseudostem, sheaths, piths, peels, and leaves are produced annually [149]. Pseudostem, sheaths, piths, peels, and leaves of the plant are agricultural waste [83, 86, 93, 119, 126–128, 134, 145, 146]. Approximately, 60% of the plant biomass is left as waste after the banana plant harvest [58, 87, 147]. These wastes are mostly disposed of by composting, aerobic separation, incineration or simply left to rot in the fields. However, these methods can result in some serious environmental and ecological problems [149]. In addition, emissions during combustion cause an increase in global warming every year [58]. It is also in the sources when every ton of banana waste emits an average of half a ton of carbon dioxide per year [58]. In fact, banana waste can be converted into renewable energy due to their rich organic content [149]. They can be used as raw materials due to the rapid growth of the so-called banana stem and its high biomass [126, 127]. Recycling, in other words, sustainability can be achieved by obtaining fiber from these plant stems and leaves, which are considered as waste [2, 58, 91, 111, 127]. Banana fibers obtained from the evaluation of waste products are being used as a component in various industry branches due to natural a fiber, easy accessibility, renewable, and biodegradable [58, 87, 90, 91, 93, 99, 105, 110, 111, 127, 134, 150].

Geographical factors are very important in the cultivation of banana plants [89, 100, 139, 148, 151]. In particular, climate and soil properties are the determining factors in plant cultivation [89, 100, 139, 148, 151]. Banana plants are grown in tropical climates conditions, mostly in hot regions [57, 58, 86, 100, 125–128, 134, 137, 139, 148, 151–154]. In these regions, precipitation rates are important as well as temperature changes. When the amount of precipitation falls below 50 mm in order to ensure that the plant grows efficiently, irrigation is done, and the amount of water required by the banana plant that requires a lot of water is met [151]. 27–28 °C is considered as the ideal temperature for banana cultivation. At temperatures below 15–16 °C, the development of the banana plant cannot be achieved as desired. At lower temperatures, the above-ground part of the plant can die completely and the body under the ground can be damaged [148, 151]. The development of the pseudostem part of the banana plant, which is used as a fiber, can be delayed in cool weather [148, 151]. Temperature is not desired to be above 35 °C for the efficient growth of the banana plant, and the humidity in the air should be 60% or more [148, 151]. Light alkaline, rich in organic matter, sandy-loamy, deep, and permeable soil types are considered as ideal soil types for the cultivation of banana plants [151]. Although this plant has a high demand for water, the high ground water level is not a desirable soil feature for banana cultivation [151]. In such cases, well-drained land should be preferred [151]. In addition, the banana plant cannot grow in tired soils because it is sensitive to salt. Banana plant, which loves humid climates, causes a lot of evaporation with its large leaves. Therefore, its interest in water is increasing. Also, soil moisture is important for nutrient uptake [100, 148, 151]. For instance, in Turkey, the cultivation of banana plant is mostly grown in microclimate fields in the southern part of the country [100, 151, 153, 155].

3   Banana Fibers and Their Production

Banana fiber is a lignocellulosic natural plant fiber. Moreover, banana fiber is renewable, sustainable, and biodegradable. As aforementioned, the pseudostem is a part of the banana plant that appears to be a trunk, that arises from a soft central core and tightly wrapped up to 25 leaf sheaths and banana fiber can be extracted from the banana tree which possesses circular leaf structure often named to be as bast [95, 96]. Therefore, banana pseudostem fibers could be considered as leaf fibers and as bast fibers. For this reason, banana fibers can be referred to as leaf fibers, pseudostem fibers, or bast fibers in different sources [82–84, 88–90, 92, 95, 97–117]. Banana stems are composed of long and strong fibers in cylindrical structure [106, 113].

The innermost parts of the stem cut after banana harvest contain thin fibers. Therefore, silk-like flashly fibers are acquired from the innermost part of the banana plant stem. Coarser fibers are obtained from the outermost layers [87, 99, 106, 127, 156].

It is important to mention that besides the fibers acquired from the pseudostem of the banana plant, fibers can also be obtained from banana inflorescence bracts [87] and from the peel of the banana fruit [124]. In addition to those, it is also possible to obtain cellulose nanofibers from banana peels, which are known to contain 29% cellulose, 18% hemicellulose, and 3% lignin [124, 149]. Banana peels are becoming a by-product for food companies that process bananas. Cellulose nanofiber production from lignocellulosic materials such as bananas includes a pretreatment, partial hydrolysis, and mechanical breakdown of the raw material, respectively. The method used industrially and traditionally to separate cellulose microfibers into nanofibers is the acid hydrolysis method. Sulfuric or hydrochloric acid is generally used in this process. In this way, it becomes possible to separate glycosidic bonds between cellulose chains [57]. It is known that cellulose nanofibers acquired by acid hydrolysis can be used as reinforcing agents in composites. Nevertheless, it should not be forgotten that acid hydrolysis efficiency depends on factors such as acid concentrations [57]. It was reported that cellulose nanofibers can be successfully obtained from banana peels, an agricultural industrial waste, and the obtained nanofibers have properties that prove they could be utilized as reinforcement material in composites [57]. In the study of Ferrante et al. [124], the mechanical characteristics of the fibers obtained from the fruit peels of the Cavendish type banana plant were investigated. In this study, banana fibers were extracted from green unripe banana peels by hand stripping method. The average length of the fibers obtained was measured between 150 and200 mm [124]. Moreover, it was reported that the stiffness values of banana fruit peel fibers were noticeably lower than from those of banana fibers acquired from stem or leaves of the banana plant [124]. Here, on the one hand, banana fruit peel fiber is a material obtained from fruit waste. On the other hand, banana fibers obtained from the stem, or the leaves of the banana plant are much larger and stiffer structures leading to higher stiffness levels when compared with banana fruit peel fibers [124]. In addition, it is possible to obtain fiber from the bracts that fall from the banana plant during the formation of the banana fruit [87]. The banana plant bunch carries a great number of spirally arranged bracts, fleshy, boat-shaped, red, and purplish, on its axis after 4 weeks of florescence (Fig. 2). Bracts that shed as the florescence development of a banana plant is complete are also an agricultural waste and can cause environmental problems, just like banana pseudostems cut off during banana harvest. Cellulosic banana fibers can be extracted from banana bracts to solve these problems and to use bracts effectively [87]. Amutha et al.[87] examined the characterization of natural fibers obtained from banana inflorescence bracts. In this study, they removed the banana inflorescence bracts fibers by hand scraping, water retting method, and retting with enzymes. In enzymatic retting process, treatment was carried out with 0.1% surface active agent and 0.2% pectinase enzyme for 45 min at 50 °C. The obtained banana inflorescence bract fibers have similar properties with banana fibers obtained from banana leaves and pseudostems. There are fine lines and voids on the surface of banana inflorescence bract fibers. This shows that it has a kind of scaly appearance and an uneven surface. The length of the banana inflorescence bract fibers varies between 22 and 33 cm leading to average of 27.6 cm. The fiber diameter was found to be 79.6 μm, and the density of the fiber to be 1.39 g/cc. The banana inflorescence bract fiber possesses 56.48% cellulose content, 28.44% lignin content, 3.44% ash content, and 10.45% moisture content akin to other lignocellulosic fibers. It was reported that banana inflorescence bracts fibers can be utilized in technical and clothing applications by mixing them properly with other natural fibers [87].

Highly durable fibers obtained from pseudostems and leaves of the banana plant grown for the food industry in the world are considered as a waste product of banana [83, 88, 102, 111, 121]. When grown in tropical regions, every banana plant, which is grown without the need for pesticides and fertilization, only carries fruit once. The most important feature that distinguishes banana fibers acquired from the superimposed leaves of the harvested banana plant from other natural fibers is that it is obtained from agricultural wastes [83, 101]. It is recorded that the fiber ratio obtained from such wastes is over 3.87 million tons [82]. On average, 1 kg of banana fiber is produced from 37 kg of banana stem [82, 106]. Each banana leaf has a pseudostem, leaf stem, and basic leaf scabbards that form the leaf section [106, 122]. Banana leaves are made up of different layers containing longitudinal fibers, and the length of the fibers obtained ranges from 60 to 110 cm [99, 156]. In addition, they are bright in color and have a stronger structure than cotton fibers [111]. The fibrous portions from the banana leaves could be removed by raspador or decorticator machines, manually scraped or extracted using chemical methods [90, 93, 101, 106, 110, 120, 122, 125].

The fibers obtained from the banana leaves are in bundles [101, 106]. The color of the obtained banana fiber had observed to be white, brownish, and yellowish [121]. Different methods have been employed since ancient times for separating the leaf and stalk layers of the banana plant from the plant tissue to obtain bright and durable fibers [121, 157]. Brindha et al. [110] evaluated four different extraction processes of banana fibers in terms of industrial application. They investigated the effects of different processes applied for extraction of banana fibers on fiber yield, surface morphology, physical, chemical, and mechanical characteristics [110]. In practice, they used banana pseudostem sheaths about 30.5 cm long, 8 cm wide, and 1 cm thick. They preferred methods including mechanical extraction, chemical, microbial, and enzymatic retting for the extraction of banana fibers. According to the results obtained, every method used to obtain fiber has its advantages and disadvantages. The chemical retting method exhibited a twofold rise in yield without considerably affecting strength in comparison with mechanical extraction [110]. If banana fibers will be used in garment production, it has been determined that it is more appropriate to obtain the fibers by chemical retting method using sodium hydroxide. Because the chemical retting process that applied using sodium hydroxide allows us to obtain fiber samples with a low tex value, which is more suitable for the textile industries. They concluded that banana fibers obtained by microbial retting method contain higher lignin; and therefore, it would be more appropriate to use them in fiber reinforced composites [110]. They stated that banana fibers obtained by enzymatic retting method contain high cellulose and therefore a good alternative to wood pulp in the paper industry [110].

3.1 Banana Fiber Production with Retting Method

The banana fibers need to be extracted and cleaned from the banana plant leaves and pseudostem for making banana fibers usable [101]. The oldest method known for this is retting method [101, 110, 158–161]. But as it is a time-consuming method, it is not preferred much today [158]. Retting is a biochemical process applied for the separation of non-fiber tissues from banana fibers in banana plant leaves [93, 158]. In the retting method, lignocellulosic banana fibers in the banana plant are ensured to be obtained with the help of microorganisms [158, 159]. In this way, banana fibers with higher amounts of cellulose are obtained compared to mechanical methods [159]. However, it is also known that insufficient or excessive retting process negatively affects banana fiber quality [110]. Microorganisms or enzymatic methods make the pectin water-soluble, which binds the fibers together and is also found in banana plants at a rate of 4% [158, 159]. Enzymatic methods allow the banana fibers to be extracted without breaking [83]. Pectinase and xylanase enzymes are commonly used in the enzymatic method [83]. It is possible to obtain high quality fibers with this method [83]. As a result of the studies on obtaining banana fibers by enzymatic methods, it has been stated that the enzymatic process is effective in the extraction of banana fiber, one of the most effective enzymes is polygalacturonase, which is specific for substrates with high specific activity and do not damage the cellulosic structure of the fibers. However, it is also stated that the effectiveness can be reduced due to enzyme deactivation in long-term processes. In addition, studies have proven that the fibers obtained by the enzymatic method can be spun with or without mixing with other fibers to produce yarn [83].

The extraction process of banana fibers can be carried out chemically or enzymatically in dew, still water, rivers, hot water [121, 158]. If the retting method is to be carried out in hot water, the temperature should be at 20–30 °C [101, 158, 159]. The easy separation of fibers from the leaves varies with the awaiting time in the water [101, 158, 159]. In the retting process, the leaf and shell parts obtained from the pseudostem of the banana fibers are sliced lengthwise and kept in water in several pieces for 15 days each [160]. In this way, pectin, hemicellulose, lignin, waxes, and fats in the structure of banana fiber can be removed. The quality of the fiber is determined by the rate at which these chemicals are removed from the fiber [121, 158, 160, 161]. After docking, the fibers were washed in standing water or tap water and suspended for 8 h to dry [121, 158, 160, 161]. In order to prevent oxidation of the banana fibers, they were treated with 5% sodium hydroxide (NaOH) solution for 4 h in total dipping time and washed until they reach neutral pH [106, 156, 158, 161]. NaOH opens the cellulose structure, allowing the hydroxyl groups to be prepared for reactions [91]. Some of the wax, cuticle layer, lignin, and hemicellulose are removed during the treatment with sodium hydroxide [91, 92]. Thus, cellulosic fibers can be extracted by sodium hydroxide treatment [92]. Generally, sodium hydroxide (NaOH) is used in banana fiber extraction with chemical method [83]. Moisture content of banana fiber, NaOH concentration, vapor pressure, and times are generally known as the most important factors affecting the method of retting of banana fiber [158].

The dew method, which is the preferred method in regions with high humidity, is used to rot banana leaves with the help of microorganisms [121]. These microorganisms disrupt the structure of lignin and pectin in the banana fiber and ensure the removal of lignin and pectin from the fiber [121, 158, 160, 161]. The banana fibers obtained after this process, which was completed in 1–1.5 months, are quite soft. However, sometimes, it may not be preferred because it is a very slow process. In addition, some microorganisms formed affect pigments and cause fiber color change [160]. Also, coarser and lower quality fibers are obtained with dew retting process than water retting process [121]. The fibrils are joined in bundles and the banana fiber surfaces are smooth owing to the existence of waxes and oils present in the structure of the banana fiber [158, 162]. Lignin is depolymerized and oxidized with hydrogen peroxide during alkaline steaming and bleaching [57, 63, 158]. Oxidation of lignin causes degradation of lignin and formation of hydroxyl carbonyl and carboxylic groups. This facilitates the dissolution of lignin in an alkaline environment [57, 63, 158]. The fiber bundles are less than the average size after these operations. This is due to the dissolution of hemicellulose and lignin. It was reported that the banana fibers possess a hard surface and most of the banana fiber bundles were completely separated [121, 158, 159, 163]. Treatment with alkaline solution removes wax and dirt from the banana fiber surface. Many studies have also revealed that impurities in banana fibers can be removed by alkali treatment [57, 63, 85, 158].

3.2 Banana Fiber Production with Mechanical Methods

It is preferred in mechanical methods as well as biological methods in obtaining banana fibers [106, 156]. Banana fibers can be acquired from the banana plant leaves that make up the pseudostem of the banana plant by hand or with the help of machines [120, 156]. Manual fiber extraction is a very primitive method and known as low-cost fiber extraction [156]. Generally, the banana fiber is located near the outer surface of the leaf sheath and could be easily peeled into strips of 5–8 cm wide and 2–4 mm thick along the entire length of the leaf scabbard [106]. In this method, it is ensured that the outer surface of the leaves is scraped manually in the direction of the fibers in the leaf by means of a material or cutting blades [106, 109, 156]. Thus, the banana fibers in the leaf are removed to the surface. It was reported that the average yield of the banana fibers thus obtained and dried is 1–2% [106, 109, 156].

Instead of this laborious and time-consuming method, simple and low-cost user-friendly extractor machines were designed, and the manual fiber extraction method was replaced by these machines [106]. The production of banana fibers by machines called decorticating or raspador is not much different from the manual process [106, 118, 164]. With these machines, which have a single-phase electric motor with roller and are fixed on a frame that works with the arrangement of belts and pulleys fiber production results in 20–25 times more fiber production than manual fiber production [93, 106, 118, 126]. These machines are more efficient than manual processing and can extract 20–30 kg of fiber per day [90]. After the banana harvest, the trunks cut to

120–180 cm are crushed with the rollers of the decorticating machine and the fibers are intended to be removed [86, 106, 118, 126, 127, 164]. Banana leaves are scraped and cleaned by these machines which has sharp rotary cylinder [86, 93, 106, 126, 134]. Decortication, i.e., paring process, allows the removal of narrow strips of the stem with a toothed blade [106, 118]. This machine, which is used in the production of banana fiber, was developed in India. Obtaining fiber from the wastes of banana harvests is done quickly and efficiently with this machine [90, 106, 118]. These machines, which also known as raspador, consist of clamping rollers, leveling rollers, and scraper rollers [118, 164]. The leaf sheaths of the banana plant fed between the rotating rollers are taken to the machine to a certain place and then withdrawn, and the non-fiber tissues are peeled with the help of scraper rollers [93, 118, 122, 125, 164]. The other part of the plant pseudostem sheaths and leaf sheaths that do not enter the machine are taken into the machine, and the same process is performed [118, 164]. In this way, banana fibers with lengths ranging from 60 to 110 cm can be obtained [99, 156]. Banana fibers contain 40–50% moisture after extraction [106, 118, 164]. The moisture content could be decreased to 10% by centrifugal extraction before evaporative drying process [106, 118, 122, 126, 164]. The extracted banana fibers are dried under semi-shade sun until they turn white [106, 118, 126, 164]. The fiber quality obtained by these machines is quite good, and it has superior quality in terms of length, softness, strength, and color, and the fiber obtained is quite high [99, 118, 120, 156]. Even though the energy consumption is high, the processing period is very short [99, 118, 156].

3.3   Degumming Process of Banana Fibers

Water soluble substances such as hemicellulose, pectin, lignin, which form the structure of banana fibers, may not be completely removed after the retting process [64, 90, 158, 160, 161, 165]. These materials that are not completely removed can lead to the banana fiber to display stiff handle [158, 160, 161]. In order to obtain brighter, softer, and better-quality fibers, the banana fibers are subjected to a gum extraction process after retting process [160]. This method applied for banana fiber can also applied to other leafs fibers [166]. The deguming removal process, which consists of chemical and biological processes, enables the separation of fiber bundles [166]. It also removes the coarse and hard polysaccharides found in the fibers [90]. This process, which can be carried out by enzymatic, chemical, or hydrolytic catalysis, enables the partition or disintegration of peptide bonds [90]. For this purpose, the fibers can be processed in an alkaline solution or in a hot soap solution [64, 162, 165, 166]. Degumming process is provided by treating the fibers for 1 h at 90–95 °C in alkaline soap solution such that pH is 10–11 [162, 165, 166]. In addition to this method, known as the conventional degumming process, enzymatic degumming methods are applied to banana fibers [106, 121, 156, 158, 160]. It is preferred because it is an easy and soft process to apply on fibers [90, 121, 160]. The gummy material densely coated around the fibers is removed by means of enzymes such as ligninases, xylanases, pectinases, cellulases and laccases [90]. Degumming process is completed by processing the banana fibers with pectinase enzymes at 45–60 °C for 1–1.5 h (at pH 7.5–8) [97, 121, 160]. In the literature, there are different studies on the extraction and degumming process of banana fibers. Studies are carried out on the biological separation of banana fibers by solid state fermentation and development of degumming processes [159, 167]. For instance, banana leaf sheaths can be left in medium which produces polygalacturonase enzyme in a wheat bran environment and the banana fiber bundles can removed by manual peeling after a fermentative process [159, 167]. Since this method is more ecological than chemical fiber extraction methods, it is thought to be more advantageous in terms of environmentally friendly application [121, 159, 167].

In another study, banana fibers were biologically degummed by using environmentally friendly xylano-pectinolytic enzymes [64]. In this study, it was intended to use ecofriendly chemicals/biocatalysts to make banana fibers smooth and well and to encourage their usage in textile products [64]. They used the xylanase and pectinase-producing bacterial isolate *Bacillus pumilus* to perform an environmentally friendly wash. Pectinases and xylanase enzymes lyse large fiber molecules into galacturonic acid and xylose. Thus, they can enhance the absorption capacity and whiteness of the fibers. The fibers were treated with pectinases and xylanase enzymes at pH 6–10 [64]. The results showed that enzymatic treatment provided 13.27% rise in whiteness, 16.14% rise in brightness, and 8.63% reduction in yellowness compared to raw banana fibers. They have also proven that the bio-degumming process reduces chemical usage by 50% compared to traditional degumming processes [64]. Thus, toxic chemicals that pollute the environment and energy consumption were also reduced. It was concluded that the process of removing non-cellulosic impurities from banana fibers with xylano-pectinolytic enzymes has proven to be an environmentally safe method [64].

Another study was carried out by Paramasivam et al. [90] to remove the gum left on banana fibers; they revealed that degumming agents are a determining factor affecting the brightness and processing characteristics of the fibers. In their study, the degumming process of banana fibers was achieved with enzymes, thus reducing the increasing environmental concerns. The use of enzymes such as pectinase and laccase in the process of degumming the banana fibers resulted in the production of higher quality fibers. They also confirmed that the laccase enzyme plays a more effective role in improving the surface quality of the fibers compared to other degumming methods [90].

3.4 Yarn Formation from Banana Fibers

Banana fibers are somewhat stiff due to their lignin content [111]. As aforementioned, banana fibers were softened by enzyme or chemical processes so that they can be used in textile products (Degumming process) [111]. Different yarn spinning methods are applied to make the fibers into yarn after the degumming process applied to banana fibers [97, 107]. The fibers are made ready for the spinning phase by keeping them moist to prevent breakage [97]. Banana fibers, which have appearance similar to bamboo and ramie, are much easier to spin than bamboo and ramie [97]. Spinning methods applied to cotton fibers can also be applied to banana fibers [107]. In this way, banana fibers become suitable for use as a mixture with other fibers (natural and/or synthetic fibers). Banana fibers can be twisted by almost all spinning methods including ring, open-end, and semi-worsted spinning [107]. In the literature, yarns used in the production of cotton/banana, viscose/banana, modal/banana blend fabrics were obtained by ring spinning method and hand spun method [111]. Although worsted spinning method is used for wool spinning, it can also be applied for banana fibers [107]. Studies have proven that banana fibers can be spun into yarn with or without mixing with other fibers [83, 111]. However, producing banana fiber and wool fiber blend yarns does not require major equipment changes in industries and is therefore said to be easier to manufacture [83]. In addition, it is also reported in the literature that banana/PP blend yarn is more homogeneous and shows higher strength than flax/PP yarn [83]. Thick and coarse banana yarns are used in bags, baskets, hats, and floor covering. However, thin banana yarns and banana yarns formed from their natural and synthetic fiber blends are used in garment [168]. In particular, it is known that banana fiber blended fabrics can be used in men's suits and these kinds of similar areas [111]. The yarns obtained from banana fibers have a natural sheen that gives an extremely durable and satin-like appearance. Chemical processing is not used during the processing of banana fibers, which are also very good at moisture absorption. Moreover, the banana fibers are completely biodegradable [82].

4   Chemical Structure of Banana Fibers

Banana fibers have a complex structure [106, 113]. Banana fibers are lignocellulosic and consist of spirally wound cellulose microfibrils [106, 109, 113, 126]. The cell wall thickness is 1.25 microns [141]. The hemicelluloses found in banana fibers, which are connected to cellulose fibrils by strong hydrogen bridges, have an amorphous structure. Hemicelluloses take charge in determining moisture absorption, thermal degradation properties, and fiber biodegradability of the fiber [169, 170]. There are different values in the literature regarding the amounts of cellulose, hemicellulose, lignin, pectin, oils, waxes, and different nitrogenous compounds in the structure of the banana fiber [58, 62–64, 83, 85, 94, 97, 109–113, 118, 119, 122, 127, 170–182]. The ratio of these fiber components also varies depending on which part of the banana plant the banana fiber is obtained from [82]. The amount of cellulose and microfibril angle in the fiber structure is an substantial factor in determining the mechanical properties of the banana fiber [58, 82, 106, 113]. Banana fiber is a good fiber with good mechanical characteristics owing to the high cellulose content and low percentage of lignin [58, 93, 125]. The lignins in the structure of the banana fiber coexist with hemicelluloses which are non-cellulosic, pectin-free cell wall polysaccharides and has an important role in the natural degradation resistance of the lignocellulosic material [58, 113, 171, 182]. Due to the pentazone gums and essential oils contained in banana fibers, it is mixed with bamboo pulp and used in papermaking [99]. Bilba et al. [82] were investigated the chemical structure of the leaf part of the banana plant by elemental analysis. According to the results, they find out to be 31–35% cellulose, 14–17% hemicellulose, 15–16% lignin in the structure of banana fiber. In addition, according to different studies on banana fibers, it was monitored that the cellulose content of banana fiber was higher than some other plant fibers (such as coconut fiber and barley bast fiber) [82]. The content and proportions of banana fiber found in different sources in the literature are given in Table 1 [58, 64, 82–85, 91, 92, 94, 98, 106, 107, 112, 113, 118, 120, 142, 156, 158, 161, 170–178, 180–184].

The main constituents in the structure of banana fiber are cellulose, hemicellulose, lignin, pectin, wax, and extracts [118]. The chemical component of banana fiber varies according to the type and environmental conditions in which it is grown. According to Table 1, while the cellulose content in banana fiber varies between 35 and 65%, hemicellulose ratio can be said to vary between 6 and 34%, lignin 5–20%, pectin 2–5% [82, 83, 98, 106, 107, 113, 142, 156, 158, 161, 170–178]. Comparison of the chemical structure of banana fibers with other natural fibers is given in Table 2 [78, 84, 85, 106, 112, 170, 183, 185–188].

It was observed that banana leaf fibers have similar comparable close cellulose content as ramie, jute sisal fibers [161, 184]. It was also stated that the amount of cellulose in the structure of banana fibers is higher than bamboo and coconut (coir)

Table 1 Chemical composition of banana fiber

Cellulose content (%)	Hemicelulose (%)	Lignin (%)	Pectin (%)	Ash (%)	Moisture content (%)	References
48–60	10–15	14–20	2–4	–	–	[82]
38–43	18	16	4	8	7–12	[113, 142, 158]
60–65	6–18	5–10	3–4	4.7	10–15	[64, 106, 112, 170–175]
43.6	14	11	–	3.8	6.5	[83, 176, 177]
62–64	19	5	–	–	10–11.5	[84, 94, 98, 120, 161, 178]
55–63	22–34	10–13	–	–	9–11	[107, 156]
54–65	–	–	–	–	–	[85]
58.5	15.4	13.2	–	–	–	[180]
50–60	–	5–10	3–5	1–3	–	[181]
35	17	16	–	9	11	[58]
60–65	19	5–10	–	–	–	[183]
60–65	5–19	5–10	2–3	–	–	[91]
60–65	6–9	5–10	3–5	1–3	–	[92]
43.6	14	11	–	–	–	[83]
60–65	19	5–15	–	–	–	[184]

Table 2 Comparison of chemical composition of banana fibers with those of other natural fibers [78, 84, 85, 106, 112, 170, 183, 185–188]

Fiber types	Cellulose	Hemicelulose	Lignin	Pectin
Banana	60–67.6	6–19	5–10	3–5
Cotton	82–96	2–6	0.5–1	5–7
Bamboo	26–43	30	21–31	–
Linen	71	18.6–20.6	2.2	2.3
Hemp	70.2–74.4	17.9–22.4	3.7–5.7	0.9
Jute	61–71.5	13.6–20,4	12–13	0.2
Ramie	68.6–76.2	13.1–16.7	0.6–0.7	1.9
Sisal	60–78	10–14.2	8–11	10
Kenaf	31–39	21.5	15–19	2
Pineapple	70–82	16–22.2	5–13	1–3
Coconut	36–43	0.15–0.25	41–45	3–4

fiber [82, 161]. The high cellulose content in the fiber structure also affects the performance characteristics of the fiber [161]. Banana fibers are one of the lignocellulosic plant fibers due to their cellulose and lignin content.

5 Physical and Chemical Properties of Banana Fibers

As aforementioned, banana fiber is a lignocellulosic fiber [84, 93, 102, 109, 113, 187]. Banana fiber has good mechanical characteristics because of their high alpha cellulose and low lignin content [84, 93, 102, 113, 187]. Various factors such as the growth conditions of the banana plant, the maturity of the fibers during harvest, the methods utilized to extract the fibers, and the transportation and storage of the fibers over time are among the factors that affect the mechanical characteristics of banana fibers and lignocellulosic fibers such as banana fibers [179]. The helical angles of cellulose crystallites (microfibrillar angle) of banana fibers are arranged in spirals of 11°–12° [94, 112, 113, 161, 178, 187]. Microfibrillar angle is also given as 10°–12° in some sources [65]. Generally lumen size is 5 μm in the literature [94, 161, 178]. However, there are sources stating that the lumen size is between 13.4 and 22.4 μm [65]. The high cellulose content, the small microfibrillar, angle and lumen size of the banana fiber give high strength and modulus to the fiber [58, 120, 178]. The low microfibrillary angle and modulus of the banana fiber increase the strength of the banana fiber [58, 178]. The effect of varying cross-sectional areas on the rupture strength of banana fibers of different lengths (10, 20, 25 and 40 mm) was examined and they observed that as the length of the banana fibers enhanced, the fiber strength decreased. It was concluded that this was due to the increasing number of defects with changing cross-sectional area [85]. Preethi and Balakrishna Murthy [82] analyzed the manufacture process, structure, properties, and suitability of biofiber for a variety of industrial applications. Accordingly, they determined that the diameter of banana fiber varies between 80 and 250 μm and the elongation percentage is 1.0–3.5% [82]. The diameter of the banana fiber determines the fineness of the fiber. The thinnest fibers are acquired from the leaf parts of the banana fiber forming the pseudostems. Banana fibers with an average fiber fineness of 2,400 nm (nm) are lightweight and have high moisture absorption [82, 122]. The breaking load was 332.33 g, the elongation at break was 2.01%, and the strength was 39.56 g/tex [82]. The properties of banana fibers obtained from various sources are given in Table 3.

Property comparison of banana fibers with other fibers were shown in Table 4. Accordingly, it was monitored that the banana fiber is lighter than the flax fiber [84]. In the literature, it was stated that bamboo fiber and ramie fibers are similar to banana fibers, but it is known that banana fiber is thinner and more easily bent than bamboo and ramie fibers [122]. Balakrishnan et al. [126] explored the influence of banana fiber

Table 3 Physical properties of banana fibers

Density (g/m^3)	Fiber diameter (μm)	Tensile strength (MPa)	Young's modulus (GPa)	Elongation at break (%)	References
1.35	120	550	20	5–6	[94, 178]
1.35	–	537	3.48	–	[161]
1.50	150	–	–	–	[175]
1.35	80–250	540	–	1–3.5	[82, 112]
1.35	–	540	3.48	–	[84]
–	50–250	711–789	27–32	2–3	[189]
–	105–210	–	6.6–13.7	1.93–3.27	[85]
0.95–0.75	–	529–914	27–32	–	[181]
	12.0–30.0	700–800	27–32	2.5–3.7	[65]

Table 4 Comparison of the properties of banana fibers with other fibers [78, 84, 98, 178]

Fiber types	Density (g/m^3)	Tensile strength (MPa)	Young's module (GPa)	Elongation at break (%)	Moisture content (%)
Banana	1.35	540–550	20	5–6	10–11
Sisal	1.41	350	12.8	6–7	–
Flax	1.50	345	27–39	–	8
Cotton	1.60	300–590	5.5–12.6	3–10	8
Bamboo	0.8	220–1000	22.8	1.3	–
Hemp	1.48	550–900	70	1.6	8–10
Rami	1.5	220–938	44–128	2–3.8	12–17
Pineapple	1.07–1.52	170–1627	6.21–82.5	1.6	11.8
Glassfiber	2.5	3000	65	3	–

extraction and process parameters on fiber fineness and suitable methods to decrease banana fiber fineness. After removing the banana fibers, they used in their studies by mechanical methods, they treated them with chemicals and enzymes to evaluate their physical characterization. They applied physico-biological, physico-chemical, and physico-enzyme and chemical combined processes [126]. In the physico-biological process, the treatment was carried out with enzymes at a concentration of 5% in acid medium at a temperature of 55 °C. In the physico-chemical process, the treatment was carried out at 95 °C with 6% hydrogen peroxide (bleach), 2% sodium silicate, 3% caustic soda, and 0.2% wetting agent. In the case of physico-enzyme and chemical combined process, the treatment was carried out with 5% concentration of pectinase enzyme at 45–55 °C. From all these applied processes, it was concluded that the fiber fineness and fiber linear density of banana fibers obtained by mechanical methods and then treated with physical-biological and chemical combination improved compared to mechanically extracted banana fibers [126].

Banana fibers possessing high cellulose content and low microfibrillar angle have high tensile characteristics [91, 178]. The tensile properties of banana fibers were observed to be better than sisal fibers compared to the sisal fibers observed to have nearly the same cellulose content (Table 4) [84, 91, 178]. It was stated that banana and sisal fibers could be utilized together to create a low-cost and user-friendly composite material with sufficient rigidity and damping behavior. The combination of banana and sisal fibers creates a highly qualified composite material [84]. As the properties of these two fibers differ, they are suitable for use in composite structures together [178]. Bending and impact strength characteristics of banana composites were found to be better than hybrid composites [84]. If banana fibers are to be utilized in composite materials, chemical modification processes such as acetylation, mercerization, methylation, cyanoethylation, benzoylation, acrylation, peroxide, and permanganate can be applied to the banana fibers to support the interface bond between fiber-matrix [88, 94, 133, 179]. Here, it was intended to diminish the amount of moisture absorbed by the fibers with these processes that is a chemical modification of the fibers [88, 94, 133]. Because banana fibers have a high-water absorption feature thanks to their hygroscopic material content such as cellulose and hemicellulose [88, 92, 94, 179, 184]. This situation causes dimensional imbalance in the fibers, creates a weak interface with the matrix material, and limits the use of banana fibers as reinforcement material in composites [88, 92, 94, 179, 184]. Relatively inefficient bonding between the fiber and the matrix is observed when hydrophilic fibers are combined with a hydrophobic thermoplastic matrix [92, 179]. The surface of banana fibers can be roughened to help improve this bond, or the surface of the fibers must be chemically modified to reduce the hydrophilicity of the banana fibers [92, 179]. Chemical modifications and chemical coupling agents applied to banana fibers are thought to optimize the interface of the fibers [92, 94]. Chemical coupling agents react with the hydroxyl groups of the cellulose and the functional groups of the matrix [94]. Silane, alkali, peroxide, permanganate, isocyanate treatment enhances the bonding between banana fibers and matrix [133, 179, 184]. Surface modification processes using chemicals affect not only the surface characteristics of fibers but also their physical and mechanical characteristics [133, 184]. Peroxide treatment is a process that increases the tensile characteristics of banana fibers. This enhancement is believed to be due to the free radical reaction initiated by the peroxide between cellulose [179]. Studies started and conducted to improve fibers are mostly focused on fiber-matrix interface adhesion and reducing water absorption [179].

Chemical, thermal, and morphological characteristics of banana fibers were investigated, and the treatment with alkali (NaOH) were explored [63]. First, dried banana leaf fibers were soaked in 1, 3, 5, 7 and 9% NaOH solution separately for 24 h at normal room temperature. It was then cleaned with distilled water to remove NaOH from the fiber surface and dried. As an outcome of the treatment of banana fibers with alkali, it has been observed that lignin and hemicellulose components and other impurities were removed from the fiber surface. Strong hydrogen bonding has been formed on the fiber, which provides better mechanical characteristics to the fiber [63]. It enabled the banana fibers to show better interfacial properties. An increase in the thermal resistance of banana fibers has also been observed with the removal of waxy layers and other impurities from the fiber surface [63].

Treatment of banana fibers with NaOH (mercerization) diminishes the spiral angle of the cellulosic molecular chain orientation of the fibers and causes an increase in modulus [184]. In addition, it was proven by many studies in the literature that the NaOH treatment effectively removes the lignin and non-cellulosic phases of banana fibers and cellulosic fibers such as banana fibers [63, 83, 91, 92, 184]. Mohan and Kanny [184] grafted nano-layered clay particles into banana fibers utilizing alkali and silane methods and evaluated the thermal, mechanical, and fiber-matrix adhesion characteristics of banana fibers. It was observed that the tensile modulus of the fibers impregnated with nanoclay increased 35.7% and the strength value increased by 12.5% compared to the unprocessed fibers. It has also been seen to provide improved fiber-matrix adhesion. The results of this study showed that a strong banana fiber could be developed with nanoclay infusion and can be used in polymeric matrices with improved thermomechanical and adhesion characteristics [184].

Shenoy et al. [133] examined the influence of chemical treatment on the physical and mechanical characteristics of banana fiber. Here, they used sodium bicarbonate, potassium permanganate, and chromium sulfate for surface modifications of the fibers. As a result of this study, it was monitered that the single stage treatment of chromium sulfate reduces the density of all fibers. They monitored that the potassium permanganate treatment reduced the tensile strength of all examined fibers. It has been turned out that the most effective treatment for banana fibers is a 72 h sodium bicarbonate treatment [133].

In another study, banana fibers processed with laccase and xylanase enzymes and their effects on the physiochemical characteristics of banana fibers were examined [88]. The laccase enzymes used in the study change the surface characteristics of lignocellulosic fibers by the removal of the lignin content. Xylanase enzymes are utilized in several applications to hydrolyze the xylan component, which is an interface between lignin and cellulose. It also facilitates lignin removal. Vardhini and Murugan [88] processed the banana fibers they obtained using an extraction machine in three different concentrations (10%, 15%, and 20%) with laccase and xylanase enzymes. According to the results, it was observed that after treatment with enzymes, lignin and hemicelluloses in banana fibers were removed, and more hydroxyl groups were released on the surface of the fibers. This helps the fibers achieve better chemical bonding with the resin and strengthens the interface in banana fiber reinforced composites. It has been observed that laccase enzyme treatment removes lignin in banana fiber up to 29%, while xylanase enzyme treatment removes it up to 27.7%. It was reported that a concentration higher than 15% does not give more positive or negative effects. Removing hemicelluloses and lignin from banana fiber increased fiber density thanks to both enzyme processes. In addition, it was concluded that the improvement in cellulose content provides the improvement of the tensile strength of the fibers [88]. It was stated that the most important property that should be known for using banana fibers in composites is tensile strength [89]. In line with the work carried out, it is known that most of the plant fibers have different dimensions and internal defects. The study showed banana fibers have medium tensile strength values in the stress–strain curve, but fibers with a low percentage of elongation are brittle [89].

6   Applications Areas of Banana Fibers

Banana fiber is different from other fibers thanks to it is biodegradable, environmentally friendly, and sustainable fiber and low-cost, light, and high mechanical properties [58, 82–84, 88, 90–92, 109–111, 126, 190]. The fact that banana fibers have different properties than other fibers increase the usability its in both textile materials and composite materials. Therefore, the use of banana fiber has a wide range of industry [58, 82–84, 88, 91, 92, 111, 126].

Generally, banana fibers are made into fabric by weaving. Many different types of fabrics can be woven from banana fibers, abaca fibers, raffia fibers, and pineapple fibers in the weaving looms (Fig. 3). Some woven fabric examples made from 100% banana fibers are shown in Fig. 4. The woven fabric containing 50% banana fiber and 50% silk fiber for use in clothing and apparel is shown in Fig. 5.

Banana fiber, which is commonly used in handicraft products, is widely used in traditional textile products in Japan, especially in the production of kimono and kamishimo, which is a formal garment worn by the Japanese samurai military [141, 168, 191]. Moreover, banana fiber, which has more usage area with increasing environmental awareness, is preferred in summer clothes in Japan with its light weight and comfortable wearable fabrics [141, 191]. Coarse fibers acquired from the outermost shells of the banana plant are generally more suitable for use in the manufacture of tablecloths and home furnishings in home decoration. India has produced and exhibited silk quality banana fiber that could be utilized in the handicraft and textile industries [126]. Carpets made of these yarns find use in the world. Carpets made of these yarns are used all over the world [168]. In addition, it is known that banana fabrics made of banana fiber can be used in combination with leather in leather goods applications [120]. In the study conducted in this field, researchers were found that most of the physical characteristics of fabrics made of banana fiber are comparable

Fig. 3 Weaving loom used in the manufacture of woven fabrics from banana fibers or other exotic natural plant fibers, such as abaca, and raffia fibers

to those of buffalo shrunken grain leather utilized for leather products application. It was also stated that the friction and light fastness characteristics of fabrics obtained from banana fibers are better than buffalo shrunken grain leather [120].

Banana fibers are also utilized in the manufacture of products such as souvenirs, bags, ornaments, wallets, purses, shoes, sanitary pads, etc. Handbags and purses produced from exotic natural plant fibers such as banana, raffia, abaca, pineapple fibers, etc., are shown in Fig. 6. Similar production efforts for different products have started in Turkey and these efforts are growing day by day. Projects and festivals have been organized in the Mediterranean region of Turkey regarding the use of banana fiber in clothing, shoes, ornaments, bags, etc. Projects related to the promotion of the use of banana fiber were carried out in Turkey [135]. For instance, different products produced within the scope of "Banana Fiber Project" prepared by Alanya Municipality for housewives were presented at the tourism fair held in Berlin (ITB) [135]. The products produced within the scope of Alanya Banana and Banana Fiber Project, which constitute the theme of the festival, were exhibited in the 15th Alanya Tourism and Art Festival organized by the Municipality of Alanya [135, 192–194]. The steps of extraction of banana fiber by machine, drying, dyeing, yarn spinning, and weaving from these yarns were shown at this festival. The t-shirt made from banana fiber and polyester blend attracted the attention of the visitors and its news appeared in the newspapers. 80% polyester fiber-20% banana fiber produced t-shirts

Fig. 4 Some woven fabric examples made from 100% banana fibers

has seen intense interest because this was the first in Turkey [135, 192–194]. At the same time, here, the traditional Alanya belt specially had produced from warp yarn silk and weft yarn banana fiber [192, 194].

Various R&D initiatives have been initiated to ensure that the solid and liquid biomass of banana pseudostem added value with the recent development of environmental awareness [128]. In India, they developed different uses of banana pseudostem in a project carried out in collaboration with the agricultural university and research institutes. They obtained environmentally friendly organic fertilizer from banana

Fig. 5 Woven fabric
containing 50% banana fiber
and 50% silk fiber

pseudostem [128]. Banana pseudostem juice (liquid obtained from the pseudostem) can be used in medicine, health drinks (food), and textiles thanks to innovative research. In addition, they reported that it could be utilized as a bio-mordant for natural dyeing of textile materials and fixing of color in the textile industry [88, 128]. In Indonesia and the Philippines, banana pseudostem is widely used for dyeing cotton fabrics. It is also used to obtain an artistic appearance with printing effects on cotton fabric [128]. Because banana pseudostem contains various pigments and colored biomolecules, they are used in making printer ink in the Philippines [128]. The bananas' pseudostem and fiber are used in shirting fabrics, shoes, bags, hats, ropes, papers, and even money papers in different countries [82, 87, 88, 90, 109–111, 120, 126, 128].

Banana fibers are an important source of raw materials other than wood, especially in paper production [88, 110, 120, 125, 127]. The use of banana fibers in paper production significantly reduces the cost of paper production [127]. It was reported that 100% banana paper is the most absorbent, abrasion resistant and has the lowest crease recovery angle [127]. In a study carried out by Sakare et al. [147], banana fibers were used in the production of greaseproof paper utilized in the packaging of butter and fatty foodstuffs. The most important feature of these papers is that they are resistant to grease, fat, and oil. It is generally obtained from finely ground wood pulp. It was concluded that banana fibers have good physical and mechanical strength and are low cost and lightweight, making them appropriate for use in the paper and pulp industry [147].

Fig. 6 Handbags and purses produced from exotic natural plant fibers such as banana, raffia, abaca, pineapple fibers, etc.

The high cellulose content of banana fibers enables it to be used in the purification of organic and inorganic water pollutants in wastewater [88, 180, 195]. Because cellulose can be used as an adsorbent for the carboxyl and hydroxyl functional group, which becomes the active bonding site of the contaminant [180, 195]. Banana fiber has become promising filler in the preparation of chitosan/banana fiber composites due to its porous structure, being environmentally friendly, low cost, and renewable [180]. Suhani et al.[180] also used chitosan adsorbed banana fiber to remove ammoniacal nitrogen from kitchen wastewater. The mixture of banana body fiber and chitosan composite has shown that it can be used as an adsorbent for removing contaminants from kitchen wastewater [180]. In another study, banana fibers obtained from banana plants were utilized to remove acid green dye from aqueous solution and were obtained effective results [195].

It is possible to find different experimental studies in the literature regarding the usage of banana fibers. One of these efforts is about the use of modified banana fiber as oil absorbent in oil spills [196]. Some researchers have reported natural fibers can be used as absorbents to clean up oil spills. Banana fibers can also be utilized as a natural oil absorbent, which can aid clean up oil spills, and especially recover some of the oil lost during transportation. Teli and Valia [196] improved the affinity of banana fiber to oil and ensured that banana fibers became hydrophobic and oleophilic by grafting with butyl acrylate monomer. Thanks to this process, the hydroxyl group in the cellulose is replaced by butyl acrylate monomer and forms esters. They were monitored that the oil absorption capacity of modified banana fibers increased with the addition of butyl acrylate. They concluded that the modification process carried out had no negative influence on the thermal resistance of the fibers up to 300 °C and that the reusability of the sorbents would be possible at least three times. As a result, it was reported that modified banana fibers can be used to clean up spilled oil [196].

Banana plant fibers are utilized in the production of technical textiles and composites with desired physical and mechanical properties [83, 90]. Therefore, banana fibers, which are utilized as reinforcement material in composite materials, are preferred because of their low cost, recyclability, and perfect tensile and impact resistance [83, 90–92, 120]. Banana fibers can be used in thermoset plastic products, construction materials, products that provide thermal insulation, automotive and transportation applications [58, 109, 120, 130]. Banana fiber is preferred in automotive and space technologies because of its characteristics such as lightness and abrasion resistance which are important in composite materials used in automotive and space technologies [91, 92, 94, 120, 197]. Banana fiber, which is among the micro and nano-based cellulose materials, is also used in the automobile industry as it is an environmentally sensitive material and has very good mechanical properties [92, 94, 120, 197]. In addition, the reduction in water intake and the improvement of dielectric properties of composites produced using banana fiber and polyurethane foam is an example of the suitability of banana fiber for use in the automotive sector [100, 197]. It is also known that banana fibers have high tensile strength, are resistant to rotting, and their specific flexural strength is close to glass fibers [94]. The fact that banana fibers are resistant to effects such as stone impact and moisture indicates that they are suitable for components utilized in the exterior of road vehicles

[94]. Mercedes uses in the Class A vehicles with floor protection product made of banana fiber composites [94, 198]. The use of banana fibers in underfloor protection for passenger cars by Daimler Chrysler is increasing the popularity of banana fiber reinforced composites in the automotive industry due to its innovative applications [94]. The use of banana fiber in natural fiber reinforced cement and concrete was also investigated. Studies on the usage of banana fibers in the construction sector are available in the literature [94, 199, 200].

The place of the fruit of the banana plant in the food sector is known [201]. But from banana peels, studies have been conducted on the extraction of fructo-oligosaccharides, a functional foodstuff obtained using different solvents and mixtures [201]. As a result of the experiment, it was concluded that fructo-olysaccharides can be produced from banana peels which is a waste material [201]. In another study, the feasibility of the acid–base behavior of the banana peels as a bioab-sorbent for the dyes had determined and the biological removal strategy of hazardous wastes containing dangerous compounds began to be developed with banana peels [202].

Banana fiber, which is a biodegradable, renewable, natural, and sustainable fiber obtained entirely from banana plant waste, has also started to be mentioned in the fashion world. Banana fiber, which has been popular lately, appears in different textile products such as dresses, sarees, and trousers in textile fashion [203–209]. It is expected that different textile products containing solely banana fibers or blended along with different other fibers will have a greater say in the near future in the fashion world.

7   Conclusion

Banana plants are mostly grown in tropical regions such as the Philippines and India. However, the banana plant can also be grown in some Mediterranean countries such as Turkey for its fruit. Banana harvest wastes accumulated in agricultural industrial areas cause serious environmental problems as they generally have not been used in industrial and commercial areas. Banana fibers obtained by evaluating the agricul-tural wastes left to the environment after harvesting of banana plant are becoming a preferred fiber type in several industrial fields because of their superior mechanical characteristics as well as being environmentally friendly and biodegradable nature. Banana fibers, which have superior mechanical, chemical, and performance proper-ties, are suitable for not only for textile use but also preferred in bio-composite struc-tures due to their biodegradability. Banana fibers play a role in numerous different applications in many different fields such as textile, composite, paper, construction, thermal insulation, automotive and transportation, etc. Banana fibers could also be used in the manufacture of clothing such as t-shirts, shirts, kaftans, kimonos, etc., and carpets, bags, handbags, wallets, purses, belts, shoes, ornaments, souvenirs, handicraft products, sanitary pads, etc. Banana fiber, whose popularity has increased recently, also appears in different textile products such as dresses, sarees, and trousers in textile fashion. Nowadays, the diversity of new produced fibers such as banana fibers is increasing with the aim of more of a naturalness, renewability, and sustainability. It is very crucial to expand the awareness of such fibers and to extend the use of these fibers obtained by evaluating agricultural wastes. The preferability of these fibers are recently increasing due to its ecofriendly nature. These fibers are important not only for the textile industry, but also for several other industrial areas. The best example of these fibers is banana fibers. In addition to the good mechanical, chemical, and performance properties exhibited by the banana fiber obtained from banana plant wastes, banana fiber finds its place not only in the textile sector but also in numerous different industrial areas due to its natural, biodegradable, sustainable, and renewable nature. It is expected that the interest in this type of environmentally friendly and sustainable fibers, which are obtained and evaluated from waste, will increase day by day.

References

1. Doğan Z (2012) Tekstil Sektöründe Atık Ekolojisi Uygulamaları. Akdeniz Sanat 4(8):24–26
2. Korkmaz M (2015) Sektörel-Ev Tekstilinde Çevre Dostu Ürünler. İzmir Ticaret Odası, https://api.izto.org.tr/storage/SectoralReport/original/mJ4zVMr23EjnoPUv.pdf, Ar&Ge Bülten, Eylül-Ekim, pp 13–18
3. Kalayci E, Avinc O, Yavas A, Coskun S (2019) Chapter 1: Responsible textile design and manufacturing: environmentally conscious material selection. Responsible manufacturing: issues pertaining to sustainability book. In: Alqahtani AY, Kongar E, Pochampally K, Gupta SM (eds) CRC Press., Taylor and Francis Group, USA, International Standard Book Number-13: 978–0–8153–7507–4, pp 1–19
4. Yıldırım FF, Yavas A, Avinc O (2020) Bacteria working to create sustainable textile materials and textile colorants leading to sustainable textile design. In: Muthu S, Gardetti M (eds) Sustainability in the textile and apparel industries: sustainable textiles, clothing design and repurposing. Print ISBN: 978–3–030–37928–5, https://doi.org/10.1007/978-3-030-379 29-2_6, Springer Nature Switzerland AG, Cham, 2020, pp 109–126
5. Unal F, Avinc O, Yavas A, Aksel Eren H, Eren S (2020) Contribution of UV technology to sustainable textile production and design. In: Muthu S, Gardetti M (eds) Sustainability in the textile and apparel industries: sustainable textiles, clothing design and repurposing. Print ISBN: 978–3–030–37928–5, https://doi.org/10.1007/978-3-030-37929-2_8, Springer Nature Switzerland AG, Cham, 2020, pp 163–187
6. Unal F, Yavas A, Avinc O (2020) Sustainability in textile design with laser technology. In: Muthu S, Gardetti M (eds) Sustainability in the textile and apparel industries: sustainable textiles, clothing design and repurposing. Print ISBN: 978–3–030–37928–5, https://doi.org/10.1007/978-3-030-37929-2_11, Springer Nature Switzerland AG, Cham, 2020, pp 263–287
7. Yıldırım FF, Avinc O, Yavas A, Sevgisunar G (2020) Sustainable antifungal and antibacterial textiles using natural resources. In: Muthu S, Gardetti M (eds) Sustainability in the textile and apparel industries: sourcing natural raw materials. Print ISBN 978–3–030–38540–8, https://doi.org/10.1007/978-3-030-38541-5_5, Springer Nature Switzerland AG, Cham, 2020, pp 111–179
8. Bakan E, Avinc O (2021) Sustainable carpet and rug hand weaving in Uşak Province of Turkey. In: Gardetti MÁ, Muthu SS (eds) Handloom sustainability and culture. Sustainable textiles: production, processing, manufacturing and chemistry. Print ISBN: 978–981–16–5271–4, https://doi.org/10.1007/978-981-16-5272-1_3, Springer, Singapore, 2021, pp 41–93

9. Fattahi FS, Khoddami A, Avinc O (2020) Sustainable, renewable, and biodegradable poly (lactic acid) fibers and their latest developments in the last decade. In: Muthu S, Gardetti M (eds) Sustainability in the textile and apparel industries: sourcing synthetic and novel alternative raw materials. Print ISBN 978–3–030–38012–0, https://doi.org/10.1007/978-3-030-38013-7_9, Springer Nature Switzerland AG, Cham, 2020, pp 173–194

10. Yıldırım FF, Yavas A, Avinc O (2020) Printing with Sustainable Natural Dyes and Pigments. In: Muthu S, Gardetti M (eds) Sustainability in the textile and apparel industries: production process sustainability. Print ISBN 978–3–030–38544–6, https://doi.org/10.1007/978-3-030-38545-3_1, Springer Nature Switzerland AG, Cham, 2020, pp 1–35

11. Eren HA, Yiğit İ, Eren S, Avinc O (2020) Ozone: an alternative oxidant for textile applications. In: Muthu S, Gardetti M (eds) Sustainability in the textile and apparel industries: production process sustainability. Print ISBN 978–3–030–38544–6, https://doi.org/10.1007/978-3-030-38545-3_3, Springer Nature Switzerland AG, Cham, 2020, pp 81–98

12. Eren HA, Yiğit İ, Eren S, Avinc O (2020) Sustainable textile processing with zero water utilization using super critical carbon dioxide technology. In: Muthu S, Gardetti M (eds) Sustainability in the textile and apparel industries: production process sustainability. Print ISBN 978–3–030–38544–6, https://doi.org/10.1007/978-3-030-38545-3_8, Springer Nature Switzerland AG, Cham, 2020, pp 179–196

13. Avinc O, Bakan E, Demirçalı A, Gedik G, Karcı F (2020) Dyeing of poly (lactic acid) fibers with synthesised novel heterocyclic disazo disperse dyes. Color Technol 136(4):356–369

14. Filiz Yıldırım F, Ozan Avinç O, Yavaş A (2014) Soya fasülyesi protein lifleri bölüm 1: soya fasülyesi protein liflerinin genel yapisi, üretimi ve çevresel etkileri. Uludağ Üniversitesi Mühendislik-Mimarlık Fakültesi Dergisi, Cilt 19(2):29–50

15. Filiz Yıldırım F, Ozan Avinç O, Yavaş A (2015) Soya fasülyesi protein lifleri bölüm 2: soya liflerinin özellikleri ve kullanim alanlari. Uludağ Üniversitesi Mühendislik-Mimarlık Fakültesi Dergisi, Cilt 20(1):1–21

16. Eren S, Avinc O, Saka Z, Aksel Eren H (2018) Waterless bleaching of knitted cotton fabric using supercritical carbon dioxide fluid technology. Cellulose 25(10):6247–6267

17. Kalayci E, Avinc O, Yavaş A (2019) Usage of horse hair as a textile fiber and evaluation of color properties. Annals of the University of Oradea Fascicle of Textiles, Leatherwork, (Print) = ISSN 1843 – 813X, (On-line) = ISSN 2457–4880, Vol 2019, No 1, pp 57–62

18. Kalayci CB, Melek G, Kalayci E, Avinc O (2019) Color strength estimation of coir fibers bleached with peracetic acid. Annals of the University of Oradea Fascicle of Textiles, Leatherwork, (Print) = ISSN 1843 – 813X, (On-line) = ISSN 2457–4880, Vol 2019, No 2, pp 65–70

19. Setthayanond J, Sodsangchan C, Suwanruji P, Tooptompong P, Avinc O (2017) Influence of MCT-β-cyclodextrin treatment on strength, reactive dyeing and third-hand cigarette smoke odor release properties of cotton fabric. Cellulose 24(11):5233–5250

20. Hasani H, Avinc O, Khoddami A (2017) Effects of different production processing stages on mechanical and surface characteristics of polylactic acid and PET fiber fabrics. Ind J fiber Textile Res (IJFTR) 42(1):31–37

21. Rahmatinejad J, Khoddami A, Mazrouei-Sebdani Z, Avinc O (2016) Polyester hydrophobicity enhancement via UV-Ozone irradiation, chemical pre-treatment and fluorocarbon finishing combination. Prog Org Coat 101:51–58

22. Avinc O, Aksel Eren H, Erismis B, Eren S (2017) Treatment of cotton with a laccase enzyme and ultrasound. Industria Textila 67(1):55–61

23. Rahmatinejad J, Khoddami A, Avinc O (2015) Innovative hybrid fluorocarbon coating on UV/Ozone surface modified wool substrate. Fibers Polym 16(11):2416–2425

24. Gedik G, Avinc O, Yavas A, Khoddami A (2014) A novel eco-friendly colorant and dyeing method for poly(ethylene terephthalate) substrate. Fibers Polym 15(2):261–272

25. Davulcu A, Aksel Eren H, Avinc O, Erismis B (2014) Ultrasound assisted biobleaching of cotton. Cellulose 21:2973–2981

26. Aksel Eren H, Avinc O, Erismis B, Eren S (2014) Ultrasound-assisted ozone bleaching of cotton. Cellulose 21:4643–4658

27. Avinc O, Aksel Eren H, Uysal P (2012) Ozone applications for after-clearing of disperse-dyed poly(lactic acid) fibers. Color Technol 128, 479–487
28. Bilal Turkoglu K, Kalayci E, Avinc O, Yavas A (2019) Oleophilic Buoyant Kapok fibers and new approaches. Düzce Üniversitesi Bilim ve Teknoloji Dergisi 7:61–89
29. Hasani H, Avinc O, Khoddami A (2013) Comparison of softened polylactic acid and polyethylene terephthalate fabrics using KES-FB. fibers Text East Eur 21, 3(99):81–88
30. Avinc O, Celik A, Gedik G, Yavas A (2013) Natural dye extraction from waste barks of Turkish red pine (pinus brutia Ten.) Timber and eco-friendly natural dyeing of various textile fibers. Fibers Polym 14(5):866–873
31. Gedik G, Yavas A, Avinc O, Simsek Ö (2013) Cationized natural dyeing of cotton fabrics with corn poppy (papaver rhoeas) and investigation of antibacterial activity. Asian J Chem 25(15):8475–8848
32. Avinc O, Aksel Eren H, Uysal P, Wilding M (2012) The effects of ozone treatment on soybean fibers. Ozone Sci Eng 34(3):143–150
33. Avinc O, Day R, Carr C, Wilding M (2012) Effect of combined flame retardent, liquid repellent and softener finishes on poly(lactic acid) (PLA) fabric performance. Text Res J 82(10):975–984
34. Avinc O, Phillips D, Wilding M, Bone J, Owens H (2006) Preferred alkaline reduction-clearing conditions for use with dyed Ingeo poly(lactic acid) fibers. Color Technol 122:157–161
35. Avinc O, Phillips D, Wilding M (2009) Influence of different finishing conditions on the wet fastness of selected disperse dyes on polylactic acid (PLA) fabrics. Color Technol 125:288–295
36. Avinc O, Khoddami A (2009) Overview of poly(lactic acid) (PLA) fiber part I: Production, properties, performance, environmental impact, and end-use applications of poly(lactic acid) fibers. fiber Chem 41(6):391–401
37. Avinc O, Wilding M, Phillips D, Farrington D (2010) Investigation of the influence of different commercial softeners on the stability of poly(lactic acid) fabrics during storage. Polym Degrad Stab 95:214–224
38. Khoddami A, Avinc O, Mallakpour S (2010) A novel durable hydrophobic surface coating of poly(lactic acid) fabric by pulsed plasma polymerization. Prog Org Coat 67:311–316
39. Avinc O, Khoddami A (2010) Overview of poly(lactic acid) (PLA) fiber part II: Wet processing; pretreatment, dyeing, clearing, finishing, and washing properties of poly(lactic acid) fibers. fiber Chem 42(1):68–78
40. Avinc O, Wilding M, Bone J, Phillips D, Farrington D (2010) Colorfastness Properties of dyed reduction cleared, and softened poly(lactic acid) fabrics. AATCC Rev 10(5):52–58
41. Avinc O, Wilding M, Bone J, Phillips D, Farrington D (2010) Evaluation of colour fastness and thermal migration in softened polylactic acid fabrics dyed with disperse dyes of differing hydrophobicity. Color Technol 126:353–364
42. Avinc O, Wilding M, Gong H, Farrington D (2010) Effects of softeners and laundering on the handle of knitted PLA filament fabrics. Fibers Polym 11(6):924–931
43. Aksel Eren H, Avinc O, Uysal P, Wilding M (2011) The effects of ozone treatment on polylactic acid (PLA) fibers. Text Res J 81(11):1091–1099
44. Avinc O (2011) Clearing of dyed poly(lactic acid) fabrics under acidic and alkaline conditions. Text Res J 81(10):1049–1074
45. Avinc O (2011) Maximizing the wash fastness of dyed poly(lactic acid) fabrics by adjusting the amount of air during conventional reduction clearing. Text Res J 81(11):1158–1170
46. Avinc O, Khoddami A, Hasani H (2011) A mathematical model to compare the handle of PLA and PET knitted fabrics after different finishing steps. Fibers Polym 12(3):405–413
47. Khoddami A, Avinc O, Ghahremanzadeh F (2011) Improvement in poly(lactic acid) fabric performance via hydrophilic coating. Prog Org Coat 72:299–304
48. Avinc O, Owens H, Bone J, Wilding M, Phillips D, Farrington D (2011) A colorimetric quantification of softened polylactic acid and polyester filament knitted fabrics to 'water-spotting.' Fibers Polym 12(7):893–903

49. Yasin Odabaşoğlu H, Ozan Avinç O, Yavaş A (2013) Susuz Boyama. Tekstil ve Mühendis 20:90, 63–79
50. Filiz Yıldırım F, Gökçin Sevgisunar H, Yavaş A, Ozan Avinç O, Çelik A (2014) UV Korumada Ekolojik Çözümler. Tekstil ve Mühendis 21:96, 37–51
51. Tungtriratanakul S, Setthayanond J, Avinc O, Suwanruji P, Saebae P (2016) Investigation of UV protection self cleaning and dyeing properties of nano TiO2 treated poly lactic acid fabric. Asian J Chem 28(11):2398–2402
52. Fattahi F, Khoddami A, Avinc O (2020) Nano-structure roughening on poly (Lactic Acid) PLA substrates: scanning electron microscopy (SEM) surface morphology characterization. J Nanostruct 10(2):206–216
53. Avinc O, Yavas A (2017) Soybean: for textile applications and its printing. Chapter 11 of Soybean-The Basis of Yield, Biomass and Productivity, DOI: https://doi.org/10.5772/66725, Dr. Minobu Kasai (Ed.), ISBN 978–953–51–3118–2, Print ISBN 978–953–51–3117–5, InTech, Rijeka, Crotia, May 2017, pp 215–247
54. Sema P, Soydan AS, Avinc O (2019) Gizem Karakan Günaydin. In: Yavas A, Niyazi Kıvılcım M, Demirtaş M (eds)Physical properties of different Turkish organic cotton fiber types depending on the cultivation area. Organic cotton, pp 25–39. Springer, Singapore
55. Günaydin GK, Yavaş A, Avinc O, Soydan AS, Palamutcu S, Şimşek MK, Dündar H, Demirtaş M, Özkan N, Niyazi Kıvılcım M (2019) Organic cotton and cotton fiber production in Turkey, recent developments. Organic cotton, pp 101–125. Springer, Singapore
56. Görkem G, Ozan Avinç O, Yavaş A (2011) Bromus Tectorum Bitkisinin Tekstilde Doğal Boyarmadde Kaynağı Olarak Kullanımı. Tekstil Teknolojileri Elektronik Dergisi, Cilt: 5(1):40–47
57. Tibolla H, Pelissari FM, Martins JT, Vicente AA, Menegalli FC (2018) Cellulose nanofibers produced from banana peel by chemical and mechanical treatments: characterization and cytotoxicity assessment. Food Hydrocolloids 75:192–201
58. Priyadarshana RWIB, Kaliyadasa PE, Ranawana SRWMCJK, Senarathna KGC (2020) Biowaste management: Banana fiber utilization for product development. J Nat Fibers 1–11
59. Kumartasli S, Avinc O (2020) Important step in sustainability: polyethylene terephthalate recycling and the recent developments. In Muthu S, Gardetti M (eds) Sustainability in the textile and apparel industries: sourcing synthetic and novel alternative raw materials, Print ISBN 978–3–030–38012–0, https://doi.org/10.1007/978-3-030-38013-7_1, Springer Nature Switzerland AG, Cham, 2020, pp 1–19
60. Kumartasli S, Avinc O (2020) Recycling of marine litter and ocean plastics: a vital sustainable solution for increasing ecology and health problem. In: Muthu S, Gardetti M (eds) Sustainability in the textile and apparel industries: sourcing synthetic and novel alternative raw materials, Print ISBN 978–3–030–38012–0, https://doi.org/10.1007/978-3-030-38013-7_6, Springer Nature Switzerland AG, Cham, 2020, pp 117–137
61. Kumartasli S, Avinc O (2021) Recycled thermoplastics: textile fiber production, scientific and recent commercial developments. In: Parameswaranpillai J, Mavinkere Rangappa S, Gulihonnehalli Rajkumar A, Siengchin S (eds) Recent developments in plastic recycling. composites science and technology, Print ISBN: 978–981–16–3626–4, https://doi.org/10.1007/978-981-16-3627-1_8, Springer, Singapore, 2021, pp 169–192
62. Üner İ, Başaran FN (2016) Tekstilde Sürdürülebilirlik İçin Yöresel Ürünlerin Yaşam Döngüsü Değerlendirmesindeki Rolü: Çaput Dokumacılığı Örneği. Akdeniz Üniversitesi, IV. Yöresel Ürünler Sempozyumu ve Uluslararası Kültür/Sanat Etkinlikleri, Antalya, Turkey, 3–5 Kasım, pp 243–252
63. Parre A, Karthikeyan B, Balaji A, Udhayasankar R (2020) Investigation of chemical, thermal and morphological properties of untreated and NaOH treated banana fiber. Materials Today: Proceedings 22:347–352
64. Kaur A, Varghese LM, Battan B, Patra AK, Mandhan RP, Mahajan R (2020) Bio-degumming of banana fibers using eco-friendly crude xylano-pectinolytic enzymes. Prep Biochem Biotechnol 50(5):521–528

65. Al-Oqla FM, Omari MA, Al-Ghraibah A (2017) Predicting the potential of biomass-based composites for sustainable automotive industry using a decision-making model. Woodhead Publishing, In Lignocellulosic fiber and Biomass-Based Composite Materials, pp 27–43

66. Gedik G, Avinc O (2018) Bleaching of hemp (Cannabis Sativa L.) fibers with peracetic acid for textiles industry purposes. Fibers Polym 19(1):82–93

67. Günaydin GK, Palamutcu S, Soydan AS, Yavas A, Avinc O, Demirtaş M (2020) Evaluation of fiber, yarn, and woven fabric properties of naturally colored and white Turkish organic cotton. J Text Inst 111(10):1436–1453

68. Arık B, Avinc O, Yavas A (2018) Crease resistance improvement of hemp biofiber fabric via sol–gel and crosslinking methods. Cellulose 25(8):4841–4858

69. Yavaş A, Avinc O, Gedik G (2017) Ultrasound and microwave aided natural dyeing of nettle biofiber (Urtica dioica L.) with madder (Rubia tinctorum L.). fibers Text East Euro 25(4, 124):111–120

70. Arık B, Yavaş A, Avinc O (2017) Antibacterial and wrinkle resistance improvement of nettle biofiber using chitosan and BTCA. fibers Text East Euro 25(3, 123):106–111

71. Kurban M, Yavas A, Avinc O, Aksel Eren H (2016) Nettle biofiber bleaching with ozonation. Ind Text 67(1):46–54

72. Günaydin GK, Avinc O, Palamutcu S, Yavas A, Soydan AS (2019) Naturally colored organic cotton and naturally colored cotton fiber production. Organic cotton, pp 81–99. Springer, Singapore

73. Soydan AS, Yavas A, Günaydin GK, Palamutcu S, Avinc O, Niyazi Kıvılcım M, Demirtaş M (2019) Colorimetric and hydrophilicity properties of white and naturally colored organic cotton fibers before and after pretreatment processes. Organic cotton, pp 1–23. Springer, Singapore

74. Unal F, Avinc O, Yavas A (2020) Sustainable textile designs made from renewable biodegradable sustainable natural abaca fibers. In: Muthu S, Gardetti M (eds) Sustainability in the textile and apparel industries: sustainable textiles, clothing design and repurposing, Print ISBN: 978–3–030–37928–5, https://doi.org/10.1007/978-3-030-37929-2_1, Springer Nature Switzerland AG, Cham, 2020, pp 1–30

75. Unal F, Yavas A, Avinc O (2020) Contributions to sustainable textile design with natural raffia palm fibers. In: Muthu S, Gardetti M (eds) Sustainability in the textile and apparel industries: sustainable textiles, clothing design and repurposing, Print ISBN: 978–3–030–37928–5, https://doi.org/10.1007/978-3-030-37929-2_4, Springer Nature Switzerland AG, Cham, 2020, pp 67–86

76. Gedik G, Avinc O (2020) Hemp fiber as a sustainable raw material source for textile industry: can we use its potential for more eco-friendly production? In: Muthu S, Gardetti M (eds) Sustainability in the textile and apparel industries: sourcing natural raw materials, Print ISBN 978–3–030–38540–8, https://doi.org/10.1007/978-3-030-38541-5_4, Springer Nature Switzerland AG, Cham, 2020, pp 87–109

77. Görkem G, Avinç O, Yavaş A (2010) Kenevir Lifinin Özellikleri ve Tekstil Endüstrisinde Kullanımıyla Sağladığı Avantajlar. Tekstil Teknolojileri Elektronik Dergisi 4(3):39–48

78. Kalayci E, Ozan Avinc O, Bozkurt A, Yavaş A (2016) Tarımsal Atıklardan Elde Edilen Sürdürülebilir Tekstil Lifleri: Ananas Yaprağı Lifleri. Sakarya Üniversitesi Fen Bilimleri Enstitüsü Dergisi 20(2):203–221

79. Soydan AS, Yavaş A, Ozan Avinç O, Günaydın GK, Niyazi Kıvılcım M, Demirtaş M, Palamutcu S (2019) The effects of hydrogen peroxide and sodium hypochlorite oxidizing treatments on the color properties of naturally colored green cotton. J EJENS-Euro J Eng Nat Sci 3(2):1

80. Kalayci E, Avinc O, Yavas A (2019) The effects of different alkali treatments with different temperatures on the colorimetric properties of lignocellulosic raffia fibers. Int J Adv Sci Eng Technol 7(1, 1):15–19. ISSN(p): 2321–8991, ISSN(e): 2321–9009

81. Kurban M, Yavaş A, Ozan Avinç O (2011) Isırgan Otu Lifi ve Özellikleri. Tekstil Teknolojileri Elektronik Dergisi 5(1):84–106

82. Preethi P, Balakrishna Murthy G (2013) Physical and chemical propertie s of banana fiber extracted from commercial banana cultivars grown i n Tamilnadu State. Agrotechnol S11 8:2
83. Ortega Z, Morón M, Monzón MD, Badalló P, Paz R (2016) Production of banana fiber yarns for technical textile reinforced composites. Materials 9(5):370
84. Srinivasan VS, Boopathy SR, Sangeetha D, Ramnath BV (2014) Evaluation of mechanical and thermal properties of banana–flax based natural fiber composite. Mater Des 60:620–627
85. Sia CV, Fernando L, Joseph A, Chua SN (2018) Modified Weibull analysis on banana fiber strength prediction. JJ Mech Eng Sci 12(1):3461–3471
86. Manohar K (2016) A comparison of banana fiber insulation with biodegradable fibrous thermal insulation. Am J Eng Res (AJER) e-ISSN, 2320–0847
87. Amutha K, Sudha A, Saravanan D (2020) Characterization of natural fibers extracted from banana inflorescence bracts. J Nat Fibers 1–10
88. Vishnu Vardhini KJ, Murugan R (2017) Effect of laccase and xylanase enzyme treatment on chemical and mechanical properties of banana fiber. J Nat Fibers 14(2):217–227
89. Subramanya R, Satyanarayana KG, Shetty Pilar B (2017) Evaluation of structural, tensile and thermal properties of banana fibers. J Nat Fibers 14(4):485–497
90. Paramasivam SK, Panneerselvam D, Sundaram D, Shiva KN, Subbaraya U (2020) Extraction, characterization and enzymatic degumming of banana fiber. J Nat Fibers 1–10
91. Waghmare PM, Bedmutha PG, Sollapur SB (2017) Review on mechanical properties of banana fiber biocomposite. Int J Res Appl Sci Eng Technol 5(10):847
92. Naeem MA, Siddiqui Q, Mushtaq M, Farooq A, Pang Z, Wei Q (2020) Insitu self-assembly of bacterial cellulose on banana fibers extracted from peels. J Nat Fibers 17(9):1317–1328
93. Vadivel K, Vijayakumar A, Solomon S, Santhoshkumar R (2017) A review paper on design and fabrication of banana fiber extraction machine and evaluation of banana fiber properties. Int J Adv Res Electr Electron Instr Eng 6(3)
94. Zaman HU, Beg MDH (2016) Banana fiber strands–reinforced polymer matrix composites. Compos Interfaces 23(4):281–295
95. Subagyo A, Chafidz A (2018) Banana pseudo-stem fiber: preparation, characteristics, and applications. Banana Nutr Funct Process Kinet
96. Different Types of Textile Fibers. https://ordnur.com/fiber/different-types-of-textile-fibers/, Accessed Sept 2021
97. Molina SMG, Pelissari FA, Vitorello C (2001) Screening and genetic improvement of pectinolytic fungi for degumming of textile fibers. Braz J Microbiol 32(4): 320–326
98. Joseph S, Sreekala MS, Oommen Z, Koshy P, Thomas S (2002) A comparison of the mechanical properties of phenol formaldehyde composites reinforced with banana fibers and glass fibers. Compos Sci Technol 62(14):1857–1868
99. Sivashankar S, Nachane RP, Kalpana S (2006) Composition and properties of fiber extracted from pseudostem of banana (Musa sp). J Hortic Sci 1:95–98
100. Ungüren E, Kaçmaz YY (2016) Alanya Bölgesinde Muz Lifinin Değerlendirilmesine Yönelik Bir Vaka Analizi: 4-K Kulüpleri. Asos J J Acad Soc Sci 4(29):574–594
101. Jarman CG (1977) Banana fiber: a review of its properties and small-scale extraction and processing. Trop Sci 19(4):173–185
102. Teli MD, Valia SP (2013) Acetylation of banana fiber to improve oil absorbency. Carbohydr Polym 92(1):328–333
103. Bledzki AK, Mamun AA, Faruk O (2007) Abaca fiber reinforced PP composites and comparison with jute and flax fiber PP composites. Express Polym Lett 1(11):755–762
104. Murali Mohan RK, Mohana Rao K (2007) Extraction and tensile properties of natural fibers: Vakka, date and bamboo. Compos Struct 77(3):288–295
105. Prabu VA, Manikandan V, Uthayakumar M, Kalirasu S (2012) Investigations on the mechanical properties of red mud filled sisal and banana fiber reinforced polyester composites. Mater Phys Mech 15:173–179
106. Ray DP, Nayak LK, Ammayappan L, Shambhu VB, Nag D (2013) Energy conservation drives for efficient extraction and utilization of banana fiber. Int J Emerg Technol Adv Eng 3(8):296–310

107. Franck RR (2005) Bast ve Other Plant fibers, cilt 39: Crc Press
108. Suhaily SS, Khalil HA, Asniza M, Fazita MN, Mohamed AR, Dungani R, Syakir MI (2017) Design of green laminated composites from agricultural biomass. Woodhead Publishing, In Lignocellulosic fiber and biomass-based composite materials, pp 291–311
109. Sengupta S, Debnath S, Ghosh P, Mustafa I (2019) Development of unconventional fabric from banana (Musa acuminata) fiber for industrial uses. J Nat Fibers
110. Brindha R, Narayana CK, Vijayalakshmi V, Nachane RP (2019) Effect of different retting processes on yield and quality of banana pseudostem fiber. J Nat Fibers 16(1):58–67
111. Doshi A (2017) Banana fiber to fabric: process optimization for improving its spinnability and hand (Doctoral dissertation, The Maharaja Sayajirao University of Baroda)
112. Franck RR (Ed) (2005) Bast and other plant fibers, vol 39. Crc Press
113. Mukhopadhyay S, Fangueiro R, Arpac Y, Şentürk Ü (2008) Banana fibers–variability and fracture behavior. J Eng Fibers Fabr 3(2):1–7; 155892500800300207
114. Suhaimi SA (2020) Physical properties of banana pseudo-Stem woven fabric treated with softening agent. J Acad 8(2):76–83
115. Balakrishnan S, Dharmasri Wickramasinghe GL, Samudrika Wijayapala UG (2020) A novel approach for banana (Musa) pseudo-stem fiber grading method: extracted fibers from Sri Lankan banana cultivars. J Eng Fibers Fabr 15:1558925020971766
116. Praveena BA, Shetty BP, Akshay AS, Kalyan B (2020) Experimental study on mechanical properties of pineapple and banana leaf fiber reinforced hybrid composites. In AIP conference proceedings, vol 2274, No 1, p 030015. AIP Publishing LLC
117. Saxena T, Chawla VK (2021) Banana leaf fiber-based green composite: an explicit review report. Mater Today Proc
118. Pappu A, Patil V, Jain S, Mahindrakar A, Haque R, Thakur VK (2015) Advances in industrial prospective of cellulosic macromolecules enriched banana biofiber resources: a review. Int J Biol Macromol 79:449–458
119. http://www.fao.org, Accessed on August 2020
120. Singaraj SP, Aaron KP, Kaliappa K, Kattaiya K, Ranganathan M (2019) Investigations on structural, mechanical and thermal properties of banana fabrics for use in leather goods application. J Nat Fibers, 1–11
121. Jacob N, Niladevi KN, Anisha GS, Prema P (2008) Hydrolysis of pectin: an enzymatic approach and its application in banana fiber processing. Microbiol Res 163(5):538–544
122. Balakrishnan S, Wickramasinghe GD, Wijayapala US (2019) Study on dyeing behavior of banana fiber with reactive dyes. J Eng Fibers Fabr 14:1558925019884478
123. TMMOB Ziraat mühendisleri Odası, (2020), Muz Raporu/2019, https://www.zmo.org.tr/genel/bizden_detay.php?kod=32505&tipi=38, Accessed on February 2021
124. Ferrante A, Santulli C, Summerscales J (2019) Evaluation of tensile strength of fibers extracted from banana peels. J Nat Fibers 17(10):1519–1531
125. Arafat KMY, Nayeem J, Quadery AH, Quaiyyum MA, Jahan MS (2018) Handmade paper from waste banana fiber. Bangladesh J Sci Ind Res 53(2):83–88
126. Balakrishnan S, Wickramasinghe GLD, Wijayapala US (2019) Investigation on improving banana fiber fineness for textile application. Text Res J 89(21–22):4398–4409
127. Ramdhonee A, Jeetah P (2017) Production of wrapping paper from banana fibers. J Environ Chem Eng 5(5):4298–4306
128. Basak S, Chattopadhyay SK, Samanta KK (2016) Banana pseudostem sap: an important agro-waste for diversified applications including textile. In Pearson VA (ed) Bananas: cultivation, consumption and crop diseases, Nova Science Publishers, New York, pp 91–106
129. Prasad AV, Mohana Rao K, Nagasrinivasulu G (209) Mechanical properties of banana empty fruit bunch fiber reinforced polyester composites
130. Sivaranjana P, Arumugaprabu V (2021) A brief review on mechanical and thermal properties of banana fiber based hybrid composites. SN Applied Sciences 3(2):1–8
131. Hon DNS, Chao WY (1993) Composites from benzylated wood and polystyrenes: their processability and viscoelastic properties. J Appl Polym Sci 50(1):7–11

132. Raj RG, Kokta BV, Grouleau G, Daneault C (1990) The influence of coupling agents on mechanical properties of composites containing cellulosic fillers. Polym-Plast Technol Eng 29(4):339–353
133. Shenoy Heckadka S, Nayak SY, Joe T, Zachariah NJ, Gupta S, Kumar NVA, Matuszewska M (2020) Comparative evaluation of chemical treatment on the physical and mechanical properties of areca frond, banana, and flax fibers. J Nat Fibers 1–13
134. Manohar K, Adeyanju AA (2016) A comparison of banana fiber thermal insulation with conventional building thermal insulation. Curr J Appl Sci Technol, 1–9
135. Alanya Adres Alanyanın İnternet Gazetesi (2017) ITB Berlin Fuarı'nda muz lifine büyük ilgi. http://www.alanyaadres.com/gundem/itb-berlin-fuari-nda-muz-lifine-buyuk-ilgi-h16148.html, Accessed in March 2021
136. Ware M (2021) Benefits and health risks of bananas. Newsletter, Medical News Today, https://www.medicalnewstoday.com/articles/271157, Accessed on October 2021
137. Guiné R, Costa DVTA (2016) Chemical composition and bioactive compounds in bananas and postharvest alterations. In: Pearson VA (ed) Bananas: cultivation, consumption and crop diseases, Nova Science Publishers, New York, pp 27–68
138. Musaceae, Wikipedia the Free Encyclopedia, Musaceae, https://en.wikipedia.org/wiki/Mus aceae, Accessed in July 2021
139. Muzgiller, Wikipedia the Free Encyclopedia, Muzgiller, https://tr.wikipedia.org/wiki/Muz giller, Accessed in June 2021
140. Morton J (1987) Banana, p 2946. In: Morton JF (ed) Fruits of warm climates, Miami, FL
141. Eichhorn S, Hearle JW, Jaffe M, Kikutani T (eds) (2009) Handbook of textile fiber structure. Volume 2: Natural, regenerated, inorganic and specialist fibers
142. Bilba K, Arsene MA, Ouensanga A (2007) Study of banana and coconut fibers: Botanical composition, thermal degradation and textural observations. Biores Technol 98(1):58–68
143. Morton JF (1987) Fruits of warm climates. Morton JF (1987) Creative Resource Systems, Inc.
144. Musa G (2021) https://en.wikipedia.org/wiki/Musa_(genus), Accessed in May 2021
145. Türkiye İK (2021) https://www.tuik.gov.tr/, Accessed in September 2021
146. Food And Agriculture Organization of The United Nations, October 2015, Banana Market Review 2013–2014, https://www.fao.org/home/en, Accessed in September 2020
147. Sakare P, Bharimalla AK, Dhakane-Lad J, Patil PG (2020) Development of greaseproof paper from banana pseudostem fiber for packaging of butter. J Nat Fibers, 1–9
148. Türkay C (2015) Örtüalti Muz Yetiştiriciliği Ve Sera Kurulmasinda Dikkat Edilecek Hususlar. Ziraat Yüksek Mühendisi-Erdemli, Türkiye Cumhuriyeti Gıda Tarım ve Hayvancılık Bakanlığı, https://arastirma.tarimorman.gov.tr/alata/Belgeler/Diger-belgeler/%C3%96rt% C3%BCalt%C4%B1MuzYeti%C5%9Ftiricili%C4%9FiCT%C3%BCrkay.pdf., Accessed in May 2021
149. Li C, Liu J, Ivo Achu N, Fang J (2016) Sustainable management of banana waste through renewable energy and bio-fertilizer generation. In: Pearson VA (ed) Bananas: cultivation, consumption and crop diseases, Nova Science Publishers, New York, pp 69–89
150. Venkateshwaran N, Elayaperumal A, Sathiya GK (2012) Prediction of tensile properties of hybrid-natural fiber composites. Compos B Eng 43(2):793–796
151. Akova SB (2012) Türkiye'de Muz Ziraatinin Coğrafi Dağilişi Ve Özellikleri. Coğrafya Dergisi, (5)
152. Kumar K, Gokuleshwar KJ, Ranganathan S, Thiagarajan C (2018) Fabrication and mechanical property study on glass/sisal/banana natural fibers. Int J Pure Appl Math 119(12):15637–15645
153. Gubbuk H, Pekmezci M (2004) Comparison of open-field and protected cultivation of banana (Musa spp. AAA) in the coastal area of Turkey. New Zealand J Crop Horticul Sci 32(4):375–378
154. Van den Bergh I, Vézina A, Picq C (2013) Where bananas are grown. ProMusa. Bioversity International

155. Emekli NY, Büyüktaş K (2009) A research over current state of the banana greenhouses in Anamur province of Mersin. Ziraat Fakültesi Dergisi, Akdeniz Üniversitesi 22(1):23–38
156. Brindha D, Vinodhini S, Alarmelumangai K, Malathy NS (2012) Physico-chemical properties of fibers from banana varieties after scouring. Indian J Fund Appl Life Sci 2(1):217–221
157. İnci B (1992) Elyaf Bilgisi. Marmara Üniversitesi Yayın No: 524, Teknik Eğitim Fakültesi Yayın No: 7, İstanbul
158. Sheng Z, Gao J, Jin Z, Dai H, Zheng L, Wang B (2014) Effect of steam explosion on degumming efficiency and physicochemical characteristics of banana fiber. J Appl Polym Sci 131(16)
159. Ganan P, Zuluaga R, Velez JM, Mondragon I (2004) Biological natural retting for determining the hierarchical structuration of banana fibers. Macromol Biosci 4(10):978–983
160. Eklahare SR (2005) Eco-friendly chemical processing of textiles and environmental management, pp 28–29
161. Sumaila M, Amber I, Bawa M (2013) Effect of fiber length on the physical and mechanical properties of ramdom oriented, nonwoven short banana (musabalbisiana) fiber/epoxy composite. Cellulose 62:64
162. Paul SA, Joseph K, Mathew GG, Pothen LA, Thomas S (2010) Influence of polarity parameters on the mechanical properties of composites from polypropylene fiber and short banana fiber. Compos A Appl Sci Manuf 41(10):1380–1387
163. Fernandez-Bolanos J, Felizon B, Heredia A, Guillen R, Jimenez A (1999) Characterization of the lignin obtained by alkaline delignification and of the cellulose residue from steam-exploded olive stones. Biores Technol 68(2):121–132
164. Banana fiber, Banana Star, 2012, http://bananafiber.com/bananafiber.html, Accessed in July 2021
165. Bisanda ETN (2000) The effect of alkali treatment on the adhesion characteristics of sisal fibers. Appl Compos Mater 7(5):331–339
166. Tarakçıoğlu I (1978) Tekstil Terbiyesi ve Makinaları. Ege Üniversitesi Matbaası, Cilt 1 İzmir
167. Jacob N, Prema P (2008) Novel process for the simultaneous extraction and degumming of banana fibers under solid-state cultivation. Braz J Microbiol 39(1):115–121
168. Kırmızı GM (2009) Japon tekstil boyama ve desenlendirme teknikleri üzerine bir araştırma (Doctoral dissertation, DEÜ Güzel Sanatlar Enstitüsü)
169. Li X, Tabil LG, Panigrahi S (2007) Chemical treatments of natural fiber for use in natural fiber-reinforced composites: a review. J Polym Environ 15:25–33
170. Bulut Y, Erdoğan ÜH (2011) Selüloz Esasli Doğal Liflerin Kompozit Üretiminde Takviye Materyali Olarak Kullanimi. Tekstil ve Mühendis 18(82)
171. Reddy N, Yang Y (2005) Biofibers from agricultural byproducts for industrial applications. Trends Biotechnol 23(1):22–27
172. Kulkarni AG, Satyanarayana KG, Rohatgi PK, Vijayan K (1983) Mechanical properties of banana fibers (Musa sepientum). J Mater Sci 18(8):2290–2296
173. Satyanarayana KG, Sukumaran K, Mukherjee PS, Pillai SGK (1986) Materials science of some lignocellulosic fibers. Metallography 19(4):389–400
174. Nolasco AM, Soffner MLAP, Nolasco AC (1998) Physical–mechanical characterization of banana fiber, Musa cavendish-nanicao variety. In Second international symposium on natural polymers and composites. Sao Carlos: Embrapa Agricultural Instrumentation, Sao Paulo University (USP), Sao Paulo State University (UNESP)
175. Satyanarayana KG, Guimarães JL, Wypych FERNANDO (2007) Studies on lignocellulosic fibers of Brazil. Part I: source, production, morphology, properties and applications. Compos Part A Appl Sci Manuf 38(7):1694–1709
176. Pothan LA, Thomas S, Neelakantan NR (1997) Short banana fiber reinforced polyester composites: mechanical, failure and aging characteristics. J Reinf Plast Compos 16(8):744–765
177. Kumar, M., and Kumar, D. (2011). Comparative study of pulping of banana stem. Int. J. fiber Text. Res, 1(15.13).

178. Idicula M, Malhotra SK, Joseph K, Thomas S (2005) Dynamic mechanical analysis of randomly oriented intimately mixed short banana/sisal hybrid fiber reinforced polyester composites. Compos Sci Technol 65(7):1077–1087
179. Jordan W, Chester P (2017) Improving the properties of banana fiber reinforced polymeric composites by treating the fibers. Proc Eng 200:283–289
180. Suhani N, Mohamed RMSR, Nasir N, Ahmad B, Oyekanmi AA, Awang H, Daud Z (2020) Removal of COD and ammoniacal nitrogen by banana trunk fiber with chitosan adsorbent. Malaysian J Fund Appl Sci 16(2):243–247
181. Das MC, Singh SP, Raja DE, Prabhuram T (2020) Evalution of mechanical properties of banana fiber & particulate reinforced epoxy composite. Stud Indian Place Names 40(3):7086–7092
182. Sadasivam S, Manickam A (1996) Biochemical methods, 2nd edn. New Age International (P) Limited Publishers, New Delhi, pp 13–19
183. Usmani MA, Khan I, Haque A, Bhat AH, Mondal D, Gazal U (2017) Biomass-based composites from different sources: Properties, characterization, and transforming biomass with ionic liquids. Woodhead Publishing, Lignocellulosic fiber and biomass-based composite materials, pp 45–76
184. Mohan TP, Kanny K (2017) Mechanical and thermal properties of nanoclay-treated banana fibers. J Nat Fibers 14(5):718–726
185. Mohanty AK, Misra MA, Hinrichsen GI (2000) Biofibers, biodegradable polymers and biocomposites: an overview. Macromol Mater Eng 276(1):1–24
186. Taj S, Munawar MA, Khan S (2007) Natural fiber-reinforced polymer composites. Proc Pakistan Acad Sci 44(2):129–144
187. Lewin M, Pearce EM (eds) (1998) Handbook of fiber chemistry, revised and expanded. Crc press
188. Kozlowski R (2012) Handbook of natural fibers: types, properties and factors affecting breeding and cultivation. Woodhead Publishing Limited, Cambridge
189. Pothan LA, Thomas S (2003) Polarity parameters and dynamic mechanical behaviour of chemically modified banana fiber reinforced polyester composites. Compos Sci Technol 63(9):1231–1240
190. Ünal F, Avinç O, Yavaş A, Gündoğan M, Liflerinin M (2018) Tekstil Yaş İşlemleri Ve Son Gelişmeler, II. Uluslararasi Bilimsel Ve Mesleki Çalişmalar Sempozyumu (Bilmes 2018), 2018, pp 944–950
191. Hendrickx K (2007) The origins of banana-fiber cloth in the Ryukyus. Leuven University Press, Japan
192. Belediyesi A (2016) Muz Lifinden Yapilan Ürünlere Yoğun İlgi, 15.03.2016, http://alanya.bel.tr/haber/17499/muz-lifinden-yapilan-urunlere-yogun-ilgi, Accessed in September 2021
193. Adres A (2021) Alanya Muzu Farkli Bir Yüzüyle Türkiye'ye Tanitildi, 06.04.2016, http://www.alanyaadres.com/m/gundem/alanya-muzu-farkli-bir-yuzuyle-turkiyeye-tanitildi-h7106.html , Accessed in September 2021.
194. Belediyesi A (2021) Alanya Muzu Farkli Bir Yüzüyle Türkiye'ye Tanitildi, 06.04.2016, http://alanya.bel.tr/haber/17552/alanya-muzu-farkli-bir-yuzuyle-turkiye-ye-tanitildi , Accessed in August 2021
195. Abdul Karim SK, Lim SF, Chua SN, Salleh SF, Law PL (2016) Banana fibers as sorbent for removal of acid green dye from water. J Chem
196. Teli MD, Valia SP (2016) Grafting of butyl acrylate on to banana fibers for improved oil absorption. J Nat Fibers 13(4):470–476
197. Çavdar AD, Boran S (2016) Doğal liflerin otomotiv sanayinde kullanımı. Kastamonu University Journal of Forestry Faculty 16(1):253–263
198. Silva JLG, Al-Qureshi HA (1999) Mechanics of wetting systems of natural fibers with polymeric resin. J Mater Process Technol 92:124–128
199. Coutts RSP (1990) Banana fibers as reinforcement for building products. J Mater Sci Lett 9(10):1235–1236

200. Deelaman W, Chaochanchaikul K, Tungsudjawong K (2018) Effect of banana fibers on mechanical and physical properties of light weight concrete blocks. Appl Mech Mater 879:151–155). Trans Tech Publications Ltd.

201. Kurtoğlu G (2011) Muz Kabuklarindan Frükto-Oligosakkarit Karişimlarinin Özütlenmesi. Thesis, 2011, SDÜ Fen Bilimleri Enstitüsü, Isparta

202. Palma C, Contreras E, Urra J, Martínez MJ (2011) Eco-friendly technologies based on banana peel use for the decolourization of the dyeing process wastewater. Waste Biomass Valorization 2(1):77–86

203. Belediyesi A (2021) Alanya tarihinin ilk moda defilesi bedesten'de gerçekleştirildi, 30.09.2017, https://www.alanya.bel.tr/haber/19721/alanya-tarihinin-ilk-moda-defilesi-bed esten-de-gerceklestirildi , Accessed in September 2021

204. Nanda ST (2020) The Indian designers turning food waste into high fashion. The Third Pole, December 4, 2020, https://www.thethirdpole.net/en/food/the-indian-designers-turning-food-waste-into-high-fashion/, Accessed in September 2021

205. Mavolu (2018) From waste to value: banana fiber for fashion and textiles, July 8, 2018, https://mavolu.com/blogs/news/from-waste-to-value-banana-fiber-for-fashion-and-tex tiles , Accessed in September 2021

206. Beyer A (2017) This shirt is bananas! banana fiber sustainable fashion. Eluxe Magazine, December 15, 2017, https://eluxemagazine.com/fashion/banana-fiber-sustainable-fashion/, Accessed in September 2021

207. Kumar A (2020) Banana fiber for Sustainable Fashion and Beyond, October 22, 2020, https://www.onlineclothingstudy.com/2020/10/banana-fiber-for-sustainable-fashion.html, Accessed in October 2021

208. Made Trade Magazine (2019) Behind the Fiber: Transforming Banana Tree Waste Into Fabric", Made Trade Magazine, August 19, 2019, https://magazine.madetrade.com/banana-fabric/, Accessed in September 2021

209. Hendriksz V (2017) Sustainable textile innovations: banana fibers. Fashion United, 28 Aug 2017, https://fashionunited.uk/news/fashion/sustainable-textile-innovations-banana-fiber/2017082825623, Accessed in July 2021

Development of Sustainable Sound and Thermal Insulation Products from Unconventional Natural Fibres for Automobile Applications

P. Ganesan and T. Karthik

Abstract An attempt has been made in this present work to develop sound and thermal insulation products from unconventional natural fibres that could be effectively used in the automotive interiors. The fibres intended for the work include century plant, common milkweed, and Indian Bowstring Hemp. The physical properties of the fibres were tested using standard methods and the morphological structure of the fibres was analysed using scanning electron microscopes (SEM), the crystallinity index nature of these fibres was evaluated by X-ray Diffraction (XRD) method. These fibres were blended with different proportion to produce the needle punched non-woven structure. The non-woven fabrics were evaluated for their sound and thermal insulation properties along with the other basic properties. The results showed that the blended non-woven structures have better sound and thermal insulation properties.

Keywords Non-woven · Natural fibre · Sound insulation · Sustainability · Thermal insulation · Milkweed

1 Introduction

Increase in electromechanical systems and technological development has led to high noise pollution. Work place and environmental noise pollution caused by automotives has become a serious threat to human ear. Now a day's specially developed non-wovens and composites are becoming one amongst the important sound absorption materials. About 50% of the automobile products are developed using non-wovens. Currently certain synthetic fibre-based non-wovens and composites are widely used in the automotive interiors for sound insulation. Automotive parts being sent to scrap after use, the synthetics pollute the external environment on a greater note. Also due to the ever decreasing petroleum resources, we have to find alternative source of raw

P. Ganesan (✉) · T. Karthik
Department of Textile Technology, PSG College of Technology, Coimbatore, India
e-mail: pgn.textile@psgtech.ac.in

© The Author(s), under exclusive license to Springer Nature Singapore Pte Ltd. 2022   195
S. S. Muthu (ed.), *Sustainable Approaches in Textiles and Fashion*, Sustainable Textiles: Production, Processing, Manufacturing & Chemistry,
https://doi.org/10.1007/978-981-19-0878-1_8

material which is eco-friendly to the environment and biodegradable, which can be best combined to use as automotive interiors.

Natural fibre has broader application ranges in the area of textile, mainly in the light of the current global tendency towards eco-friendly textiles. The emerging sustainable and green economy should be based on renewable and energy efficiency feed stocks in the polymer products, processes that might reduce emission of carbon and recyclable materials. Natural fibres are renewable and have less weight, good strength, and also low cost, which make them attractive for the industry. The utilization of fibres from natural resources for reinforcement in composites is a promising research area. In recent times, attempt has been carried out to minimize the use of synthetic fibres and other expensive glass, carbon or aramid fibres and also reduce the weight of the car's body considerably by utilizing the advantage of the low density and cost of natural fibres.

Automotive textile includes all textile components e.g. filaments, fibres, yarns, and fabrics which are used in automobile industry. Few of these components can be visible on the outside of the product and some of them are concealed. The components such as carpets, upholstery, roof liners, and seat belts are visible and on the other hand components such as composites, tyre cords and rubber reinforced components (hose and filters) and airbags are concealed. In general, acoustic porous materials can have porosity greater than 90%. Common sound absorption materials are open cell foam and fibre. It is generally an energy conversion process. Due to the excitement of sound, air molecules in the pores undergo periodic compression and relaxation, this results in change of temperature. All thermal insulation materials work on a single basic principle: heat moves from warmer to colder areas. Therefore, on cold days, heat from inside a building seeks to get outside. And on warmer days, the heat from outside the building seeks to get inside. Insulation is the material which slows this process. When a hot surface is surrounded by an area that is colder, heat will be transferred and the process will continue until both are at the same temperature.

The fibres which are used presently in automobiles are rock wool, fibre glass, Polypropylene used in Carpets and Sun visors and Polyurethane—Headliners and Insulation felts. Inhaled slivers of mineral wool irritate the alveoli and can cause lung disease. Scientists found isocyanates, a compound that can bring potential harm to one's lungs, in materials made up of polyurethane. Exposure to the said product can cause lung irritation and asthma attacks. The main fibre properties contributing the sound and thermal insulation properties of non-woven fabrics are fineness, length, and hollowness. Based on the properties required for the sound and thermal insulation, the unconventional fibres which are less investigated and explored were selected for the study. Hence in this project attempt is made to develop sound and thermal insulation interiors for automotive application that are completely eco-friendly and biodegradable. The fibres used for the intended purpose include century plant (*Agave americana*) [1], common milkweed (*Asclepias Syriaca*), and Indian Bowstring Hemp (*Sansevieria Roxburghiana*) [2]. The fibre properties have been characterized and analysed and accordingly blended with each other in different proportions to produce non-woven fabric. The blended needle punched non-wovens were analysed for their sound insulation and thermal insulation properties and are reported.

2 Material and Methods

2.1 Selection and Extraction Fibres

The common milkweed fibres, and stems of century plant, and Indian Bowstring Hemp were collected from local area Coimbatore, Tamilnadu, India. The common milkweed fibres were obtained from the pods. The full-grown pods were collected from the plant and the floss is extracted from the pods and partly dried. The milkweed fibres, with their seeds attached, were removed from the opened pod by hand. The century plant and Indian Bowstring Hemp were extracted by the process similar to extraction of sisal and other bast fibres. Initially the fibres were done retting in process for 7 days. This is followed by the extraction process of fibres by crushing them. The Century plant fibres were done manual extraction by beating the stem against rough surface followed by cleaning. The fibres were extracted using mechanical extraction process. The portions of the bast plant were inserted through the rollers of the machine, which is crushed the bast plant to separate the fibres. The fibres thus produced were washed in running water to clean then fibres were dried at room temperature. The plants and extracted fibres from them are shown in Fig. 1.

3 Characterization of Fibres

3.1 Physical Properties of Fibres

Physical properties of the fibres were characterized as per standard methods; the standards are listed in the following Table 1.

3.2 Surface Morphology Using Scanning Electron Microscope

The surfaces of fibres were observed using scanning electron microscope. Prior to the test, the samples were coated with a thin layer of gold by a plasma sputtering apparatus. The observation was performed in high vacuum mode with secondary electron detector and accelerating voltage between 5 and 10 kV.

(a) Century plant and extracted fibre

(b) Common milkweed pod and fibre

(c) Indian Bowstring Hemp plant and extracted fibre

Fig. 1 Plants and extracted fibres

Table 1 Standard testing method for physical properties of fibre

S. No.	Parameter	Method of testing	Standard
1	Single fibre length	Baer sorter	ASTM D 1447
2	Single fibre strength	Instron	ASTM D 3822
3	Single fibre fineness	Gravimetric method	ASTM D 1577
4	Density of fibre	Gradient column	ASTM 1505-03
5	Moisture regain	Moisture analyser	ASTM D 2654-89a

3.3 Crystallinity of Fibres Using X-Ray Diffraction (XRD)

X-ray diffractograms of the samples were obtained with Lab-X, wide angle X-ray diffractometer, having an X-ray tube, producing monochromatic radiation ($\lambda = 1.54$ Å) at 30 kV and 20 mA. The crystallinity index of the fibre was calculated according to the Segal empirical method as given in the formula below.

$$CrI(\%) = \frac{(I_{002} - I_{am})}{I_{002}} * 100$$
$$Cr \text{ ratio} = \frac{i002}{(i002 + i101)} * 100$$

where I_{002} is the maximum intensity of the I_{002} lattice reflection and I_{am} is the height of the minimum (I_{am}) between the 002 and the 101 peaks which is due to amorphous region of the sample. Fibres were cut with a knife crusher to about 500 μm and pressed into a pellet using a cylindrical steel mould (diameter = 16 mm) with an applied pressure of 35 MPa in a laboratory press.

3.4 Analysis of Chemical Groups by FTIR

FT-IR spectra of the fibres (raw, alkali treated, dyed) were recorded using Shimadzu FT-IR in KBr matrix with a scan rate of 32 scans per minute with a resolution of 4 cm^{-1} in the wave number region of 400–4000 cm^{-1}. The fibre samples were chopped into smallest particles using a scissor and ground well to a fine powder using mortar and pestle. Then it was mixed with KBr and pelletized by pressurization to record the FT-IR spectra under standard conditions.

3.5 Thermal Properties of Fibres

Thermo Gravimetric Analysis: TGA can be used to evaluate the thermal stability of a material. In a desired temperature range, if a species is thermally stable, there will be no observed mass change. Negligible mass loss corresponds to little or no slope in the TGA trace. TGA also gives the upper use temperature of a material. Beyond this temperature the material will begin to degrade.

   Differential Scanning Calorimetry: The DSC thermograms of the conditioned (25 °C, RH 75%) raw and treated fibre samples were recorded on Perkin Elmer DSC 7 as well as in Netzsch DSC 204 instruments from room temperature to 400 °C and 200 °C respectively at a heating rate of 10 °C/min, in nitrogen atmosphere.

3.6 Preparation of Fibre and Web Formation

The seed from common milkweed fibres was manually removed prior to web formation in card. The fibres of *Indian Bowstring Hemp* and *Century plant* were cut to the length of 50 mm and then opened using Shirley trash analyser. It reduces the length and to open it further. It consists of a feed table, feed roll, licker-in, baffles, air blast, and condenser. The licker-in is very similar to those used in commercial cards. The fibre is fed slowly to the licker-in and as it is broken up, the air blast carries the lint around the bottom of the flow plate and to the condenser. Trash and the heavy particles drop into the waste camber. The clean fibres are deposited into the lint chamber. Thus opened fibres were taken to the carding machine for web formation.

   The raw fibres were mixed well. Each lot weighing 30, 45, and 75 g was prepared for each fibre. The samples were then fed into a miniature card (Trytex, India) equipped with stationary flat arrangement. It is a microprocessor-based carding machine having possibility of processing up to 50–100 g per batch. It has metallic wire clothing individual drive to the rollers. The minimum amount of fibre required for web formation was found to be 30 g. Hence in order to identify the contribution of the three fibres individually towards the final properties of the non-woven they were blended in the above ratios. Also to find the contribution significantly 100% pure webs of all the three fibres were also produced.

3.7 Needle Punched Non-woven Production

The webs were blended in different proportions to obtain the best result. Totally 9 samples were produced and six samples from blending of common milkweed, century plant and into different proportions and three samples were 100% single fibres non-wovens were produced.

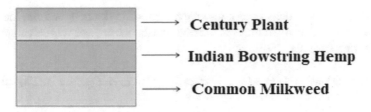

Fig. 2 Arrangements of web layers for the non-woven

The webs were placed in such a way that the coarser fibre was on the receiving end of the sound wave, followed by finer fibres. Thus Century plant was made as the top layer, followed by Indian Bowstring Hemp and then Common milkweed as shown in Fig. 2 for needle punching process. This arrangement is done so that when the sound wave reaches the top surface, due to the more open structure, it will initially enter into the product without reflecting back. Now when it reaches the second layer some amount of the sound waves are absorbed by the little closed structure and finally when it reaches the third layer, there is a complete absorption of the sound. The non-woven fabric was formed in needle punching machine (Dilo machines, Germany) with 10 mm needle penetration depth, 150 strokes/min and stitch density of 96.

4 Characterization of Developed Non-woven

4.1 Physical Characterization

The physical properties of the non-wovens such as the areal density (ASTM D1059), thickness (ASTM D1777-96), air permeability (TS 391 EN ISO 9237) were tested as per standard methods. The bulk density was calculated using the following relationship

$$pb = \frac{W}{t} \text{ Kg/m}$$

W—Weight of sample per unit area (g/m^2), determined the following standard method ASTM D 3776.

t—Thickness of sample (m).

The porosity has a direct relationship with the sound and thermal insulation of the fabric. The porosity is calculated using this formula,

$$H = 1 - (pa/pb)$$

where ρb is the bulk density of non-woven composite (kg/m^3); and ρa, the weight average absolute density of fibres in the non-woven (kg/m^3). ρa was calculated using the following formula:

$$\rho a = (P_{CP}D_{CP} + P_{CM}D_{CM} + P_{IBH}D_{IBH})/(P_{CP} + P_{CM} + P_{IBH}) \text{ kg/m}^3$$

where P_{CP}, P_{CM}, and P_{IBH} are the % blend proportion of century plant, common milkweed, and Indian Bowstring Hemp respectively; and D_{CP}, D_{CM}, and D_{IBH}, the absolute densities of century plant, common milkweed, and Indian Bowstring Hemp respectively fibres respectively.

4.2 Thermal Conductivity by Lee's Disc Method

The thermal conductivity of samples was determined using Lee's Disc method using following formula

$$k = \frac{ms\left(\frac{dT}{dt}\right)x}{A(T_2 - T_1)}$$

where,

k = thermal conductivity of fabric ($\times 10^{-3}$ WM^{-1} K^{-1})

T_1 = Temperature of C in steady state

T_2 = Temperature of B in steady state

s = Specific heat of the material of the disc (370 JKg^{-1} K^{-1})

m = mass of disc (C)

Surface area of fabric $G = A$ (mm^2)

Thickness of fabric $G = x$ (mm).

4.3 Sound Absorption Coefficient by Impedance Tube Method

The developed non-wovens were tested for their sound absorption coefficient by Impedance Tube Method (ASTM E 1050) as shown in Fig. 3. An impedance tube was used with a sound source (loudspeaker) connected to one end, and the test sample mounted to another end. The loudspeaker generates the broadband random sound waves. Sound waves propagating as plane waves in the tube hit the sample,

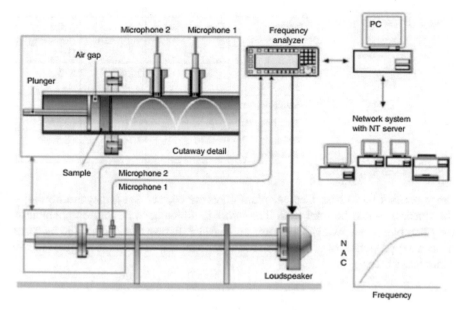

Fig. 3 Impedance tube

get partially absorbed, and subsequently reflected. The acoustical properties of the test sample were tested in the frequency range of 63–6300 Hz. This system tests a sound absorptive material, processes the results, and reports the results in a graph of the absorption coefficient in various frequencies. Thus, the absorption coefficient of each sample was obtained. In present research work, the acoustic characteristics of the non-woven fabric were expressed as a NRC (Noise Reduction Coefficient and was calculated using the relationship:

$$NRC = \frac{\alpha_{250} + \alpha_{500} + \alpha_{1000} + \alpha_{2000}}{4}$$

5 Results and Discussions

5.1 Physical Properties of Fibres

Table 2 represented the physical properties of selected fibres. Tenacity of the Indian Bowstring Hemp fibres is the highest compared to the three fibres that were selected for the study which is in the range of the hemp fibres. Century plant being a stiff fibre the tenacity is low, even less than that of milkweed. Milkweed being seed hair fibre, it is very finer whereas Indian Bowstring Hemp has optimum fineness in the

Table 2 Physical properties
of fibres

Property	Century plant	Common milkweed	Indian Bowstring Hemp
Tenacity (g/tex)	13.423	16.575	53.58
Fineness (den)	235.98	0.99	47.97
Density (g/cc)	1.38	0.9	1.35
Moisture Content (%)	9.15	10.4	8.5

range of other bast fibres. Century plant fibres are coarser, but if they are fibrillated the fineness would be improved. The two bast fibres have density comparable to the other bast fibres, whilst Common milkweed is lighter than water. The moisture absorption properties of all the three fibres were almost in the same range as that of other bast fibres.

5.2   Surface Morphology of Fibres Using SEM

From Figs. 4a and 5c it is clear that the surface morphology of the Century plant and Indian Bowstring Hemp fibres is similar to that of the other bast fibres. They have a multicellular structure and very rough surface morphology with many striations. These rough surfaces are as a result of the frictional contact of the fibres between each other. From Fig. 4b, the Common milkweed fibres on the other hand have a very smooth surface morphology. This smooth and lustrous effect on the fibre surface could be observed as a result of the higher wax content in the fibres.

5.3   FTIR Analysis of Fibres

The Century plant, common milkweed, and Indian Bowstring Hemp fibres showed common absorptions (Fig. 5) around 3400, 2925, 1470, 1356, 1169, and 1038 cm^{-1} and they were identified as reported in other lignocellulosic fibres [3, 4]. The assignment of the characteristic IR peaks and their common relative sources are given in Table 3 [5]. Both the bast fibres (Century plant and Indian Bowstring Hemp) have functional groups like lignin, pectin, and hemicellulose that are easily soluble which are identified by the presence of certain carbon linkages. In case of the milkweed fibres the polysaccharides form the major portion of the fibre which is due to the presence of O–H and C–H linkage.

It can be noted that the absorption band at ~1730 cm^{-1} and 1240^{cm-1} seen in all the fibres. These bands which are attributed to the stretching vibrations of C=O

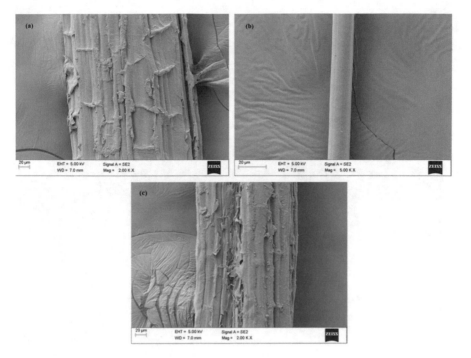

Fig. 4 Scanning electron microscope images of fibres. **a** Century plant **b** common milkweed and **c** Indian Bowstring Hemp

and C–O groups and are prevalent in lignin and hemi-cellulosic structures. These absorption bands which are recognized to the vibrations of C=O and C–O groups and are established in hemi-cellulosic and lignin structures.

5.4 Crystallinity of Fibres Using X-Ray Diffraction

The XRD study reveals the crystallinity index and crystallinity ratio which measures the orientation of cellulose crystals in a fibre to the fibre axis and the patterns are shown in Fig. 6 and the values are given in Table 4.

The XRD results as shown in Fig. 6 show two main peaks, representing the planes (002) and (101) at 2θ around 22.46° and 16.3° respectively, which are assigned to Cellulose I. The peak intensity at 22.46° is said to represent the total intensity (crystalline + amorphous) of material and peak intensity at 16° correspond to the amorphous material in the cellulose. The Indian Bowstring Hemp fibre has the highest crystallinity values which may be due to the relative fineness of the fibre and the chemical groups present in it. In case of the milkweed fibres the crystallinity values are comparable with the other seed fibres like cotton. From Fig. 6, it can be inferred

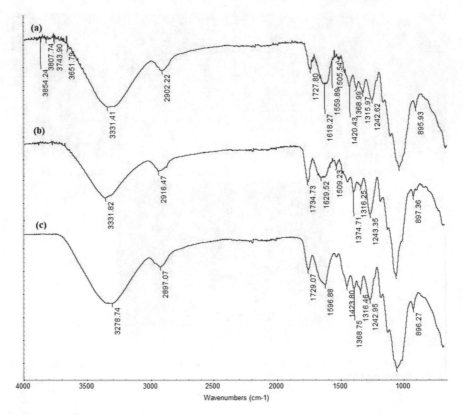

Fig. 5 FTIR graphs of **a** century plant **b** common milkweed (milkweed) and **c** Indian Bowstring Hemp

that in the case of Century plant, the crystalline regions are scattered across the different lattices.

5.5 Thermal Properties of Fibres

Since majority of natural fibres are having lower thermal stability compared to synthetic fibres, thermal characteristics were determined for all the three fibres. The thermal degradation of natural cellulosic fibres depends on their polymer morphology, molecular weight, and crystallinity (Rosen 1993). TGA and DTG curves of fibres are shown in Fig. 7.

From the Thermogravimetric Analysis curves of all the three fibres, three main distinctive regions were noticed during the thermal degradation process of fibre. The first region of degradation starts with weight loss from 30 to 90 °C, which is commonly attributed to evaporation of moisture from the fibre. In the second

Table 3 Assignment of FT-IR peaks and their relative sources [6]

Wave number (cm^{-1})	Vibration	Source
3300	O–H linked shearing	Polysaccharides
2885	C–H symmetrical stretching	Polysaccharides
2850	CH_2 symmetrical stretching	Wax
1732	C=O unconjugated	Hemicellulose
1650–1630	OH (Water)	Water
1505	C=C aromatic symmetrical stretching	Lignin
1425	CH_2 symmetrical bending C=C stretching in aromatic groups	Pectins, lignins, hemicelluloses, calcium pectates
1370	In-the-plane CH bending	Polysaccharides
1335	C–O aromatic ring	Cellulose
1240	C–O aryl group	Lignin
1162	C–O–C asymmetrical stretching	Cellulose, hemicellulose
895	Glycosidic bonds	Polysaccharides
670	C–OH out-of-plane bending	Cellulose

degradation region, which is a major one, happened in the temperature region of 150–360 °C, which may be attributed to the lignin and hemicelluloses degradation in the fibre. The final degradation occurred with weight loss in the range of 360–530 °C due to the degradation of α-cellulose and other non-cellulosic materials in the fibre which are in agreement with Dollimore and Holt [7].

In all the curves, the first weight loss attributed to the evaporation of moisture/water in the fibre sample were around 9.5%, 11% and 8% respectively for Century plant, Common milkweed and Indian Bowstring Hemp. From the Table 5, it is noticed that the degradation for the Century plant and Indian Bowstring Hemp starts at 230 and 260 °C, whereas for Common milkweed the degradation starts at 150 °C after the evaporation of 11% moisture. Also with the milkweed fibres the abrupt weight loss is noticed but in case of bast fibres, degradation starts at a particular point and then remains steady over a period of temperature range and then continues to degradation [8].

Figure 8 shows the DSC of three fibres. In DSC, numerous processes related to desorption of water and decomposition of polymer were observed. The DSC of fibres showed exothermic peak in the temperature range between 80 and 90 °C in all the samples, which are mainly due to the evaporation of moisture/water from the fibre. In cellulosic natural fibres, degradation of lignin begins around 200 °C whereas in the other polysaccharide like cellulose degrades at much higher temperature. Hence, the higher peaks which are at temperature range above 200 °C indicate the cellulose decomposition in the fibres. Thermal degradation in milkweed fibres starts above 200 °C with breaking of bonds and development of components which are volatile. With continuous supply of energy, degradation of fibre continues to happen which

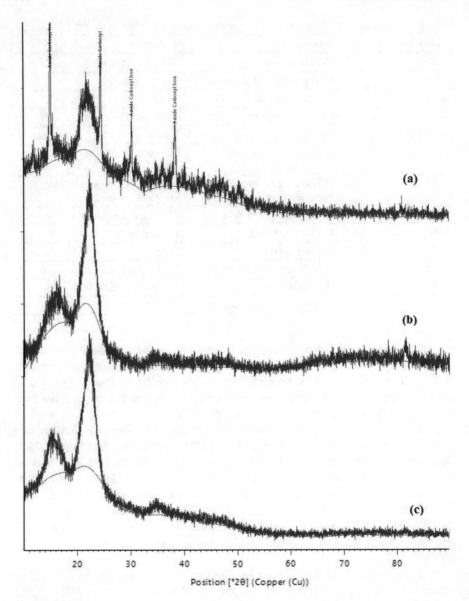

Fig. 6 XRD patterns of **a** century plant; **b** common milkweed and **c** Indian Bowstring Hemp

further results in breaking of cellulosic chains and emission of volatile by-products. The very strong endothermic peak corresponding to degradation of cellulose was noticed around 360 °C in all fibres [9, 10].

Table 4 Crystallinty index and crystallinity ratio of the fibres	Fibres	Crystallinity index (%)	Crystallinity ratio (%)
	Century plant	54.0	68.5
	Common milkweed	41.3	63
	Indian Bowstring Hemp	64.8	73.9

5.6 Characterization of Non-woven Fabrics

The non-woven samples were produced from webs of Century plant, Common milkweed, and Indian Bowstring Hemp using needle punching technique, the following Table 6 shows the different proportion of fibres used for non-woven fabric production.

The physical properties of developed non-woven samples were characterized based on the standards and the results were shown in Table 7.

The areal density and thickness of developed non-woven samples are represented in Fig. 9. From the results, from the figure, it is clear that whilst increase in thickness of non-woven samples areal density (GSM) of samples were increased. The non-woven samples having higher proportion of milkweed showed higher thickness and areal density owing to their less fibre density. The bulk density and porosity of developed non-woven samples are represented in Fig. 10. The 100% Century plant non-woven is having higher bulk density than other non-wovens due to its high fibre density. 100% Indian Bowstring Hemp sample is a bit less bulky than Century plant [11]. The coarser and longer fibres will produce bulkier structure [3]. Common milkweed contains shorter and finer fibres hence its structure is less bulky.

Amongst non-wovens from different blends, it is observed that 50% Century plant, 20% Indian Bowstring Hemp, and 30% Milkweed non-woven is bulkier than all other non-woven blends. The Milkweed possesses higher lignin content and it is smooth and brittle. Thus during opening in carding, they break and turn into shorter fibre. Thus they are having lesser cohesiveness between fibres. Hence it is not compacted easily during needling. In Century plant/Indian Bowstring Hemp/Common milkweed non-wovens, the blend ratio is found to have significant impact on the variation of bulk density. It is observed that bulk density and thickness of non-wovens increases as the percentage of Century plant increases.

The non-woven fabric porosity was calculated theoretically. The fabric porosity depends on fabric bulk density and fibre density. To allow sound dissipation by friction, the sound wave has to enter the porous material. This means, there should be enough pores on the surface of the material for the sound to pass through and get dampened. From the Fig. 10 it is seen that 30% Century plant/50% Indian Bowstring Hemp/20% Common milkweed and 100% Common milkweed sample is having higher porosity. 100% Common milkweed is having lesser bulk density and higher porosity. This can result in higher tortuosity of fabric.

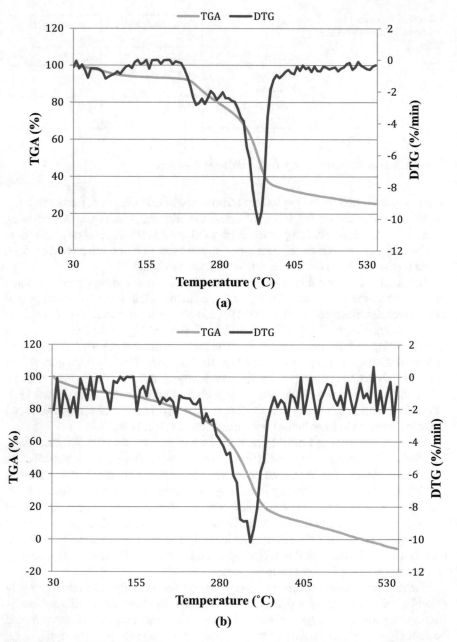

Fig. 7 TGA and DTG graphs of fibres **a** century plant **b** common milkweed and **c** Indian Bowstring Hemp

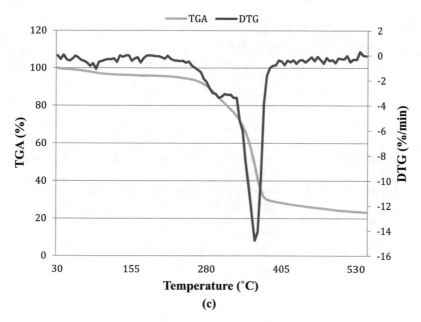

Fig. 7 (continued)

Table 5 Thermal degradation values of fibres

Fibre	Degradation temperature (°C) Td	Maximum degradation temperature (°C) Td max
Century plant	230	360
Common milkweed	150	350
Indian Bowstring Hemp	260	375

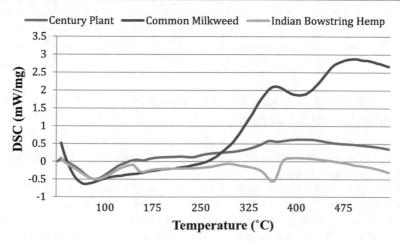

Fig. 8 DSC curves of unconventional fibres

Table 6 Sample code of samples

S. No.	Sample code	Composition
1	A-20/S-30/M-50	Century plant—20; Indian Bowstring Hemp—30; Common milkweed—50
2	A-20/S-50/M-30	Century plant—20; Indian Bowstring Hemp—50; Common milkweed—30
3	A-30/S-20/M-50	Century plant—30; Indian Bowstring Hemp—20; Common milkweed—50
4	A-30/S-50/M-20	Century plant—30; Indian Bowstring Hemp—50; Common milkweed—20
5	A-50/S-20/M-30	Century plant—50; Indian Bowstring Hemp—20; Common milkweed—30
6	A-50/S-30/M-20	Century plant—50; Indian Bowstring Hemp—30; Common milkweed – 20
7	100% A	100% century plant
8	100% M	100% common milkweed
9	100% S	100% Indian Bowstring Hemp

Table 7 Physical properties of non-woven

Sample code	Thickness (mm)	GSM	Bulk density (Kg/m^3)	Porosity (%)
A-20/S-30/M-50	8.66	430	49.67	95.61
A-20/S-50/M-30	6.08	310	50.98	95.83
A-30/S-20/M-50	8.08	420	51.98	95.42
A-30/S-50/M-20	6.24	290	46.47	96.34
A-50/S-20/M-30	7.48	410	54.81	95.54
A-50/S-30/M-20	8.56	300	45.07	95.75
100% A	5.57	390	70.04	94.92
100% M	4.65	160	34.39	96.18
100% S	4.85	310	63.89	95.27

5.7   Air Permeability and Thermal Conductivity of Non-woven Samples

The following Fig. 11 shows the air permeability and thermal conductivity of developed non-woven samples.

As the permeability through the non-woven fabric was increased, the sound absorption coefficient was found to be decreased. Good airflow resistance for certain type of non-woven material was noticed. When the density of fabric was lower, more number of air voids as well as larger voids inside the non-woven fabric but lesser acoustic wave reflection and refraction. Hence it was easy for acoustic waves to propagate through the material. In addition, the internal surface was small and the

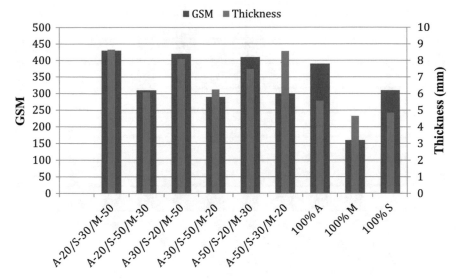

Fig. 9 Areal density and thickness of non-woven samples

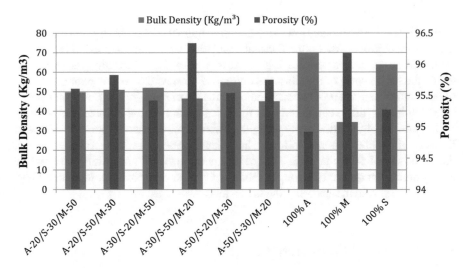

Fig. 10 Porosity and bulk density of non-woven samples

viscous and frictional resistance between the fibre and vibrating air was low. So the sound absorption property was poor.

Further, when the density was higher and the flow permeability was less, maximum of the acoustic energy was reflected at the non-woven fabric surface rather than transmission so that the sound absorption characteristics were decreased, which corresponds to 50% Century plant/20% Indian Bowstring Hemp/30% Common milkweed. For thermal conductivity. The samples 100% Common milkweed (milkweed) sample

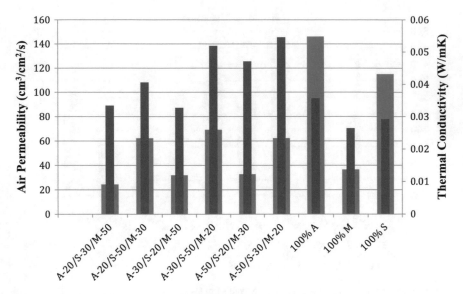

Fig. 11 Air permeability and thermal conductivity of non-woven samples

(0.02644) and 100% Indian Bowstring Hemp sample (0.02921) have given maximum thermal insulation. The better thermal insulation of Common milkweed can be attributed to its hollowness for higher thermal insulation. The air gets entrapped inside the hollowness of the fibre. As entrapped unmovable air is a poor conductor of heat, milkweed can provide maximum insulation of heat. Indian Bowstring Hemp is having higher fibre length and thus the cohesion between the fibres is absolutely high. Thus the high bulk density is higher. This caused decreased porosity of fabric. Because of decreased porosity, higher bulk density and better fibre to fibre friction, the tortuosity of fabric is high. This caused entrapment of air within the structure giving higher thermal insulation. In fabrics of similar composition, thickness plays a predominant role in insulating the heat waves.

5.8 Acoustical Characteristics

Research works carried out influence of several non-woven fabric parameters which influence the acoustic performance of non-woven materials are fibre fineness, the percentage porosity of the non-woven, air permeability, fabric thickness, and the non-woven bulk density. The results of sound absorption coefficient for non-woven fabric samples were measured separately at five different sound frequencies. From the results it is observed that all the non-woven samples have a low sound absorption coefficient (SAC) in low frequency range (250–500 Hz) and high sound absorption coefficient at medium and high frequency ranges (1–4 kHz). Figure 12 depicts the sound absorption coefficient values of non-woven samples at six different frequency

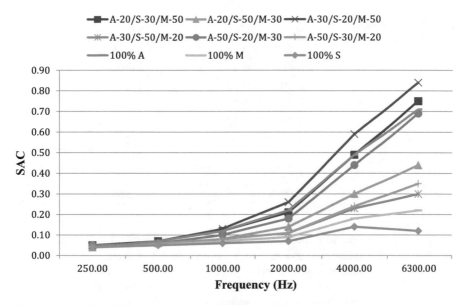

Fig. 12 SAC value of developed non-woven samples

ranges. Using the given Figures, the sound absorption coefficient differences between the non-woven fabric samples could be recognized. In the non-woven structures, air space divides at random into small sections and is distributed in the non-woven fabric sample. Hence, repeated enlargements and contractions of the sound wave during travelling in a non-woven structures were occurring. This incident, as well as the friction between the vibrating air and the fibre surface, resulting in the damping of the sound wave.

The experimental results show that the 30% Century plant/20% Indian Bowstring Hemp/50% Common milkweed non-woven samples have higher sound absorption coefficient than all other non-woven at both low frequency (250–500 Hz) and high frequency (1–4 kHz). On the other hand, 20% Century plant/30% Indian Bowstring Hemp/50% Common milkweed blend exhibits the highest acoustic performance. The contribution of thickness is appreciable. 100% Common milkweed has second highest sound absorption coefficient. Surface characteristics of fibres and their cross-sectional views play an important role. The cross-section of synthetic fibres is dependent on the cross-sectional shape of the spinneret. The Common milkweed fibres used in this research work have a hollow cross-section; whereas the cross-sectional shape of century plant and Indian Bowstring Hemp fibres has a solid cross section. The researchers show that different fibre's shapes result in various surface areas and different surface areas lead to different viscous and thermal effects [4]. The large fibre surface area results in great sound absorption.

The NRC values of different non-woven samples are given in Table 8. The NRC values of Century plant/Indian Bowstring Hemp/Common milkweed non-woven samples are higher at higher frequency, with an increase in Common milkweed

Table 8 NRC values of non-woven samples

Sample code	NRC value
A-20/S-30/M-50	0.1125
A-20/S-50/M-30	0.0800
A-30/S-20/M-50	0.1275
A-30/S-50/M-20	0.0725
A-50/S-20/M-30	0.0975
A-50/S-30/M-20	0.0725
100% A	0.0625
100% M	0.1175
100% S	0.0550

component. This could be due to the lower air permeability and hollowness led to higher frequency sound absorption. As it is known, the entanglement of fibres within the non-woven structure is influenced by the fibres flexibility. According to the milkweed fibre properties, the brittleness nature of this fibre made them to show relatively inflexible nature. Therefore, the carded web layers taken from the common milkweed fibres are anticipated to have a bigger pore size and low airflow resistance.

On the other hand, the density of bast fibre is higher than common milkweed fibres by about 52%, implying that the bast fibres will have more weight compared to milkweed fibres by considering the fibre volume. Therefore, the two non-woven samples with the same areal densities but different fibre blends, would have different numbers of fibres within their structures. Increasing the numbers of Common milkweed fibres within a fibrous material leads to creating less-dense structure, which in turn should have resulted in lower sound absorbency especially at high frequencies. But Common milkweed has given maximum sound absorption. Than that means the influence of Hollowness is higher than density in SAC. Hollow fibres due to entrapment of air in their central axis which effectively increase the boundaries within the fibrous assembly are expected to induce a two-fold increase in the sound absorbency of the fibrous material in comparison to identical fibrous material compose of solid fibres.

6 Conclusion

The tenacity of Indian Bowstring Hemp is higher when compared to that of Century plant and Common milkweed. Its value is 53.58 g/tex, which is very much higher than jute and almost similar to hemp. Milkweed is very finer whereas Indian Bowstring Hemp have most favourable fineness in the variety of other bast fibres. Century plant fibres are very much coarser than most bast fibres. When the cross-section of fibres are analysed, it is found that milkweed is a hollow fibre with smooth surface and the bast fibres Century plant and Indian Bowstring Hemp are of solid structure with serrated rough surface. When crystallinity is studied using x-ray diffraction, Indian Bowstring Hemp is highly crystalline than Milkweed and Agave fibres. The degradation of the Century plant starts at 230 °C which may be due to the reason of near to the ground moisture regain values compared milkweed and Indian Bowstring Hemp, the degradation begins at 260 °C and 280 °C respectively after the evaporation of 10% moisture.

Amongst non-wovens from different blends, it is observed that 50% Century Plant/20% Indian Bowstring Hemp/30% Common Milkweed non-woven is bulkier than all other non-woven blends. It is seen 30% Century Plant/50% Indian Bowstring Hemp/20% Common Milkweed fabric and 100% Milkweed fabric is having higher porosity. 100% Milkweed is having lesser bulk density and higher porosity. Thus tortuosity is less.

The porosity is found to be high in case of finer fibre Milkweed100% coarser fibres. Milkweed fibres are having poor air permeability property when compared with very high value of two bast fibres. This air resistivity property contributed considerably to higher SAC and thermal insulation value for 100% Milkweed fabric. The noise reduction co-efficient is higher for 30% Century Plant, 20% and 50% common Milkweed non-woven fabric and100% Milkweed non-woven fabric, of value 0.1275 and 0.1175 respectively. As the milkweed content is increased in the fabric, the sound absorption co-efficient gets increased. This can be attributed to the fineness, hollowness, and air resistivity of the fibre. Thus it is evident that finer fibres possess high Sound absorption co-efficient than coarser fibres. The Thermal insulation property is higher for 100% milkweed non-woven fabric with the thermal conductivity of 0.02644 W/mK. The products developed from unconventional fibres promote the sustainability aspects of various automotive components due to its properties and recyclability compared to synthetic fibre components.

References

1. Crawford GH, Eickhorst KM, McGovern TW (2003) Botanical briefs: the century plant—*Agave americana* L. Cutis 72:188–190
2. Kanimozhi M (2011) Extraction, fabrication and evaluation of Sansevieria Trifasciata Fiber. Indian J Appl Res 1:97–98
3. Nelson ML, Robert TO (1964) Relation of certain infrared bands to cellulose crystallinity and crystal lattice type. Part II. A new infrared ratio for estimation of crystallinity in celluloses I and II†. J Appl Polym Sci 8:1325–1341
4. Marchessault RH, Liang CY (1960) Infrared spectra of crystalline polysaccharides. III. Mercerized cellulose. J Polym Sci 43:71–84
5. Subramanian K, Kumar SP, Jeyapal P, Venkatesh N (2005) Characterization of ligno-cellulosic seed fibre from Wrightia tinctoriaplant for textile applications–An exploratory investigation. Eur Polym J 41:853–861
6. Karthik T, Ganesan P (2016) Characterization and analysis of ridge gourd (Luffa acutangula) fibres and its potential application in sound insulation. J Text Inst 107:1412–1425
7. Dollimore H (1973) Thermal degradation of cellulose in nitrogen. J Polym Sci Polym Phys Ed 11:1703–1711

8. Ganesan P, Karthik T (2016) Development of acoustic nonwoven materials from kapok and milkweed fibres. J Text Inst 107:477–482
9. Karthik T, Murugan R (2013) Characterization and analysis of ligno-cellulosic seed fiber from Pergularia daemia plant for textile applications. Fibers Polym 14:465–472
10. Rosen SL (1993) Fundamental principles of polymeric materials, vol 30, 2nd edn. John Wiley, New York
11. Sharma HS, Kernaghan K (1988) Thermogravimetric analysis of flax fibres. Thermochim Acta 132:101–109

Design and Development of Under Arm Sweat Pad

M. R. Srikrishnan

Abstract Sweat is a natural body mechanism where the body cools itself by releasing clear and salty sweat that is produced by the glands in the skin. A normal person is born with about 2–4 million sweat glands. There are two scenarios that stimulate our sweat glands causing sweat: physical heat and emotional stress. Emotional sweating typically occurs in the palms of our hands, the soles of our feet, our armpits, and sometimes our foreheads. This sweat consists of bacteria which breaks the sweat into acids causing a bad odour in the due course of time. Sweat in the underarm causes discomfort as the primary clothing we wear gets wet. The wet level due to sweat causes the primary clothing to get soggy in the underarm area. Wetness in the underarm for a longer duration causes itchiness and also bad odour. Sweat is mostly water with trace amounts of minerals like sodium, potassium, calcium, and magnesium, lactic acid, and urea. Exposure to sweat for a longer period of time can cause the fabric to become discolored and eventually weaken the fibre in the clothing. We have various options available in the market to prevent clothing from sweat and odour. Antiperspirants such as deodorants are common solution used by the people. But deodorants are temporary solutions where they give fragrance for a limited period of time. Most antiperspirants contain aluminum salts, a product that is designed to block sweat glands from producing sweat. The aluminum salts combine with the minerals in sweat which eventually causes de-coloration and weaken the fibres in the fabric. Other remedy we have is sweat-pads. These sweat-pads prevent the primary clothing from wetness. But the con is people use these only on occasions due to high cost. Disposable sweat-pads are not sustainable solution. Reusable underarm sweat pads can be the sustainable solution. Reusable sweat-pads can be skin friendly and pocket friendly. We can enhance the sweat dissipation and the value of the existing product by reducing the thickness of the sweat pad and also by constructing three fabric layers having desired property. The issue of bad odour can be overcome by finishing the sweat pad with lemongrass and lavender oil which will give a fragrance finish and antimicrobial property.

M. R. Srikrishnan (✉)
Department of Fashion Technology, PSG College of Technology, Avinashi Road, Peelamedu, Coimbatore 641004, India
e-mail: mrs.fashion@psgtech.ac.in

© The Author(s), under exclusive license to Springer Nature Singapore Pte Ltd. 2022   219
S. S. Muthu (ed.), *Sustainable Approaches in Textiles and Fashion*, Sustainable Textiles: Production, Processing, Manufacturing & Chemistry,
https://doi.org/10.1007/978-981-19-0878-1_9

Keywords Sweat · Micro denier polyester · Bamboo · Wickability · Odour · Pad · Microencapsulation · Etc

1   Introduction

In recent times, there is great concern about the body and the attire that is worn by the people. People nowadays are very much protective about their skin and the clothing. The hot summer days are really detested part of their lives. Not only these days make them sweaty, sticky and itchy, it also makes them call for sweat pads for celebrative occasions. The clothing that is preferred for parties, functions and celebrations are a way different from the customary ones. Most of them are very heavily designed and worked, which makes people drip already. And to save fellows from that, the sweat pads are mapped out. There is quite a lot of difference from the existing market pieces and one that can be developed from the following study. The sweat pad itself shouldn't be a hindrance and the cause of sweat. The pads that are developed in this study are very slimy and can go well with all kind of attires. The thickness of sweat pad matters and shouts out for the product's necessity along with the fabrics chosen. In this study, the bamboo and micro denier polyester fabrics paved the way. The physical and chemical properties of the cherry-picked fabrics were well studied and swayed the design of the product.

In this study, the design was layered with three layers, with top and bottom as micro denier polyester and the intermediate layer as bamboo. The couple of fabrics were sourced and subjected to finishing processes, which includes the drip dry and microencapsulation for hydrophilicity and antibacterial effects. The lavender and lemon grass essential oils which are used with certain percentage greatly imparts the fragrance into them. Thus, it helps the user to be odour free as well. The fabrics were then subjected to various kinds of testing. The testing includes the wickability, antimicrobial test, air permeability, moisture vapor transmission rate, and wash durability. The test result showed a positive wave towards the product development. The two creative designs developed were created into products.

The first design was very simple and had only three layers shaped to the armscye region along with a strap of elastic to hold on for the shoulder. The second design contains fissure type sweat pads for both arms that are connected by straps. The product was designed and developed with aim of giving great comfortability to the users. The product developed was thus given for users for a trail test. The survey was prepared with questions that ranked the product among the users. Most responses received were in a huge favour of the product. Therefore, the product was designed, developed and analyzed for the commercial and hale needs.

1.1 Sweat

Sweat is a clear and salty liquid which is produced by glands in our skin. Sweating is how the body cools itself. We sweat mainly under our arms and on our feet and palms. When sweat mixes with bacteria on our skin, it can cause a smell. Bathing regularly and using antiperspirants or deodorants or sweat pad can help control the odour [1]. Sweating is the sign of healthiness but the sweating causes a feeling of wetness and discoloration. It is the release of liquid from our body's sweat glands. This liquid contains salt and this process is also called perspiration. Sweating helps our body to stay cool. Sweat is commonly found under the arms, on the palms of the hands, and or on the feet. The amount we sweat depends on how many sweat glands we have. A normal person is born with about 2–4 million sweat glands. During puberty, it will begin to become fully active. Men's sweat glands tend to be more active than others. The autonomic nervous system controls sweating. This is the part of the nervous system that is not under our control. Sweating is the way of regulating the temperature naturally in our body [3].

1.2 Sweat—Common Causes

Sweating helps our body to stay cool. Mostly in all cases, it is perfectly natural. People sweat more when they exercise, when they are in warm temperatures and or in response to situations that make them to get nervous, angry, embarrassed, or afraid. Excessive sweating will occur without such triggers. People with hyperhidrosis will have overactive sweat glands. The uncontrollable sweating can also lead to significant discomfort as both physical and emotional manner. When the hands, feet and/or armpits are affected by excessive sweating, it is called focal hyperhidrosis. Sweating that is not caused by another disease is called as primary hyperhidrosis. The secondary hyperhidrosis is when the sweating occurs as a result of another medical condition. The sweating may occur in one area (focal) or all over the body (generalized) [4]. Sweating a lot is normal when it is hot or when we exercise, are anxious, or have a fever. It may also happens during menopause. If we often sweat too much, it is called hyperhidrosis. The causes of hyperhidrosis are low blood sugar, thyroid or nervous system disorders or another health problem. Sweating too little, anhidrosis, can be life-threatening because our body can overheat. The causes of anhidrosis are dehydration, burns and nerve disorders [1].

1.3 Body Odour

Skin is slightly acidic and does a good job of protecting our body from bacteria. When we sweat from our apocrine glands, however, bacteria metabolize sweat and produce odor. It's not our sweat that actually smells, it is the bacteria breaking down our sweat. Antiperspirants essentially starve bacteria and antibacterial and alkaline products make the environment (our underarm) an inhospitable place. Other factors that contribute to the causes of our body odor include, genetics, hormones, activity level, hygiene, diet and underlying disease [5].

1.4 Problems due to Sweating

Excessive sweating on our palms of the hand or in the armpits that is not caused by emotional or physical activity is called diaphoresis or hyperhydrosis. It can be an embarrassing condition. Prof. Dr. Ufuk Emekli considers sweating to be excessive (hyperhydrosis) if the body produces more than 1 mm of sweat per square meter of body surface in one minute. Hyperhydrosis can cause social problems. According to research conducted in the USA, 40% of hyperhydrosis patients have problems with their private and professional social lives, 50% lack confidence, 38% feel that they are limited in their daily activities, and 50% feel unhappy as a result of their hyperhydrosis problem [5]. According to Assoc. Prof. Dr. Nahide ONSUN, not all people with sweating problems will go to the doctor. In Turkey, many people suffer from excessive sweating because they are Mediterranean people and they mostly have fatty skin. Certain medications and surgical procedures are costly and are not covered by governmental subsidies. Thus, our proposed product is an alternative solution for people with sweating Problems [6].

1.5 Sweat Pad

Sweating is a healthy bodily activity and whereas wetness, unpleasant odours and stained garments, which are the negative effects of sweating are undesirable. The use of deodorants will prevent us from these effects. Anti-perspirant deodorant prevents some of the effects of sweating only. Surgery is another method whereas it is expensive but it is not a permanent solution. All these methods are not convenient for everyone. Washable underarm pads are used with dancing costumes and nightwear. The comfort and aesthetic properties of a final product depend on its protectiveness, softness and ease of fit. The products must have an appropriate form that fits the body and the clothing for the comfort of the user. Underarm pads should fit the body lines to ensure comfort for the user, and they should fit the clothing [7].

The most important property of the top sheet in an underarm pad is the speed at which it transfers liquid to the next layer. The top sheet prevents the liquid from escaping the absorbent layer is it's another important function.

The underarm region is active during the daytime and the pads are deformed and damaged during use. Therefore, for a certain amount of time, the pad material should be strong enough to provide resistance. The absorbent layer must absorb and contain excess sweat. A pad's absorption capacity should last for a determined amount of time i.e., 6–9 h [8].

2 Materials Used for Microencapsulation

2.1 Core Material

It is the active ingredient or also called the core material which is to be coated. The core material can be in powder or liquid form. The core material provides flexibility ensuring good design features and development of capsule.

2.2 Coating Material

It is defined as layer of substance which forms a cover over the core material for production of microcapsules. The coating material should also have some desired properties such as,

- It should be aqueous media/solvent. Both should be soluble and must also provide controlled release under specific condition.
- It should have other properties like flexibility, strength, stability, impermeability and even optical properties.
- It should be chemically compatible.
- It should be pliable, tasteless and should not have high viscosity.

Various coating materials used for microencapsulation process are:

1. Vegetable gums—Gum Arabic, agar, carrageenan, dextran sulphate and sodium alginate.
2. Cellulose—Cellulose acetate phthalate, ethyl acetate, nitrocellulose, cellulose butyrate acetate phthalate, carboxymethyl cellulose.
3. Homo polymer—Polyvinyl acetate, polystyrene, polyethylene, polyvinyl alcohol, polyvinyl chloride.
4. Copolymer—Acrylic acid copolymer, methacrylic copolymer, maleic anhydride polymers.
5. Curable polymers—Nitrated polystyrene, epoxy resin, nitro paraffin.
6. Condensation polymers—Polycarbonate, amino resins, nylon, Teflon, silicone resins, poly methane.
7. Proteins—Fibrinogen, haemoglobin, collagen, casein, gelatine, poly amino acids.
8. Waxes, paraffin, bees wax, oils, fats, rosin shellac mono glyceride (Fig. 1).

Fig. 1 Structure of
microcapsule

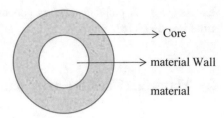

> Core

> material Wall

material

3 Materials and Methods

See Fig. 2.

Fig. 2 Methodology flow
chart

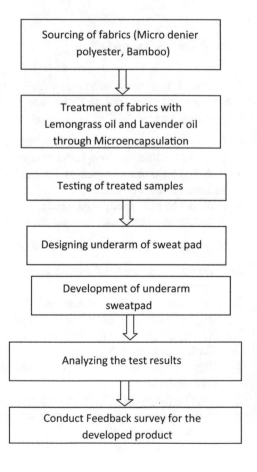

3.1 Fabric Selection—Polyester

Polyester is defined as a manufactured fiber in which the fiber forming substance is any long-chain synthetic polymer made of at least 85% by weight of an ester of a sub-aromatic carboxylic acid compromising, but not confined to di-methyl terephthalate units and parasols hydroxyl benzoate units.

3.1.1 Physical Properties

- Denier   : 0.5–15
- Tenacity   : dry 3.5–7.0: wet 3.5–7.0
- %Elongation at break   : dry 15–45: wet 15–45
- %Moisture regain   : 0.4
- Shrinkage in boiling water   : 0–3
- Crimps per inch   : 12–14%
- Dry heat shrinkage   : 5–8 (at 180 C for 20 min)
- Specific gravity   : 1.36–1.41%
- Elastic recovery @2% = 98   : @5% = 65
- Glass transition temp   : 80 °C
- Softening temp   : 230–240 °C
- Melting point   : 260–270 °C

3.2 Bamboo

Bamboo is a good source of raw material for both in the purpose of economic growth, as a food source, or used for building materials in many parts of south east Asia, south Asia and east Asia. They are light weight, high tensile, flexible, cheap. Its inheritable characteristics of kun present inside fights pesticides by default without the use of chemicals.

3.2.1 Physical Methods

- %Moisture regains   :   13
- %Absorbency rate   :   90–120
- %Elongation at break   :   23.8

(continued)

(continued)

• %Moisture regains	:	13
• %Whiteness	:	69.6
• %Oil content	:	0.17
• Dry tensile strength (cN/dtex)	:	2.2–2.5
• Wet tensile strength (cN/dtex)	:	1.3–1.7
• Resiliency	:	Excellent
• Hand feel	:	soft and lustrous

3.3 Material Details and Layers

3.3.1 Material Specifications

Bamboo	Micro-polyster
• Count—34's	Count—154 denier
• GSM—194	GSM—176
• Colour—white	Colour—white
• Qty—3 m	Qty—5 m
• Fabric—100% knit fabric	Fabric—100% knit fabric

3.3.2 Three Layer Composite

- OUTER LAYER—Micro polyester.

  The micro polyester is kept next to the skin. Micro polyester has tiny micro holes which helps in wicking the sweat away from the skin and evaporation takes place. It also makes the user feel dry because of its high drying capability.
- INNER LAYER—Bamboo.

  The bamboo is kept next second to micro polyester. It has several micro gaps which has good air flow. The wicked sweat will be collected by the bamboo layer and will evaporate quickly within fraction of seconds.
- OUTER LAYER—Micro polyester.

The final layer next to bamboo fabric is also again micro polyester. This layer will provide as backup layer, if the sweat come to the third layer micro polyester will wick away the sweat and evaporation takes place soon (Fig. 3).

Fig. 3 Three-layer
composite concept

Micro polyester
Bamboo
Micro polyester
Underarm skin

3.4 Microencapsulation

Microencapsulation is done for micro denier and bamboo fabrics as per coacervation method using lavender essential oil and lemon grass essential oil which have very good anti- bacterial property and disinfecting property.

3.4.1 Flow Process

See Fig. 4.

Fig. 4 Microencapsulation process

Materials required for microencapsulation process

Preparation of micro-capsules using essential oil

Padding process

↓

Drying

↓

SEM analysis and Light microscopy analysis

3.4.2 Materials Required

- Gum Acacia (or) Gum Arabic powder—it's an effective encapsulating agent which is used for the formation of wall material of the micro-capsules and has high water solubility.
- Lavender essential oil and Lemongrass essential oil—these two are used for the formation of core material of the micro-capsules which possess higher antibacterial properties and used to treat allergies.
- Beaker—500 ml, Conical flask, Pipette.
- Magnetic stirrer.
- Hot air oven.
- Sodium Sulphate (20%).
- Formaldehyde (17%)—for stabilization.
- Distilled water.
- Citric acid (5%)—binder in padding process.
- Centrifugal tube.
- Padding mangle (Figs. 5 and 6).

Fig. 5 Magneticstirrer

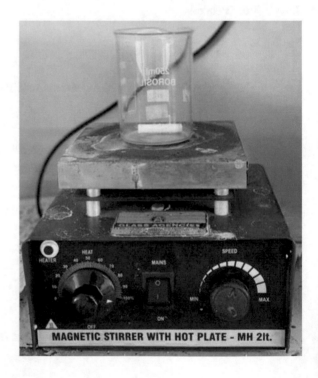

Fig. 6 Lemongrass and
Lavender essential oil

4 Preparation of Micro-Capsules

The standard procedure for preparing the micro-capsule solution is found, and the process was carried out, respectively, for the required amount of fabric. 10 g of gum acacia powder is taken with 100 ml of distilled water in a beaker and it is heated at 45 °C for half an hour until the solution swells. Take 50 ml hot distilled water and keep it in magnetic stirrer for 15 min at 40–50 °C. Take lemongrass essential oil and lavender essential oil in the ratio of 8:2 (8 ml and 2 ml, respectively) and add to the beaker. Simultaneously, add 10 ml of sodium sulphate (20%) in drops to the solution mixture in the beaker. Leave it to stir under 300–450 rpm. Then, add 5 ml of formaldehyde (17%) and allow it for 15 min. Allow it to cool for 10 min and we can see small micro-capsules formed. Finally, it is freeze dried. After that the bamboo and micro denier polyester fabrics are immersed in plate form in the freeze dried micro-capsulated liquid and kept in hot air over for 24 h (Figs. 7 and 8).

Fig. 7 Essential oils in 8:2
ratio

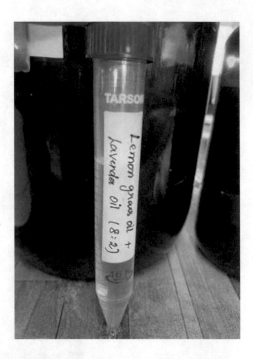

Fig. 8 Micro-capsule liquid

4.1 Padding Process

The width of the horizontal padding mangle is 300 mm. After the fabric is taken from hot air oven, it is kept out for 2–5 min and taken for pad-dry-cure process. After the padding mangle machine is switched on, pressure is set. Micro denier polyester and bamboo fabrics were treated in the padding machine one by one. The excess liquid from the fabric is squeezed out. Simultaneously, the micro-capsules were applied to

Fig. 9 Horizontal padding
machine

the two fabrics by treating it with binder. The binder used for printing micro-capsules
on the fabric is citric acid (5%) (Fig. 9).

4.1.1 Drying

After padding process, the bamboo fabric and micro denier polyester fabric is dried
in hot air oven at 37 °C for 24 h. After that to remove stiffness of fabric, fabric is
dipped once in neutral soap solution and then left to flat dry in shade (Fig. 10).

4.2 Light Microscopy and SEM Analysis

The freeze dried micro-capsule liquid is poured into 15 ml centrifugal tube and kept
in centrifugation machine for 15 min under 10,000 rpm such that micro capsules get
separated from the solution and it is ready for conventional light microscopy and
scanning electron microscopy analysis. The light/optical microscope visualizes the
fine details by magnifying it through a light beam and a series of glass lens. Scanning
electron microscope (SEM) scans and interacts with the surface of the fabric with
focused beam of electrons. Both these techniques are used here to determine the outer
structure, shape of the micro-capsules and the availability of the micro-capsules on
the bamboo and micro denier polyester fabrics, respectively.

The white dots in Fig. 11 shows the micro-capsules formed. Figures 13 and 14
show the SEM photograph of untreated bamboo fabric and bamboo fabric treated
with micro-capsules. It is seen that the micro-capsules are present in the interstices of
the fibre of the bamboo fabric. Similarly, Figs. 15 and 16 show the SEM photograph of
untreated micro denier polyester fabric and microencapsulated micro denier polyester

Fig. 10 Hotairoven

Fig. 11 SEM photograph of
micro capsules

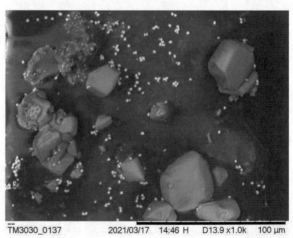

fabric. Here, the micro-capsules are stuck above the fibre surface in good amounts
(Fig. 12).

The amount of lemongrass and lavender essential oil is also important in formation
of micro capsules. The SEM analysis give us a good perspective way of breakthrough
of how fibres are intensively merged together. The microcapsules size was 70 mm.
Light microscopy already gave us a good look upon that but when compared with
SEM it gives better understanding study about the capsule formed. Resolution is the

Fig. 12 Light microscope photograph

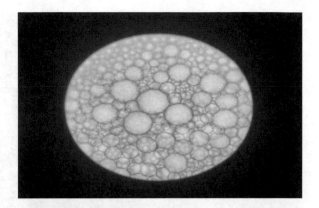

Fig. 13 Untreated bamboo fabric

Fig. 14 Microencapsulated bamboo fabric

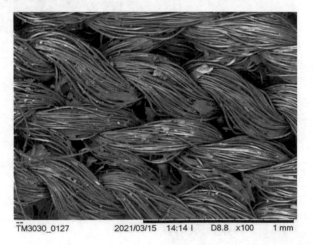

Fig. 15 Untreated micro denier polyester fabric

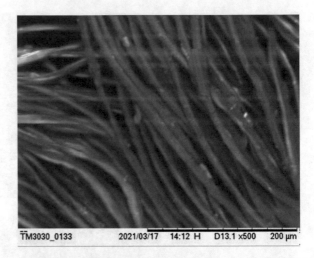

Fig. 16 Microencapsulated micro denier polyester fabric

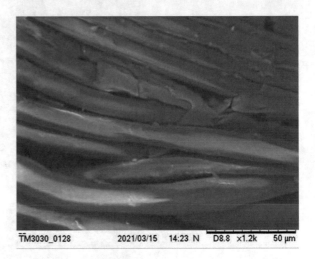

biggest advantage in SEM. It provides upto as low as 15 nm. Any imperfections can be identified with the scanning electron microscope.

5 Design and Development of Underarm Sweat Pad

The sweat pad we are developing consists of three layers. The top layer will be micro denier polyester and the middle layer is bamboo fabric and the bottom layer is Micro denier polyester (Fig. 17).

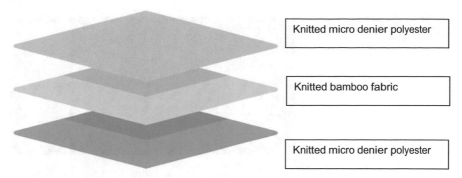

Fig. 17 Layers underarm sweat pad

Fig. 18 Design of under
arm sweat pad

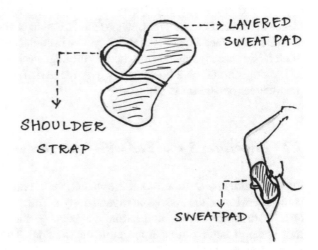

Design of Underarm Sweat Pad

The design of the sweat pad should give maximum comfort to the wearer. The wearer should not feel bulkiness when they wear it. So, the final design consists of a bulb shaped top and bottom end. The middle part of the sweat pad should have a peripheral curvilinear edge that are concave from the top and bottom end. The thinner middle portion of the sweat pad will lie in the mid of the underarm. As the thickness of the underarm sweat pad is less, there will not be any mark on the primary garment (Fig. 18).

5.1 Construction of Underarm Sweat Pad

The standard length of 14 cm and width of 7 cm are used as a standard dimension. Then the cut length is shaped according to the shape of the desired design of the sweat

Fig. 19 Prototype of under arm sweat pad

pad. The pattern is marked in the pattern sheet and cut according to the shape of the underarm sweat pad. Then, the pattern is placed on the microencapsulated fabrics of bamboo and micro denier polyester and cut according to the pattern. Then, the cut pieces are placed on the order, where the face side of the fabrics are placed facing outwards. The face side of the fabrics will be in contact with the skin of the user. The three fabrics are layered and stitched on the edges using overlock stitch. The length of the strap should be adjustable to cover the variations in the body dimensions of the different person (Fig. 19).

5.2 Underarm Sweat Pad with Body Straps

Further to enhance the comfort of the wearer, we have developed a new design which is the extension of the above-mentioned design. The new design will have sweat pad attached to the adjustable straps around the ribcage and the shoulder which provides more support and reduces slippage off the shoulder. This design can be used if the primary garment is loose fit.

Design 1:

The design 1 is normal design that is shaped for armscye. It has three layers, the top and the bottom layers being the micro denier polyester and the middle one being bamboo. The shoulder strap is attached along with it to give users the support. The edges are finished by overlock (Figs. 20, 21, 22 and 23).

Design 2:

This design has been developed in a newer version to provide more support for the users. The circumferential and shoulder straps are being used for this purpose. The shoulder straps are chosen to be transparent, so that it can be worn with any kind of clothing. The fissure design is developed for both arms and are being joined together by straps. The single needle machine has been used for attaching straps and overlock machine for edge covering in the fissure design. Bartack stitches are used to give strength to certain parts for the need of tight stitch.

Fig. 20 Under arm inclined view

Fig. 21 Under arm side view

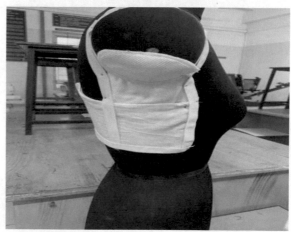

Fig. 22 Front view of design 1
Design 1 (front)

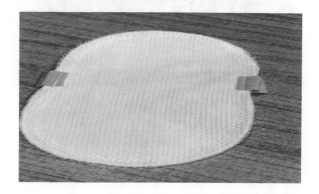

Fig. 23 Design 1 (back)

Fig. 24 Design 2 (front)

Fig. 25 Design 2 (Back)

5.3 Feedback Survey

The questionnaire for the feedback survey is created based upon the required parameters of an under arm sweat pad. The feedback survey form consists of questions that addresses the expected qualities of the sweat pad like dryness, thickness, fit and odour control. The thickness of a commercially available products ranges between 1 and 1.2 cm whereas our product has achieved a thickness of only 3.2 mm. We also added a question where we ensure that our product is not seen under the primary garment as we reduced the thickness to the most extent. The questions in the questionnaire is as follows. (1) On which occasional clothing, do you prefer using sweat pad? (2) Does it fit on the clothing? (3) During usage, is there any slippage from the garment or shoulder? (4) How many hours do you use our sweat pad? (5) Is there any wetness in the clothing? (6) How many hours do you feel dry? (7) Rate the dryness of the product? (8) Does it cause an unpleasant odour? (9) How many hours do you feel dry? (10) Is it seen under the primary clothing? (11) Do you feel any bulkiness while using the product? (12) Rate the product basis overall performance. Basis the responses we will interpret the improvements required in our product.

6   Results and Discussion

6.1   Air Permeability Test

See Table 1 and Fig. 26.

Inference:

Table 1 Result of air permeability test

Sample	Control sample (untreated)	Treated sample (microencapsulated)
Micro denier polyester	250	242
Bamboo	199	125
Cotton single jersey	123	–

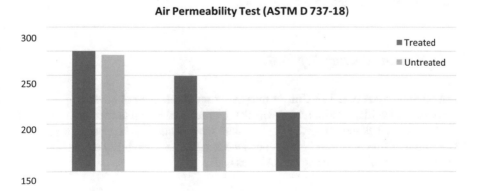

Fig. 26 Air permeability test

Table 2 Result of moisture vapour transmission test

Samples		MVTR (gm/m²/day)
Micro polyester	Control	2093
	Treated	1953
Bamboo	Control	2873
	Treated	2150
3 layers (MDP-bamboo-MDP)	Control	2613
	Treated	2084
Cotton	Control	1912

Air permeability is found to be higher for both untreated micro denier polyester ($250\ cm^3/cm^2/s$) and bamboo ($199\ cm^3/cm^2/s$) fabrics than untreated cotton ($123\ cm^3/cm^2/s$) which is used in commercial sweat pads. Air permeability increases with increase in porosity (i.e.) inter yarn pores and air gaps present in the fabric. The porosity and air gap are decided by the knit structure of the fabric. Loosely knit fabrics will exhibit more air permeability than tightly knit fabrics because of its higher amount of air being entrapped in it.

6.2 Moisture Vapour Transmission Test

Test standard: ASTME96.
    Unit of measure: $gm/m^2/day$ (Table 2 and Fig. 27).

Inference:

Moisture can diffuse through air space between fibres and yarns. Diffusion will more likely occur when the fabric has open pores within the structure and also when the fabric held more weight it will transport more moisture and quickly evaporation process takes place keeping the fabric dry.

6.3 Vertical Wickability Test

Test Standard: AATCC 197 Unit of measure: mm (Table 3 and Fig. 28)
    Thus, wickability also increases with the required finishing treatments. The treated bamboo fabric has more wickability rate than treated micro denier polyester fabric. The nature of fibre has clearly influenced the result. The wickability rate is higher in bamboo fabric due its knitting structure as well.

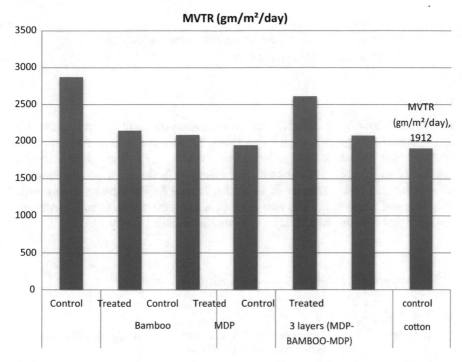

Fig. 27 moisture vapour transmission test

Table 3 Result of vertical wickability Test

Vertical wicking	Treated MDP	Treated Bamboo
Wickability in mm, after 5 min	80	100
Wickability in mm, after 30 min	95	130
Wickability in mm, after 1 h	106	150
Wickability in mm, after 24 h	110	-

6.4 Antimicrobial Activity Test

6.4.1 Antimicrobial Activity—Qualitative Analysis

Test Name: Antibacterial activity evaluation (Qualitative test)

Test Standard: AATCC 147-2004

Test Organism used: *Escherichia coli* and *Staphylococcus aureus*

(continued)

(continued)

Test Name:	Antibacterial activity evaluation (Qualitative test)
Unit of measure:	mm (Table 4)

Inference:

AATCC 147-2004 is a qualitative method to measure the ability of an antimicrobial textile to inhibit the growth of microorganisms. The specimen will have antimicrobial activity when there is a zone of inhibition formed. The antimicrobial activity is considered as nil only when the zone of inhibition is 0 mm. It is clear from the observation that there is a zone of inhibition formed for both treated micro denier polyester and bamboo fabrics, and the values are greater than 0 mm. Hence, we conclude that both the treated fabrics show antimicrobial activity.

6.5   Antimicrobial Activity—Quantitative Analysis

Test Name:	Antibacterial activity evaluation (Quantitative test) Test Standard:	ASTME E 2149-13a
Test Organism used:	*Escherichia coli* and *Staphylococcus aureus* (24 h Culture) (Table 5)	

Inference:

From the observation, the percentage reduction of CFU/mm for treated bamboo fabric are 99.9% and 97.29% for *Escherichia coli* and *Staphylococcus aureus*, respectively. Also, the percentage reduction of CFU/mm for treated micro denier polyester are 97.98% and 94.82% for *Escherichia coli* and *Staphylococcus aureus*, respectively. These percentage reduction is considered as the percentage of antimicrobial activity imparted on the surface of the fabric. This is due to the micro capsules consisting of antimicrobial agent gets attached to the surface of the fabric.

The attached antimicrobial agents disrupt the cell membrane of the microorganism that comes in contact and prevents functioning and reproducing. Hence, we conclude that final product will give an average antimicrobial inhibition of 97.49%. This will inhibit the bad odour and skin irritation caused due to the microbes in sweat.

6.6   Washing Durability Test

Test standard: ASTM E3162-18
    Unit of measure: Number of washes (Table 6 and Fig. 29)

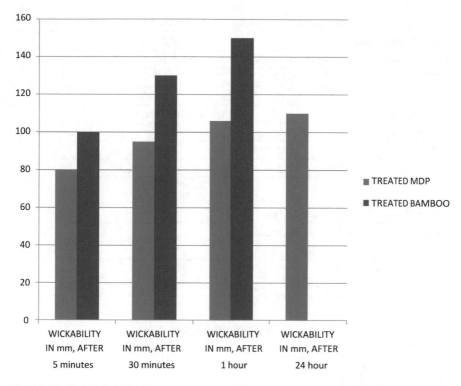

Fig. 28 Vertical wickability test

Table 4 Result of antimicrobial activity test (qualitative test)

Control bamboo fabric		Treated bamboo fabric	
Escherichia coli	*Staphylococcus aureus*	*Escherichia coli*	*Staphylococcus aureus*
(mm)		Zone of inhibition (mm)	
–	–	23	21
Control MDP fabric		Treated MDP fabric	
Escherichia coli	*Staphylococcus aureus*	*Escherichia coli*	*Staphylococcus aureus*
(mm)		Zone of inhibition (mm)	
–	–	18	16

Inference:

The treated bamboo and micro denier polyester fabrics can withstand 15 washes.

The washing durability of antimicrobial property of the bamboo and micro denier polyester fabric is 60.18% and 52.15%, respectively, after 15 washing. Thus, the treated fabrics show good washing durability.

Table 5 Result of antimicrobial activity test (quantitative test) colony count (CFU/ML)

Control bamboo fabric				Treated bamboo fabric			
Escherichia coli		*Staphylococcus aureus*		*Escherichia coli*		*Staphylococcus aureus*	
10^{-6}	10^{-7}	10^{-6}	10^{-7}	10^{-6}	10^{-7}	10^{-6}	10^{-7}
521	447	764	490	3	-	24	10
Control MDP fabric				Treated MDP fabric			
Escherichia coli		*Staphylococcus aureus*		*Escherichia coli*		*Staphylococcus aureus*	
10^{-6}	10^{-7}	10^{-6}	10^{-7}	10^{-6}	10^{-7}	10^{-6}	10^{-7}
933	744	643	321	24	10	28	22

Table 6 Result of washing durability test

Sample	Bacterial reduction %			
	No wash	5 washes	10 washes	15 washes
Bamboo fabric	99.42	89.65	74.69	60.18
MDP fabric	96.35	81.77	78.32	52.15

Fig. 29 Washing durability test

6.7 Survey Results

6.7.1 Questions Asked in the Survey

(1) On which occasional clothing do you prefer using sweat pad?
(2) Does it fit on the clothing?
(3) During usage, is there any slippage from the garment or shoulder?
(4) How many hours do you use our sweat pad?
(5) Is there any wetness in the clothing?

(6)   How many hours do you feel dry?
(7)   Rate the dryness of the product?
(8)   Does it cause an unpleasant odour?
(9)   Is it seen under the primary clothing?
(10) Do you feel any bulkiness while using the product?
(11) Rate the product basis overall performance.
(12) Please mention the improvements required in our product.

In India, sweat pads are most commonly used for party wears and ethnic wears. In second place, people wear it along with formal wears also. Common underarm sweat pads use elastic straps which wear out after washes. This causes slippage from shoulder. Our underarm sweat pad uses adjustable invisible straps. Most of the commercially available sweat pads have thicker fabrics which may cause bulkiness while usage. We came up with the thinner fabrics where we have reduced the thickness around 3mm. Sweat pad is given to 6 persons to know the feedback of the product, and they use this sweat pad for 3–6 h normally. Then, the survey is taken to know the difficulties they faced and the improvement that they see while using the product. According to feedback from survey, our sweat pad is more durable than other commercial sweat pads, since it can withstand 15 washes. This product is used especially while wearing party wear and sometimes during the usage of formal wear. Our product is a reusable underarm sweat pad consisting of bamboo and micro denier polyester layers and it of course fits well to the wearer. And they felt there is no slippage of sweat pad from the shoulder or garment. And also, there is no wetness in the clothing, and they felt the dryness for nearly 3–6 h. There is no unpleasant odour during the usage of sweat pad. And they didn't feel any bulkiness while using the product and the sweat pad is not seen under the primary clothing. Some suggested that this product must be made with standard size in order to make it readily available for everyone. And may have the variations in the colour is required (Fig. 30).

1.   On which occasional clothing do you prefer using sweat pad?

(1)   Please mention the improvements required in our product.

Fig. 30 Clothing preference   ■ Party wear   ■ formal wear   ■ casual wear

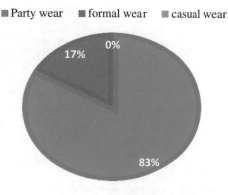

In India, sweat pads are most commonly used for party wears and ethnic wears. In second place, people wear it along with formal wears also. Common underarm sweat pads use elastic straps which wear out after washes. This causes slippage from shoulder. Our underarm sweat pad uses adjustable invisible straps. Most of the commercially available sweat pads have thicker fabrics which may cause bulkiness while usage. We came up with the thinner fabrics where we have reduced the thickness around 3 mm (Figs. 31, 32, 33, 34, 35, 36, 37, 38 and 39).

1. Does it fit on the clothing?

2. During usage, is there any slippage from the garment or shoulder?

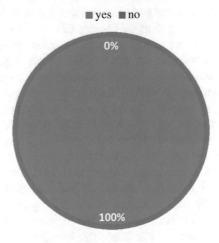

Fig. 31 Fit on clothing

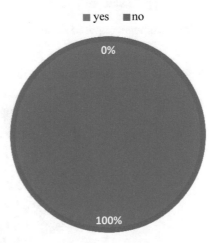

Fig. 32 Slippage from garment/shoulder

■ 0-3 hours   ■ 3-6 hours   ■ 6-9 hours   ■ more than 8 hours

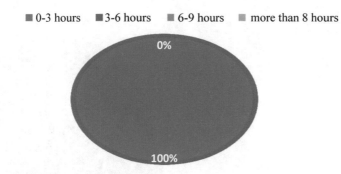

Fig. 33 Hours of usage

■ yes ■ no

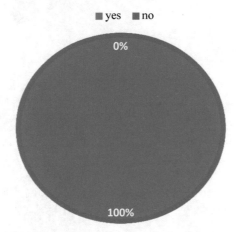

Fig. 34 Wetness in clothing

■ 0-3 hours   ■ 3-6 hours   ■ 6-9 hours   ■ more than 8 hours

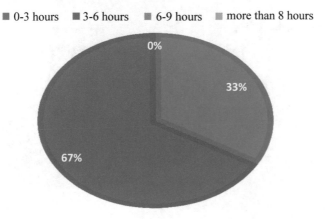

Fig. 35 Hours of dryness

Fig. 36 Unpleasant odour

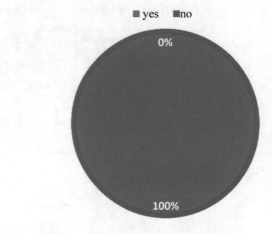

Fig. 37 Seen under primary clothing

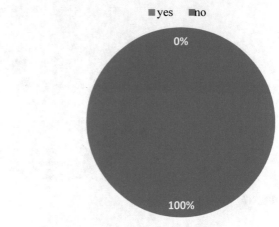

Fig. 38 Product bulkiness

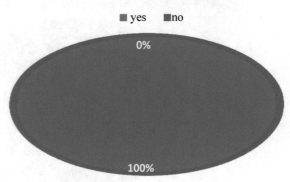

Fig. 39 Overall performance

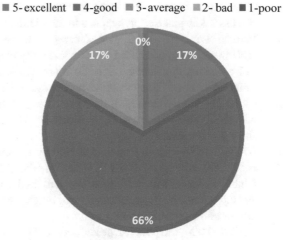

■ 5- excellent ■ 4-good ■ 3- average ■ 2- bad ■ 1-poor

3. How many hours do you use our sweat pad?

1. Is there any wetness in the clothing?
2. How many hours do you feel dry?
3. Does it cause an unpleasant odor?
4. Is it seen under the primary clothing?
5. Do you feel any bulkiness while using the product?
6. Rate the product basis overall performance of the product

7 Conclusion

Based on objective evaluation the following conclusions have been made:

1. Air permeability—In treated fabrics with micro capsules, both bamboo (125 cm^3/cm^2/s) and MDP (242 cm^3/cm^2/s) fabrics exhibit better air permeability. This is because of its higher amount of entrapped air in it due to its loose knit structure with larger air gaps. This leads to increased size of voids, hence exhibiting better air permeability. Both bamboo and MDP fabrics have more air permeability because both fabrics were interlock knitted with lower machine gauge, and this gives a loose structure with larger air gaps than the cotton single jersey fabric.
2. Moisture vapour transmission rate—The treated fabrics, Bamboo (2150 gm/m/day) and MDP (1953 gm/m/day), show a lesser result when compared with the control. This is due to the micro capsules being attached to the porous surface of the fabric. But in comparison with cotton, the treated fabrics exhibits higher amount of moisture vapour transmission rate. Thus, as the weight of micro polyester and bamboo has also an effect on transport

of moisture, sweat diffused in between the inter space yarns and evaporation process takes place soon keeping the fabric dry.

3. Vertical wicking—In treated fabrics with micro capsules, bamboo fabric (150 mm in 50 min) exhibit better wickability than treated MDP fabric(110 mm in 24 h).This is because of its fibre properties and more number of filament forms in bamboo fabric it shows better results. The knit structures help in wickability of both the fabrics. Both bamboo and MDP fabrics have good amount of wickability. Since there is higher linear density and thread density, both fabrics show good wickability.

4. AATCC 147—The antimicrobial activity is considered as nil only when the zone of inhibition is 0 mm. It is clear from the observation that there is a zone of inhibition formed for both treated micro denier polyester and bamboo fabrics, and the values are greater than 0 mm. Hence, we conclude that both the treated fabrics show antimicrobial activity.

5. AATCC 100—the percentage reduction of CFU/mm for treated bamboo fabric are 99.9% and 97.29% for *Escherichia coli* and *Staphylococcus aureus*, respectively. Also, the percentage reduction of CFU/mm for treated micro denier polyester are 97.98% and 94.82% for *Escherichia coli* and *Staphylococcus aureus*, respectively. This is due to the micro capsules consisting of antimicrobial agent gets attached to the surface of the fabric. Hence, we conclude that final product will give an average antimicrobial inhibition of 97.49%.

6. Washing durability—The washing durability of antimicrobial property of the bamboo and micro denier polyester fabric is 60.18% and 52.15%, respectively, after 15 washing. The treated bamboo and micro denier polyester fabrics can withstand 15 washes.

References

1. Onsun N (2005) Personal conversation
2. Discardable underarm garment shield issued to Pulsifer on May 29, 1956
3. Sadikoglu TG, Disposable underarm pad design
4. Anand SC, Brown KSM, Conard D (2002) Effect of laundering on the dimensional stability and distortion of knitted fabrics. Autex Res J 2
5. Fangueiro R, Filgueiras A, Wicking behavior and drying capability of functional knitted fabrics
6. Jamir MRM, Khasri A, Sustainable composites for aerospace applications, Natural hybrid composites for aircraft structural applications
7. Panigrahi NR (2007) Microfibers, microfilaments and their applications. AUTEX Research Journal
8. Xi L, Qin D, An X, Wang G (2013) Resistance of natural bamboo fibre to micro-organisms, and factors that may affect such resistance

Product Development Using Cornhusk Fibres—A Sustainable Initiative

Archana Jain, Deepali Rastogi, and Bhawana Chanana

Abstract Textile industry is striving hard to adopt sustainable practices for various aspects of production like consumption of water and electricity, new sustainable raw materials, new technologies, use of natural or non- hazardous chemical agents for wet processing and better working conditions for labour. Development of new fibres from alternative sources is gaining momentum and various agro-wastes have been explored for use in textiles. Cornhusk fibre is an addition to the same. Fibres were extracted from the husk of corn by applying chemical and enzyme treatment. The cornhusk fibres were comparable to fibres from other bast or leaf fibres. A range of products comprising of yarn, fabric, rope, hand-made mats and hygiene products were developed. Cornhusk fibres were blended with coir fibres to compensate for their lower strength. Since both the fibres had similar texture, blending could be carried out easily. The cornhusk/coir (50/50) blended yarns had similar appearance as 100% coir yarn but lower strength and elongation at break. Hence it was used for making products like mats where strength is not the primary requirement. The rope made using cornhusk blended yarn could also be used for some low end applications. The different types of mats like hand woven, sinnet and corridor mats prepared using cornhusk blended yarn were softer in feel and similar in appearance to the mats made by using 100% coir fibres. Utilization of a waste like cornhusk to manufacture home furnishings would mean conservation of resources like land and water as cornhusk is a by-product of a food crop, maize.

Keywords Sustainability · Agro waste · Cornhusk fibres · Value addition · Blended yarn · Home furnishings

A. Jain
Vivekananda College, University of Delhi, New Delhi, India

D. Rastogi (✉)
Department of Fabric and Apparel Science, Lady Irwin College, New Delhi, India
e-mail: deepali.rastogi@lic.du.ac.in

B. Chanana
Amity School of Fashion Design and Technology, Amity University, Mumbai, India

1 Introduction

Sustainability is the key mantra presently in all the industries across the world, especially textile sector as it is one of the largest industries of the world. Sustainability concerns with each and every aspect of industries, our lives and continuation of human life on the earth. The production of textiles has always added to the environmental problems, particularly due to huge water consumption and pollution caused by processing of textiles. The other significant environmental aspect of textile production is the emission of greenhouse gases as more than half of textiles produced these days are oil based. Synthetic fibres like acrylic, polyester and nylon all require use of fossil fuels and lead to emission of carbon dioxide [1]. Sustainability in textiles refers to employment of those means which make the production of textiles environment friendly. Some of the environmental benefits of using recycled raw materials rather than virgin materials for making textile products are; conservation of natural resources, reduced energy consumption, less carbon dioxide and other emissions and waste going to landfills. Utilizing renewable natural resources through value addition is an ideal transition from a petroleum-based economy to bio-based economy. Organic wastes, which are otherwise a challenge to the environment, could prove to be a potential alternative for production of value added products [2].

There is an increase in worldwide demand of man's basic needs for food and fibre according to the growing population. At the same time, communities are striving to achieve sustainable management of natural resources. Agriculturists need to strike a balance between the demand for increased output of agricultural products and sustainability. To make this possible, it is mandatory for farming industries to measure and understand their current sustainability trends and adapt practices as required for the same [3]. Keeping in pace with these growing demands and giving consumers access to their fibres of choice, the textile industry will have to take an initiative to develop a new, sustainable path for clothing the world and ending the cycle of environmental harm. The major aim should be production of fibre from wood or non-wood-based biomass to replace both cotton and polyester production, which burdens the environment, and consumes oil resulting in depletion of natural resources. It requires a fundamental change in attitude for developing a sustainable global economy, which permits improved purchasing power and living standards without exhaustion of natural resources for future generations. The products that are based on photosynthetic carbon dioxide (CO_2) fixation should be preferred due to ecological grounds. The benefit of such sustainable resources is that they can be re-grown within the foreseeable future, without having any negative side-effects on global bio-diversity in the long term [4].

Natural fibres are one of such major renewable resources for bio-based economic developments which had been used since the dawn of civilization; but had gone temporarily overshadowed by the superficial benefits of synthetic fibres that did not last long. As soon as the textile industry realized this fact, it started creating global awareness regarding benefits of using natural fibres in textile applications.

This resurgence in use of natural plant fibres is gaining momentum all over the world.

In the present scenario, technical textile is the area where application of bast and leaf fibres is being explored ingenuously. Technical textile is a high technology upcoming sector which is steadily gaining ground in India. Technical textiles are those functional fabrics which are used in various industries including automobiles, civil engineering and construction, agriculture, healthcare, industrial safety, personal protection etc. Based on usage, there are twelve segments of technical textiles; Clothtech, Agrotech, Meditech, Buildtech, Mobiltech, Oekotech, Geotech, Protech, Packtech, Hometech, Indutech and Sportech [5].

The application of jute, kenaf and other allied fibres (all bast, leaf and husk fibres), called as hard fibres by Food and Agriculture Organization (FAO), United Nations, in various areas of technical textiles is continuously increasing. Application of these fibres for tying ropes and binder twines; braided articles; stuffing material in upholstery, mattresses; reinforcing materials like plastics, wallboards; packing material like seams in vessels, barrels, piping and all kinds of paper is on a rise day by day. Hard fibres production in 2012–13 reached 3.2 million tons which was slightly less as compared to 3.34 million tons in the year 2011–12. Despite this fall, output was still significantly higher than the 2.6 million tons and 2.9 million tons reached in 2008–09 and 2009–10, respectively [6]. India is a significant player in this market as it ranked first in world production of hard fibres as per the statistics of 2012 with 58 percent share [7].

India has a rich variety of unexplored natural fibres which, if exploited, has potential to generate employment for a large section of the population. Moreover, India being an agriculture-based nation has a lot of biomass available which could be converted to a renewable resource. Utilization of the so-called waste to generate wealth will provide an additional avenue for livelihood generation and lead to green economy, which would also add to the green cover of the country [8].

Natural resources play an important role in the economy of any country, thus contributing substantially to the Gross Domestic Product (GDP). In the case of developing and underdeveloped countries, social development also gets enhanced along with economic development. Hence, it is not surprising to find an increasing world trend towards the utilization of such resources through new processes and products to their maximum. In turn, this does not help only in preventing environmental pollution, which is caused by throwing these materials in the environment without proper disposal but also in creating employment, particularly in rural/semi-urban areas which ultimately contributes to the upliftment of people's living standards. Amongst such natural resources, about 2.5 billion tons of lignocellulosic materials are available, which have been used since 6000 BC. Fibres are available from many lignocellulosic sources and they are also called plant fibres, natural fibres or vegetable fibres [9].

Globally, natural fibres industry provides employment to millions of people, especially small-scale farmers and processors. Income from this industry adds significantly to the earning and food security of the poor farmers and workers involved in fibre industries. In many countries, the fibre industry is of major importance to the national economy, like cotton in some African countries; cotton is grown on two million farms employing over 10 million people in Sub-Saharan Africa, whilst in Asia, cotton is the major source of income for about 100 million families. The jute industry in Bangladesh and sisal industry in Tanzania have great economic significance to the respective country's economy. In some countries, the fibre industry is of lesser significance at the national level but holds great local importance within the country. For example, alpaca fibre in the Andes, sisal in North-East Brazil and jute in West Bengal and India holds quite significant status [10].

This chapter gives a brief overview of various natural fibres which are used in manufacturing of different technical textiles and use of cornhusk fibres for product development. Cornhusk which is an agro waste could be utilized for a variety of applications like twines, composites, paper and pulp etc.

The study discussed in the chapter involved extraction of fibres from cornhusk, determination of composition, assessment of various physico-chemical properties and morphology of extracted fibres. The fibres were put to different end uses as per their suitability and prototypes of different products were created.

The cornhusk was collected, cleaned, dried and stored in a well-ventilated room. It was treated with alkali. The alkali treatment was optimized with respect to alkali concentration, temperature and treatment time. This was followed by enzyme treatment which was optimized on the basis of enzyme concentration and treatment time. Enzyme treated cornhusk fibres were bleached using hydrogen peroxide. A softening treatment was given to impart a smooth handle to the fibres. The extracted cornhusk fibres were analysed for their composition and physico-chemical properties using standard test methods. Fibre morphology was studied using scanning electron microscopy and X-ray diffraction. Finally, the cornhusk fibres were blended with different fibres to make various products. The cornhusk fibres were blended with cotton and polyester to make yarns and fabrics. They were blended with coir to make coarse yarn, ropes and hand-made mats of different types.

The cornhusk fibres obtained were yellowish in colour, coarse and had lower strength but higher elongation at break. The moisture regain and water retention of the fibres was found to be high. The cornhusk fibres could not be spun into yarn alone hence they were blended with cotton and polyester in different proportions and fabric was constructed using them. The cornhusk fibres were further blended with coir fibres to make coarse yarn. The cornhusk-coir blended yarn was comparable to 100% coir yarn in appearance. It was used to develop rope and different types of hand-made mats. Cornhusk fibres could be a fruitful addition to broaden the basket of natural fibres.

1.1 Natural Plant Fibres

Natural plant fibres are the renewable fibres taken from different parts of the plant like bast, leaf, seed, fruit and husk. Geometrical dimensions of plant fibres, especially the length is dependent mainly on location of fibres in the plant. For example, fibres from fruits and seeds are few centimetres long whilst those from stems or leaves are much longer [11]. Common bast fibres include flax, jute, hemp, ramie and kenaf; leaf fibres are obtained from banana, pineapple and sisal etc. whereas cotton is the most important fibre obtained from seed and coir is extracted from fruit husk i.e. coconut. Bast fibres form an important category as most of the cellulosic fibres other than cotton belong to this group. Flax fibres hold the distinction of being used in first place by human beings for clothing.

Due to the increasing concerns regarding sustainability of natural resources, textile industry is banking upon the use of natural plant fibres for various applications. Natural plant fibres have been successful due to their positive impact on our ecological balance. Various plant fibres can be used for a number of textile and technical applications, e.g. bast or stem fibres, leaf fibres and fibres of seeds and fruits. Flax, hemp, jute, ramie, sisal and coir are mainly used for technical purposes. The physical properties of the fibres are the key factor in deciding the end use to which fibres could be put and utilized exploiting their potential to maximum. Various applications for which fibres can be used are broadly categorized as textile and non-textile applications.

1.2 Textile and Non-textile Applications of Natural Cellulosic Fibres

Textile applications comprise of textiles used for apparel and aesthetic purposes. Bast and leaf fibres are a challenge to process and weave into cloth, in spite of this fact, weavers all over the world have learned to overcome this difficulty whilst maintaining the materials' subtle natural beauty and also adapt to complicated and demanding dye procedures to great effect. In a world that is completely engaged in global trade of industrially produced cottons and synthetic fabrics, no one realizes that there was a time when all the cloth needed in any community had to be woven by hand and moreover, it was primarily made from bast or leaf fibres. In present scenario, even the word 'bast' might not be familiar to many people. During the twentieth century, hand weaving was not considered to be an economically significant activity in most of the world primarily because of the industrialization of textile production, the invention of man-made fibres and the emergence of global cloth trade on a colossal scale. Even if handlooms were being used, weavers totally left bast and leaf fibres produced from local fields and forests for industrially produced cotton or synthetic yarns. But remarkably, during the first decade of the twenty-first century, there has been a revival of hand-weaving of bast and leaf fibres. It's not surviving only in few communities, but flourishing at some places even under newly globalized conditions [12].

Non-textile or technical textile is the term given to textile products used for industrial purposes. It is a growing sector which has large potential to contribute in economic growth and employment generation. It would substantially lead to increase in exports; which is still an untapped market in our country as compared to traditional textiles where it ranks high globally. In the recent past, the progress of different segments of technical textile industry has been on a gradual rise in India. Technical Textile products acquire their demand from development and industrialization in a country. Looking at the pace at which emerging nations are industrializing, the market for technical textiles can also be expected to grow proportionately with industrial growth in different parts of the world. In India, Technical Textile sector has registered compounded annual growth rate of 11% during the 11th Five Year Plan and as per the 12th Five Year Plan estimates by the sub-group on technical textiles, technical textile market size is expected to grow at compounded annual growth rate (CAGR) of 20% and reach Rs. 1, 58,540 crores by 2016–17 from the market size of Rs 70,151 crores in 2012–13. This positive trend is a strong indication that technical textiles would provide new opportunities to the Indian textile industry to have long-term sustainable future. Despite achieving high growth rate, the per capita consumption of technical textiles in India is 1.7 kg per person as compared to 10–12 kg in developed countries. This low per capita consumption needs to be improved by taking appropriate measures. Globally, the technical textiles contribute to about 27% of textile industry, in some of the western countries its share is even 50% whilst in India it is just 11% [13]. Though the textile industry fulfills the physiological needs of mankind, the technical textiles fulfill the high technical and quality requirements (mechanical, thermal, electrical, durability etc.) giving products the ability to offer technical functions. Technical textiles are those textile materials and products which are used for technical performance and functional properties and are not concerned only with traditional or decorative characteristics. There are many terms often used in place of technical textile like industrial textiles, functional textiles, performance textiles, engineering textiles, invisible textiles and hi-tech textiles. Depending on the product characteristics, functional requirement and end use application, the highly diversified range of technical textiles products have been divided into 12 sectors.

Application of natural fibres in all the areas of technical textiles is being contemplated by the manufacturers, as it makes their products biodegradable and more desirable in the industry, which is the need of the hour. For instance, jute fibre is being used for various technical textiles like canvas cloth and tarpaulins in agrotech and oekotech; laminates in buildtech; jute and its blended fabrics in clothtech; soil savers in geotech; curtains and wall coverings in hometech; non-wovens for sound and heat insulation in mobitech; soft luggage fabrics, laminated jute hessian cloth and sackings in packtech and flameproof and mildew proof fabrics made using jute yarn as core and other conventional/ high performance fibres as sheath in protech [14]. Coir fibres have been also applied for many technical applications like biodegradable erosion control mats, floor mats, mattresses etc. Europeans are extensively using car interiors made of natural fibres. Automotive industry in Germany too, has been using flax, sisal and jute for car interiors in large quantities since last 20 years. Non-wovens made from kenaf exhibit good sound insulation properties for automobile interiors [15].

1.3   Unconventional Natural Cellulosic Fibres

Researchers are continuously engaged in exploring all the possible renewable natural resources which could become a potential source of textile fibres. Many pilot studies have been conducted experimenting with a number of plant sources for extracting fibres for varied purposes. Some of the studies are reported below.

Varshney and Bhoi extracted bast fibres from the stem of *Calotropis procera* (aak) plant by water retting. They found that aak fibre's strength and fineness was comparable to that of cotton but it was difficult to spin due to small staple length, high percentage of small fibres and lack of convolutions. They concluded that aak fibres could produce good quality cloth if evenness and fineness of aak yarn got improved [16].

Date palm leaf fibres were extracted by combined mechanical and chemical action and it was found that the fibres were comparable to various bast and leaf fibre in many respects like fineness, strength and extension at break etc. The authors further contemplated the use of fibres in preparation of ropes and handicrafts successfully. They suggested setting up small-scale industries in rural areas to exploit this renewable agro-waste [17].

Collier and Arora used water pre-treatment and alkaline treatment to extract fibres from sugarcane rind. They found that the fibres were stiffer as compared to other textile fibres but could be used in preparation of non-woven geo-textile mats. They also reported that water treatment has considerable influence on the amount of lignin in fibres and the physical and mechanical properties of fibres [18].

Borah and Kalita extracted fibres from the snake plant leaves using sodium hydroxide and sodium carbonate. They determined the tensile properties of the extracted fibres and concluded that sodium hydroxide extracted fibres had better physical properties and could be used in preparation of yarn, fabric and other products [19].

Kavitha extracted fibres from areca nut husks by water retting and beating of husks. The softer fibres were selected manually for open end spinning and coarse yarn of count 10 s was prepared in blend with cotton fibres. It was concluded that well cleaned areca nut husk fibre-cotton blended yarn could be used for making fabrics used in home furnishings. It was also reported that the extracted fibre possesses some natural antimicrobial property without application of any finish [20].

India has a rich variety of unexplored natural fibres which, if exploited, has potential to generate employment for a large section of the population. Moreover, India being an agriculture-based nation has a lot of biomass available which could be converted to a renewable resource. Utilization of the so-called waste to generate wealth will provide an additional avenue for livelihood generation and lead to green economy, which would also add to the green cover of the country [8].

Many other such fibres have been extracted from different plants like milkweed stems, water hyacinth, korai grass etc. which are being applied in various fields of textiles.

Agro waste like sugarcane bagasse, rice husk and cornhusk etc. has traditionally been used as fuels but their burning adds to air pollution. They need to be utilized more efficiently to add to our existing renewable resources. One such agro waste that can be employed to broaden our textile resource basket is; cornhusk, the outer covering of the corn cobs.

1.4 Corn and Cornhusk Fibres

Corn is the 3rd most important food crop after rice and wheat in India. In India, maize is cultivated in all the states except Kerala. It is cultivated throughout the year in different parts of the country for various purposes. Corn production in India has grown at a compounded annual growth rate of 5.5% over the last ten years from 14 million metric tons in 2004–05 to 23 million metric tons in 2013–14 [21].

Most of the corn produced is processed for its grains to be used for various purposes like flour, snacks and animal feed. During processing cornhusk is removed from the cob and burnt in case of most of the varieties for corn. Only husk of baby corn is used as animal feed as it is believed to enhance the yield of milk in cattle.

The cornhusk which is a bio-waste can be utilized for extraction of fibres which can be used for various textile applications. If 23 million metric tons of corn is produced in India annually, it could generate approximately 5 million metric tons of husks which could in turn lead to production of about 0.5 million metric tons of fibre. The major advantage of using cornhusk fibre is; fulfilling the requirement of food and clothing from same land without using any additional resources. It is in tandem with current motto of economic sustainability which emphasizes on developing such means of generating income and wealth, which do not cause hazards to the environment.

By-products of food sources like pineapple and banana leaves, sugarcane rind and coconut husks have been already used for extracting fibres for use in textiles though the fibres obtained from these also have some limitations in quality, availability and geographical requirements necessary to grow these crops. On the contrary, cornhusk is easily available, with no geographical limitations, and does not have any commercial value till date. Utilization of a bio waste like cornhusk will also help in solving the problem of its disposal and reducing the cost of waste treatment.

Cornhusk fibres are being looked upon as an addition to the existing unconventional plant fibres. Cornhusk fibres are extracted from outer covering of corn cobs by chemical and enzymatic retting. The fibres from cornhusk are coarse in nature and are similar in appearance to other bast or leaf fibres.

1.5 Various Applications of Corn in Textiles

Corn fibre: Corn fibres are the fibres made from PLA (poly lactic acid) obtained from corn. Corn fibre products are quite similar in look and feel to regular fibre, but they are

100% biodegradable and compostable. PLA fibre is however biologically degradable only when subjected to the right conditions. It very well suits for waste treatment through composting. As corn is a renewable resource, it makes corn fibre much more sustainable than regular fibre [22]. In the 1980s, PLA was synthesized from corn starch for the first time. Production of fibres from corn PLA started at a large scale around the year 2000 whilst it was introduced globally in 2003 on a commercial scale with the trade name, Ingeo. Ingeo is the first melt-processable natural fibre manufactured with the PLA resin using polyester type fibre manufacturing processes [23].

PLA is aliphatic polyester, having relatively small pendant methyl groups to hinder rotation and easy access to the oxygen atoms in the ester linkage. The PLA molecule tends to assume a helical structure. Polymer characteristics may be unique due to the fact that the lactide dimer occurs in three forms which are; the L form, which rotates polarized light in a clockwise direction; the D form, which rotates polarized light in a counter-clockwise direction and the meso form, which is optically inactive. The relative proportions of these forms can be controlled during polymerization which results in relatively broad control over important polymer properties. Ingeo fibre is a novel fibre that combines the best of both the worlds i.e. the performance of a synthetic fibre and the advantages of a natural material. This fibre has good strength and the fabric thus formed has many properties such as gentle bright lustre, good air permeability, good crease recovery, moisture absorption, shrink resistance and high resistance to ultra violet rays [23].

Sisodia and Parmar (2014) evaluated the dyeing behaviour of corn (PLA) fibre using disperse dyes and reported that colour strength of the fibres increased with increase in temperature but the strength decreased due to the degradation of fibre structure [24].

Cornhusk fibres: Very limited work has been done on cornhusk fibres across the world. Some of the studies reporting varied applications of cornhusk fibres are as follows,

- *Textiles*: Reddy and Yang examined the properties and potential applications of natural cellulose fibres from cornhusk. They extracted natural cellulose fibres from cornhusk and claimed that the extracted fibres have properties between cotton and linen. They blended the fibres with cotton in various proportions and processed on the ring and rotor spinning machines and concluded that fibres extracted from cornhusk were suitable for high value textile applications and had better processability than natural fibres from other agricultural by products [25].

In another study by Salam et al. kenaf and cornhusk fibres were bleached to CIE whiteness indices of 66. Bleaching was used to partially remove the lignin from the fibres without affecting other fibre properties and variables such as the concentration of hydrogen peroxide, time, temperature and pH were optimized for both the kenaf and cornhusk fibres. The effects of various bleaching parameters on the whiteness index and the breaking tenacity of the fibres were reported [26].

According to a study conducted by Reddy et al. (2011) natural cellulose fibres extracted from cornhusk had better dyeing properties for direct and sulphur dyes and similar dyeing properties for reactive and vat dyes when compared to cotton fibres dyed under similar dyeing conditions. They quoted short single cells, higher amounts of lignin and hemi cellulose, lower percent crystallinity and relatively coarse fibres of cornhusk as the reasons which make common cellulose fibre dyeing conditions unsuitable to dye cornhusk fibres. They also concluded that higher dye sorption on cornhusk fibres too was due to the above-mentioned reasons [27].

Yılmaz (2013) studied the effect of chemical parameters like alkali concentration and treatment time on the properties of extracted cornhusk fibres. The study proved that highest strength properties were achieved with 5–10 g/l sodium hydroxide treatment for 60–90 min. and also reported that average length, linear density and moisture content of extracted fibres decreased with increase in alkali concentration and duration. It was also deduced from FTIR spectrum analysis that harsher treatments led to increase in cellulose content whilst lignin and hemicelluloses content were lowered [28].

Yılmaz et al. (2014) investigated the effect of xylanase enzyme on the mechanical properties of fibres extracted from fresh and dried cornhusk. They extracted the fibres using alkali followed by xylanase treatment at different concentrations. Their results showed that drying did not have any negative impact on the fibre properties but the colour of fibres obtained was dull as compared to those extracted from fresh ones. They reported that increase in concentration of enzyme and drying decreased the linear density of fibres. Increasing enzyme concentration also led to increase in the breaking tenacity and initial modulus up to a point after which it decreased.

Zhuanzhuan et al. (2015) obtained cellulosic fibres with high aspect ratio from cornhusk by controlled swelling in organic solvent and simultaneous tetra methyl ammonium hydroxide post treatment. They concluded that cornhusk fibres obtained had qualities comparable to cotton and linen and thus could be successfully used in industrial applications.

- *Twines*: Bridgehouse and Hawthorne (1982) carried out a study to form twine of cornhusk and leaves. The extraction of fibres was done from cornhusk and leaves or cornhusk such that it became free for any binding vegetable material. The fibrous material extracted was combed and formed in to a tow of required thickness. This bundle was then twisted and drafted to form the twine. He claimed that the cornhusk twine was comparable to jute/sisal twines of similar thickness with respect to strength and durability.
- *Reinforcement in composites*: Another research was conducted to extract cornhusk fibres with chemical treatment and use the fibres as reinforcement in light weight polypropylene (PP) composites. The composites from cornhusk fibre (CHF) and PP were evaluated for flexural and impact resistance, tensile and sound absorption properties. The authors compared jute/PP composites and CHF/PP composites and found that CHF/PP composites were similar in impact resistance, 33% higher in flexural strength, 71% lower in flexural modulus, 43% higher in tensile strength, 54% lower in tensile modulus, and had slightly higher noise reduction coefficient. They also reported an increase in mechanical and sound absorption properties after enzyme treatment of cornhusk fibres (Huda and Yang 2008).

Youssef, El-Gendy and Kamel (2014) evaluated cornhusk fibres reinforced recycled low density polyethylene composites and suggested their applicability in packaging.

- *Pulping fraction*: Byrd et al. explored the feasibility of pulping the three fractions of corn stalk i.e. stalk, leaves and husks. Each (unmilled) fraction was pulped, using a mild soda-AQ process. A portion of each fraction was milled and tested for ash content, hot water solubles, NaOH solubles, solvent extractives, and klason lignin content. They concluded that pulping of corn residue should be explored further as it showed the prospects for being used as a source of pulp [29].

Ekhuemelo et al. also found cornhusk to be suitable for application in paper and pulp industry [30].

- *Cellulosic nano fibres*: Cardoso, Teixeira et al. (2008) extracted and characterized cellulose nano fibres from cornhusk. The extraction of nano fibres was done by acid hydrolysis (as given in Orts et al. 2005 and Moran et al. 2008). Nanostructures of cellulose were characterized by thermo gravimetric analysis (TGA), atomic force microscopy (AFM), Fourier transformed infrared spectroscopy (FTIR) and X-Ray Diffraction (XRD). Their results showed a possibility to extract cellulose nano fibres from cornhusk, with potential application in polymer nano composites [31].
- *Composite biodegradable film*: Norashikin and Ibrahim (2009) carried out a study to prepare a biodegradable film from cornhusks which was characterized by Fourier transform infrared spectroscopy (FTIR), differential scanning calorimeter (DSC), thermal gravimetric analysis (TGA) and atomic force microscope (AFM) observation. The fabricated film showed a high degradability rate as it readily degraded within 7–9 months under controlled soil conditions. The film was used in making biodegradable pot for seedlings plantation. The authors concluded that the pot produced from cornhusk composite biodegradable film was appropriate for planting a seedling as it would give plant more comfortable growing conditions without harming the environment [32].

2 Materials and Methods

This study involved extraction of fibres from cornhusk, evaluation of their physico-chemical properties and suitability of extracted cornhusk fibres for construction of various products.

The extraction of fibres from cornhusk was done using alkali and enzyme treatment and various parameters were optimized for these treatments. As the extracted fibres had inherent yellow colour, bleaching followed the enzyme treatment and various parameters for bleaching were also optimized. As the fibres obtained were harsh, softening treatment was given to make them soft. The composition of cornhusk fibres was determined using chemical methods. The extracted cornhusk fibres were also assessed for various physico-chemical properties and their morphology was studied using scanning electron microscopy and X-ray diffraction. The fibres were finally used for developing diversified products. To overcome the limitation of fibre strength, coarseness and brittleness, cornhusk fibres were blended in varying proportions with cotton, polyester and coir fibres to spin different types of yarn and make a variety of products including fabric, rope and hand-made mats.

The experiments for the study were conducted at Lady Irwin College, New Delhi and North India Textile Research Association, Ghaziabad. Yarn blends with cotton and polyester fibres and fabric were prepared at the Pilot Spinning Plant, NITRA; blended yarn with coir and its products were made at Central Coir Research Institute, Alapuzzha.

Extraction of fibres from cornhusk: The husk was collected from fully mature cobs of Syngenta Sugar-75, a variety of sweet corn, from Aterna village in Panipat (Haryana) which was sourced with the help of Directorate of Maize Research, Pusa Institute (DMR). The cornhusk was cleaned and dried to prevent it from deterioration during storage. Dried cornhusk was stored in properly ventilated room.

- The fibre extraction process was initiated with alkali treatment of cornhusk using sodium hydroxide. The alkali treatment was optimized with respect to concentration of sodium hydroxide, temperature and time of treatment. The bundle strength and fineness of the fibres obtained were the parameters to be considered for optimizing the treatment conditions.
- As the fibres obtained with alkali treatment were quite coarse and harsh in nature, they were further treated with Pulzyme HC, a xylanase enzyme to make it softer and finer. For optimizing this treatment, two variables i.e. concentration of enzyme and treatment time were taken. Temperature and pH were specified by the supplier itself. Enzyme treatment did not have much impact on strength of fibres but improved the fineness considerably which also made the fibres more pliable. In fact, strength increased on treatment with enzyme probably due to increased flexibility after removal of hemicelluloses.
- As the fibres extracted from cornhusk were yellow in colour, they were bleached using hydrogen peroxide. Bleaching treatment was optimized in terms of whiteness index and bundle strength of cornhusk fibres. The treatment led to improvement in fineness of fibres and some decrease in the strength. To impart a smooth handle to the extracted cornhusk fibres, they were applied a cationic, silicone-based softener. Application of softener made opening of fibres quite easy and their brittleness was also reduced.

Physical and chemical analysis of extracted cornhusk fibres: After extraction of fibres from cornhusk, the composition of fibres was determined, their physico-chemical properties were analysed and fibre morphology was studied.

- The cornhusk fibres being lignocellulosic in nature primarily contain cellulose, hemicelluloses and lignin; and small amount of wax and ash. Bleached cornhusk fibres were mainly composed of 76% cellulose, 11.43% hemicelluloses, 7.5% lignin, 0.25% and 0.34% ash content.

- Some of the physical properties of the extracted cornhusk fibres studied included length, fineness, bundle strength, elongation at break, moisture content and regain, absorbency, water retention, whiteness and yellowness index and pH.
- The extracted cornhusk fibres were coarse with linear density of 86 deniers. The reason for coarseness is that the fibres are actually bundles of fibres and not individual fibres.
- The cornhusk fibres obtained were long staple fibres with a length of 10–11 cm which could be used for various textile applications in spite of their coarseness.
- Bundle strength of cornhusk fibres was found to be 1.3 g/d which is lower than other lignocellulosic fibres. Their low strength can be compensated by blending with other suitable fibres. Elongation at break of cornhusk fibres was determined to be 18.5%.
- The moisture regain of extracted fibres was quite high, i.e. 11.3% whilst moisture content was estimated to be 10%.
- The drop penetration time of cornhusk fibres was less than one second which prophecies their application for various absorbent materials. Water retention of cornhusk fibres was also found to be 199.99%, which is very high.
- Cornhusk fibre did not leave its yellow colour even after bleaching. The whiteness and yellowness indices of the bleached fibre were found to be 60 and 73, respectively as compared to unbleached fibres having whiteness index of 44 and yellowness index, 76. The extracted cornhusk fibres had a pH of 6.81, which is almost neutral.
- Surface characteristics of cornhusk fibres were studied using scanning electron microscopy. It was found that fibres became cleaner from alkali treatment to enzymatic treatment and finally to bleaching as cementing material between the fibre bundles got removed at each step.
- X-ray diffraction of cornhusk fibres was carried at all the three stages of fibre extraction i.e. alkali, enzyme and bleaching treatment to determine their crystallinity. The percentage crystallinity increased from 58.54% after alkali treatment, to 60.33% after enzymatic treatment and finally to 62.39% after bleaching; this was probably due to the removal of amorphous impurities during the successive treatments [33–36].

Product development: After extraction of fibre, its various applications were explored and prototypes of various products were developed.

- Yarn and Fabric Formation

This section of the study focused on using the cornhusk fibres for spinning of yarns by blending with other fibres like cotton and polyester and manufacturing of fabrics from these yarns.

Yarn spinning: Due to the harsh and coarse nature of extracted fibres, 100% cornhusk yarn could not be spun. The extracted fibres were processed for spinning cornhusk blended yarns. Cornhusk fibres were blended with cotton and polyester fibres separately in different ratios. Hand and ring spinning, both were carried out as the fibres were coarse in nature. Cornhusk fibres were blended in two ratios with cotton and polyester fibres i.e. 30% cornhusk: 70% cotton/polyester and 50% cornhusk: 50% cotton/polyester.

Component fibres were first mixed with hand and then fed into the blow room for a more uniform and homogenous blend. Thereafter, card slivers and draw frame slivers were made. Roving was then made on Simplex machine. Finally, the yarns were prepared on a ring frame and by hand spinning on Amber *Charkha*. The yarns prepared with 30% cornhusk fibres were discarded and not used further as most of the cornhusk fibres got removed from blend whilst processing. The yarns having 50% cornhusk fibres were further processed, evaluated and used for preparation of fabric samples.

Evaluation of yarn properties: The manufactured yarns were tested for physical properties like yarn count, yarn strength and yarn elongation using standard test procedures under standard atmospheric conditions.

Yarn count: The yarn count expresses the thickness of the yarn. Higher the indirect yarn count, finer is the yarn. The yarn count number indicates the length of a yarn in relation to its weight. Yarn count was determined as per IS: 1315 using a wrap reel and electronic balance. A lea of 120 yards was made using a wrap reel and then weighed using electronic balance. Yarn count was calculated using the following formula,

Yarn count (Ne) = 64.8/wt of the lea, where, 64.8 = constant.

Yarn strength and elongation: Yarn strength and elongation was assessed as per ASTM standard D: 2256 using Uster Tensorapid instrument. This is an automatic electronic instrument used to measure single yarn strength and elongation percentage. It is based on the principle of constant rate of elongation (CRE). It can test 20 samples one after the other in a continuous manner. It is self-threading and runs for specified number of tests. The count and number of tests are entered and the machine automatically calculates the strength, elongation etc. by built in processor. Yarn strength was expressed as g/tex.

As the yarns made on Amber *Charkha* were very uneven and weak, they were not processed further to form fabrics.

Fabric construction: The fabric was woven with ring spun cornhusk/polyester (50/50) blended yarns on shuttle less loom. As the yarns prepared were very fuzzy and had many protruding ends, they could not be used as warp. 100% polyester yarn was used as warp and cornhusk blended yarn was used as weft. Some difficulty was faced during fabric construction as the cornhusk fibres were continuously shedding from blended yarns. The weaving got interrupted many times due to this reason. As the fabrics prepared were insufficient in quantity to conduct all the tests, their GSM, ends per inch and picks per inch were determined.

2.1   Manufacturing of Rope/Mats

As the fibres extracted were coarse and manufacturing of yarn and fabric did not seem feasible for apparel purposes, they were used to make a coarse yarn by blending with coir fibres. Blending had to be done with coir fibres as the coarse yarn using 100% cornhusk fibres could not be prepared. The yarn was prepared at Central Coir Research Institute, Alappuzha in Kerala. The yarn was constructed by hand with a blend ratio of 50/50 (cornhusk fibre/coir fibre). The yarn formed was then tested according to test method IS 14596: 1998, for properties like breaking load, elongation at break, runnage (m/kg) and twists per meter and compared with 100% coir yarn of same thickness. The cornhusk-coir yarn thus prepared was then used to manufacture prototypes of the following articles.

Rope: A rope is a material of cordage construction with a circumference of 1 in. or more. It is made up of 3 or more strands. The strands themselves being an assemblage of yarns twisted together. A rope with four strands was constructed using the cornhusk-coir yarn on hand operated rope-making machine (Fig. 1). The rope manufactured was tested as per specifications of IS 1410: 2001.

Woven Mat: The cornhusk-coir blended yarn was used to weave a mat on handloom. The mat was evaluated for breaking load and elongation at break as per the test method IS 13162: 1992.

Sinnet/Chain/Braided Mat: The cornhusk-coir yarn was hand braided and then a sample mat was made. This type of mat is made by braids/plaits placed in a zigzag manner with inter stitching on a flat table upon which nails without heads are fixed according to size and pattern of the mat to be prepared (Fig. 2).

Fig. 1 Hand operated rope-making machine

Fig. 2 Sinnet mat being
constructed on table

Fig. 3 Corridor mats being
constructed on frames

Corridor/Hollander/Dutch Mat: It is a mat in which both warp and weft strands are continuous without tucking or binding. The pattern effect is produced by the weft strands only and has ribbed effect on both sides. It is prepared with the help of a wooden frame and pressing device (Fig. 3).

3 Results and Discussion

3.1 Yarn and Fabric Formation

The fibre was spun into yarn for manufacturing fabric which could possibly be used in apparel. The extracted cornhusk fibres exhibited many desirable properties as per the tests conducted like high elongation, moisture regain and absorbency, which enhance the comfort properties of any fabric. However, the major limitations of cornhusk fibres were their brittleness, coarseness and harsh texture.

Yarn Spinning: To overcome the limitations of cornhusk fibre and make it machine spinnable, yarn was spun by blending cornhusk fibres with cotton and polyester fibres. Yarn was spun in two ratios with cotton and polyester both, and the combinations which were used to spin yarn were; cornhusk: cotton—50:50, cornhusk: polyester—50:50, cornhusk: cotton—30:70 and cornhusk: polyester—30:70.

After the fibres were being opened and blended in blow room; they were fed in carding machine. The cornhusk fibre was getting entangled in the wires of the card due to which lot of cornhusk fibre was lost during carding. The yarn was spun on ring frame and also hand spun using Amber *Charkha*. Cornhusk fibre showed poor cohesiveness in both the yarn blends as it was shedding during all the stages of processing. Due to the reasons mentioned above, negligible amount of cornhusk fibre was left by the end of processing in yarn having 30% cornhusk fibre.

Hence, the yarn blend having 30% cornhusk fibre was discarded. Moreover, the yarn having cornhusk: cotton in 50:50 ratios could not be spun on ring frame due to lack of strength and cohesiveness. Finally, blended yarns of three types were constructed; all having 50% cornhusk fibre and 50% other component fibre (cotton or polyester). The yarns constructed are shown in Fig. 4. The yarns prepared were tested for physical properties like yarn count, yarn strength and yarn elongation which are given in Table 1.

All types of constructed yarns were coarse and low in strength. But the hand spun yarn was too weak and irregular, so it could not be used for construction of fabrics (Fig. 4). Fabric sample was made using ring spun cornhusk/polyester yarn, though even this yarn was very uneven and hairy.

Fabric construction: The fabric was constructed with ring spun polyester/cornhusk (50/50) yarn on rapier loom. Since cornhusk blended yarn was fuzzy; it was getting entangled and thus, sizing was applied to make its surface regular (Fig. 5). In spite of applying sizing, the cornhusk blended yarn remained quite hairy and caused breakage whilst weaving on the loom. Hence, it could not be used as warp. Instead 100% polyester yarn was used as warp and cornhusk blended yarn was used only as weft for fabric construction. As the fabric could not be produced in large quantity, only few characteristics could be determined. The parameters of the constructed fabric are given in Table 2.

The constructed fabric had a rough texture and ends of cornhusk fibres were protruding from fabric surface. The fabric was not suitable for apparel purposes due to its harsh feel. Even application of softener did not improve the feel of the fabric.

Table 1 Physical properties of blended cornhusk yarns

S. No.	Yarn type	Count (Ne)	Tenacity (g/tex)	Elongation (%)
1	Hand spun cornhusk/cotton	3	5.09	6.14
2	Hand spun cornhusk/polyester	3	5.4	7.9
3	Ring spun cornhusk/polyester	6.7	7.23	12.47

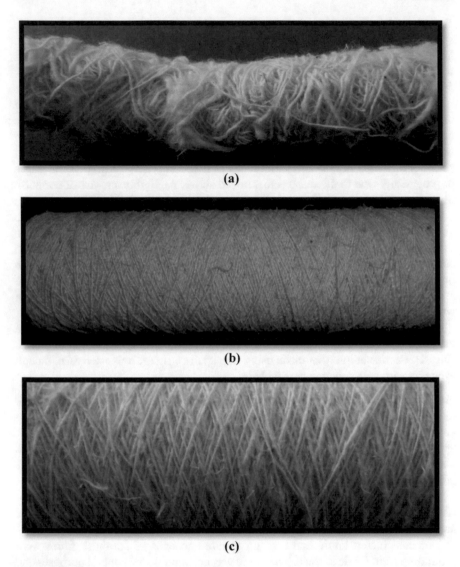

Fig. 4 a Hand spun cornhusk/cotton yarn. **b** Hand spun cornhusk/polyester yarn and **c** ring spun cornhusk/polyester yarn

3.2 Manufacturing of Rope/Mats

As blending of cornhusk fibres with cotton and polyester fibres did not give satisfactory results, the cornhusk fibre was blended with coir fibre as it is also a very coarse fibre. It was thought that both the fibres might blend well; as they were quite similar in texture. Moreover, coir is a strong fibre which could compensate for low strength of cornhusk fibres.

Fig. 5 Fabric with
polyester/cornhusk (50/50)
weft and 100% polyester
warp yarn

Table 2 Parameters of the fabric constructed

S. No.	Fabric sample	EPI	PPI	GSM
1	Ring spun polyester/cornhusk (50/50) yarn weft; 100% Polyester warp	41	21	531

EPI ends per inch; *PPI* picks per inch; *GSM* grams per square meter

Cornhusk/Coir Blended Yarn: The yarn was constructed by hand with a blend ratio of 50/50 (cornhusk fibre/coir fibre). It was formed by twisting two single loosely twisted threads. The cornhusk blended yarn was quite similar in appearance to 100% coir yarn (Fig. 6).

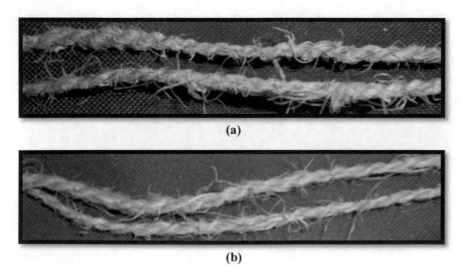

(a)

(b)

Fig. 6 **a** Cornhusk/coir (50/50) blended yarn. **b** 100% coir yarn

Table 3 Properties of cornhusk/coir (50/50) yarn and 100% coir yarn

S. No.	Parameter	Cornhusk/coir (50/50) Yarn	100% coir Yarn
1	Breaking load (N)	92.13	144.06
2	Elongation at break (%)	12.35	20.97
3	Runnage or linear density (m/kg)	291.31	366.54
4	Twist/m	66	59

The properties of the yarn constructed were tested as per test method IS 14596:1998, which is used for coir yarn. Table 3 shows the properties of cornhusk/coir (50/50) blended yarn and 100% coir yarn.

As evident from the results in Table, cornhusk/coir (50/50) blended yarn was weaker and had lower elongation at break in comparison to 100% coir yarn but its properties were comparable to some varieties of the soft twisted coir yarns as per the test standards. The cornhusk/coir (50/50) blended yarn, thus prepared, was then used to manufacture samples of various articles.

Rope: A four strand rope was constructed using the cornhusk-coir yarn on hand operated rope-making machine and each strand had four yarns (Fig. 7).

The constructed rope was tested for the required parameters and the results are reported in Table 4.

The rope constructed using cornhusk/coir (50/50) yarn had moderate strength but could be used for some low end applications where strength does not play a key role.

Woven Mat: A mat was hand woven using cornhusk-coir yarn and tested for breaking load and elongation at break (Table 5).

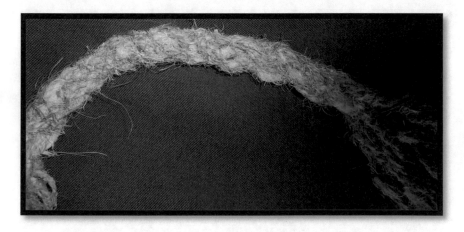

Fig. 7 Cornhusk/coir (50/50) yarn rope

Table 4 Properties of cornhusk/coir (50/50) yarn rope

S. No.	Parameter	Cornhusk/coir (50/50) Yarn rope
1	Breaking load (N)	1775.91
2	Elongation at break (%)	120.84
3	Runnage (m/kg)	12.36
4	Twist/m	24

Table 5 Properties of woven mat using cornhusk/coir (50/50) yarn

S. No.	Parameter	Cornhusk/coir (50/50) Yarn mat
1	Breaking load (N)	3284.36
2	Elongation at break (%)	17.83

The woven mat (Fig. 8) made by cornhusk/coir (50/50) yarn could be used in home furnishing textiles or in geo-textiles as cornhusk fibre is also 100% bio-degradable like coir fibres.

Sinnet/Chain/Braided and Dutch/Hollander/Corridor Mat: The samples of these mats were constructed using the cornhusk-coir blended yarn as full size mats could not be made due to shortage of cornhusk fibres. They were better in appearance and soft in feel as compared to the mats constructed with 100% coir yarn (Fig. 9).

Fig. 8 Cornhusk-coir yarn woven mat

Fig. 9 Prototypes of **a**
sinnet mat, **b** corridor mat

(a)

(b)

Thus, cornhusk/coir (50/50) yarn could be used to produce different types of hand-made home furnishing items and employment could be generated at cottage level, especially in rural areas.

4 Conclusion

Disposal of agro waste has become a challenge in present scenario which needs to be addressed urgently. Various researches are being carried out for finding applications for such agro wastes like sugarcane bagasse, rice husk, corn stalks etc. Utilizing cornhusk for textile applications could become a major breakthrough in this direction. As the fibres obtained from cornhusk were not suitable for apparel purposes due to their coarse and harsh nature, their application for different segments of technical textiles could be explored. The manufacturing of coarse yarn from cornhusk fibres by blending with fibres like coir and its use for developing a range of hand-made products has ample scope. This would lead to employment generation for a large section of rural population and would lead to value addition to an agro waste like cornhusk.

References

1. The future of textiles (2013) Advantage environment. http://advantage-environment.com/fra mtid/the-future-of-textiles. Accessed on 14 Sept 2015
2. Kivaisi A, Assefa B, Hashim S, Mshandete A (2010) Sustainable utilization of agro-industrial wastes through integration of bio-energy and mushroom production. https://cgspace.cgiar.org/bitstream/handle/10568/10816/project4_biogas.pdf?sequence=6. Accessed on 14 Sept 2015
3. Food and Agriculture Organization (n.d.) The future of food and agriculture: trends and challenges. https://www.fao.org/3/i6583e/i6583e.pdf. Accessed on 10 Apr 2021
4. Van Dam JEG (2008) Natural fibres and the environment: environmental benefits of natural fibre production and use. In: Paper presented at the symposium on natural fibres: common fund for commodities, Rome
5. Ministry of Textiles (2014) Material on technical textiles. texmin.nic.in/sector/note_technical_textiles_ammt.pdf. Accessed on 12 Jan 2015
6. Food and Agriculture Organisation (2013) Current market situation for jute and kenaf; sisal and henequen; abaca and coir. www.fao.org/fileadmin/.../est/...Hard_Fibres/.../CurrentSitua tion.docx. Accessed on 13 Sept 2013
7. Arumugam S (2013) India's textile and apparel industry. http://www.citiindia.com. Accessed on 12 Nov 2013
8. G.B. Pant Institute of Himalayan Environment & Development (2010) Promising fibre-yielding plants of the Indian Himalayan region. Retrieved from http://gbpihed.gov.in/PDF/Publication/FiberPlants.pdf
9. Satyanarayana KG, Guimarães JL, Wypych F (2007) Studies on lignocellulosic fibers of Brazil. Part I: source, production, morphology, properties and applications. Composites Part A Appl Sci Manuf 38(7):1694–1709
10. (Textile Fibre Policy, 2010)
11. Smole MS, Hribernik S, Kleinschek KS, Kreže T (2013) Plant fibres for textile and technical applications. Adv Agrophys Res 10:52372
12. Abarbanel SR (n.d.) Material choices: bast and leaf fibre textiles. https://fowler.ucla.edu/wp-content/uploads/2021/05/MaterialChoices-release.pdf. Accessed on 7 July 2021
13. Material on Technical Textiles (2021) http://texmin.nic.in/sites/default/files/scheme_techni cal_textile_070116.pdf. Accessed on 13 Dec 2021
14. Samanta AK (2016) Advantages of jute as natural bast fibre for different technical textiles. https://www.fibre2fashion.com/industry-article/7722/advantages-of-jute-as-nat ural-bast-fibre-for-different-technical-textiles. Accessed on 10 Apr 2021

15. Chapman (2010) Applications of nonwovens in technical textiles, edited by Chapman R. Elsevier, 15 June 2010, 58
16. Varshney AC, Bhoi KL (1988) Cloth from bast fibre of the Calotropis procera (Aak) plant. Biol Wastes 26(3):229–232
17. Ghosh SK, Saha SC, Bhattacharyya SK (2003) Chemical modification of date-palm leaf fibre. Textile Trends 46:23–27
18. Collier BJ, Arora MS (1996) Water pretreatment and alkaline treatment for extraction of fibers from sugar cane rind. Cloth Text Res J 14(1):1–6
19. Borah MP, Kalita B (2004) Physical properties of snake plant leaf fibre (S NPLF). Textile Trends 29–31
20. Kavitha S (2012) Application of areca fibre in home textiles. Home Textile Views 6(01):44–46
21. Federation of Indian Chambers of Commerce and Industry (FICCI) (2014) Indian maize summit, 2014. Retrieved from http://www.ficci.com/spdocument/20386/India-Maize-2014_v2.pdf
22. Tejaswini RL (2016) Fibre from corn. Fibre2Fashion. https://www.fibre2fashion.com/industry-article/7676/fibre-from-corn. Accessed on 3 Apr 2021
23. Farrington DW, Lunt J, Davies S, Blackburn RS (n.d.) Polylactic acid fibres. https://www.natureworksllc.com/~/media/Technical_Resources/Ingeo_Technical_Bulletins/TechnicalBulletin_BiodegradableSustainableFibers_Chap6_2005_pdf.pdf. Accessed on 15 May 2021
24. Sisodia N, Parmar MS (2014) Dyeing behavior and fastness properties of corn (pla) fiber. J Polym Textile Eng 1(2):01–07
25. Reddy N, Yang Y (2005) Properties and potential applications of natural cellulose fibers from cornhusks. Green Chem 7(4):190–195
26. Salam A, Reddy N, Yang Y (2007) Bleaching of kenaf and cornhusk fibers. Ind Eng Chem Res 46(5):1452–1458
27. Reddy N, Thillainayagam VA, Yang Y (2011) Dyeing natural cellulose fibers from cornhusks: a comparative study with cotton fibers. Ind Eng Chem Res 50(9):5642–5650
28. Yılmaz ND (2013) Effect of chemical extraction parameters on corn husk fibres characteristics. Indian J Fibre Text Res 38(1):29–34
29. Yılmaz ND, Çalışkan E, Yılmaz K (2014) Effect of xylanase enzyme on mechanical properties of fibres extracted from undried and dried corn husks. Indian J Fibre Text Res 39(1):60–64
30. Ekhuemelo DO, Oluwalana SA, Adetogun AC (2013) Potentials of agricultural waste and grasses in pulp and papermaking. J Res Forest Wildlife Environ 4(2):79–91
31. Cardoso RFO, Teixeira EM, Paes MCD, Marconcini JM, Mattoso LHC (2008) Cellulose nanofibers from corn husks: extraction and characterization. Retrieved from http://ainfo.cnptia.embrapa.br/digital/bitstream/item/87431/1/Proci-08.00079.PDF
32. Norashikin MZ, Ibrahim MZ (2009) The potential of natural waste (corn husk) for production of environmental friendly biodegradable film for seedling. World Acad Sci Eng Technol 58:176–180
33. Jain A, Rastogi D, Chanana B (2016) Corn—a vital crop for our economy. Res J Human Social Sci 7(3):185–192. https://doi.org/10.5958/2321-5828.2016.00030.9
34. Jain A, Rastogi D, Chanana B (2018) Utilization of cornhusk for textile usages. J Basic Appl Eng Res 5(5):405–408
35. Jain A, Rastogi D, Chanana B et al (2017) Extraction of cornhusk fibres for textile usages. IOSR J Polym Text Eng 4(1):29–34
36. Jain A (2015) Extraction and application of cornhusk fibres in textiles. PhD Thesis, University of Delhi, New Delhi

Organic Cotton: Fibre to Fashion

V. K. Dhange, S. M. Landage, and G. M. Moog

Abstract Cotton cultivated without using genetically modified (GMO) seeds, harmful pesticides, synthetic fertilisers, and chemicals is termed as "organic cotton." Organic cotton farming maintains healthy working environment for farmers, clean freshwater sources near farms and reduces soil erosion and other negative impacts on the ecosystem. Organic cotton is gaining wide acceptance in manufacturing of premium fabrics, due to its zero toxicity and being completely natural origin. During the chemical wet processing, various hazardous chemicals are used. So, during processing of organic cotton, manufacturers must ensure avoiding toxic substances such as petroleum scours, formaldehyde, chlorine bleaches, heavy metal-containing chemicals, etc. The use of natural alternatives and enzymatic treatments as partial substitutes to chemical wet processing is the need of the hour. These eco-friendly options are used to minimise and remove the hazardous effects occurring in conventional cotton fabric processing. The garment industry plays a vital role in the production of fashion products from organic cotton. Currently, the contribution of organic cotton to overall cotton production is less than 1%. Fashion consumers can help change that. If the consumers call for more organic fashion apparel, it will compel the apparel manufacturers to use more organic cotton. Many National and International brands are opting for production of fashion products from organic cotton. This chapter covers various sustainable approaches and ideas for the cultivation, processing and garment manufacturing of organic cotton.

Keywords Cultivation · Fashion · Fibre · Garment · Organic cotton · Sustainability · Textiles · Wet-processing

V. K. Dhange (✉) · S. M. Landage · G. M. Moog
Department of Textiles, DKTE Society's Textile and Engineering Institute, Ichalkaranji, Maharashtra, India
e-mail: vkdhange@dkte.ac.in

S. M. Landage
e-mail: smlandage@dkte.ac.in

© The Author(s), under exclusive license to Springer Nature Singapore Pte Ltd. 2022   275
S. S. Muthu (ed.), *Sustainable Approaches in Textiles and Fashion*, Sustainable Textiles: Production, Processing, Manufacturing & Chemistry,
https://doi.org/10.1007/978-981-19-0878-1_11

1 Introduction

According to modern agriculture science and technology, Immunoglobulin (IG) nutrients are required for plant growth. They are H, C, O, P, N, K, Mg, Ca, B, Cu, S, Fe, etc. Out of these, the first three are readily available from air and water, whilst the next 6 are called macronutrients which are subdivided into main (N, P, K) and minor (Ca, Mg, S), whilst the last 7 are called micronutrients. All these 13 are required to be provided to plants through soil only. Every plant uses all of these 13 nutrients for its growth from the soil and depletes the soil to that extent. If another crop must be grown the following year, the nutrients used must be replenished by using outside manures and fertilisers. This is how chemical fertilisers came into existence to be used in agriculture.

Before the invention of chemical fertilisers around 200 years ago, agricultural arrangements for the supply of nutrients must have existed centuries ago. These plants existed thousands of years before the first man was born. Therefore, it is obvious that the plants have some natural mechanism to exist and grow without any human support. Moreover, the new science does not explain how plants on the roadside or in forests grow without assistance from chemical fertilisers. Similarly, weeds also grow well without the application of fertilisers if a farm is kept without ploughing. Thus, the theory of NPK for plant nutrition is not applicable everywhere. This observation made to think of some other theories of plant nutrition which will explain how crops grew before the advent of chemical fertilisers and how they continue to grow in forests.

Nature has classified living organisms into two types: moveable and non-movable or static. Humans, animals, birds, and insects belong to the moveable class, and therefore they are required to search for their food, whilst immovable plants are provided with their nutrition at the place where they are born and grow. How do plants get the thirteen nutrients from the soil even if one does not add them to the soil? New Science explains this by showing that there are organisms such as rhizobia that are capable of taking nitrogen from the air and fixing it in the soil for use by the plant, as well as storing it for future crops. The supply of nitrogen is unlimited in the air. Similarly, there are other organisms such as phosphorus-solubilising bacteria and pseudoplus, which provide iron to the plant. There is a chance that the other organisms in the soil can provide the remaining 10 nutrients in some way [1, 2].

Natural fibres are divided into three categories: vegetable fibres, animal protein fibres, and mineral fibres. Cellulose is the principal component of all plant fibres. The most commonly used natural fibres for manufacturing textile goods are cotton, jute, banana, ramie, flax, wool, and silk. In general, the terms "natural fibre clothing" and "organic clothing" are used interchangeably. Organic clothing is made without using any toxic chemicals in their entire supply chain. Natural fibre clothing, on the other hand, refers to garments produced from natural fibres like cotton, banana, flax, or wool that may not be cultivated or produced under organic conditions [3]. Cotton is considered as the most significant natural fibre in the world, owing to its exceptional qualities like wearing comfort, natural appearance, breathability, and moisture absorbency. Cotton and cotton-based products are produced in large quantities in China, India, and the United States [4]. Cotton accounts for nearly a quarter of the global textile fibre market. [5].

Due to the rapid rise in cancer risk, textile customers are becoming increasingly concerned about the raw materials used in finished products, which are supposed to be free of chemical residues. Organic cotton growing is a novel approach to achieving long-term textile sustainability. Because organic cotton is grown without the use of toxic pesticides or synthetic fertilisers, it is less chemically contaminated than conventional cotton.

2 Cotton Fibre

Cotton is a white, feathery fibre, and individuals consider cotton to be a pure, natural cellulosic fibre. Cotton is a fantastically multipurpose and worldwide significant fibre that is employed for a wide range of fabrics. As a result, it is one of the most extensively traded commodities on the market. Cotton fibres have a high market value due to their versatility, absorbency, softness, breathability, and durability. Cotton is not only the most communal and successful fabric, but it also has the highest marketable price of any cloth. Cotton fibre is a vital component of many countries' foreign exchange earnings. However, there is a dark side to the international cotton industry. When most of us shop for cotton garments in stores, we are unaware of the dangers. Fibre is used for the manufacturing of one T-shirt, which requires approximately 1 lbs of cotton. In such cases, a massive and overwhelming load is placed on the water, air, and soil of the earth, affecting global health.

Cotton is a wonderful fibre for clothing, but the way most cotton is grown causes serious health concerns. A lot of people are experiencing different types of health issues such as rashes, allergies, respiratory issues, and difficulty focusing mentally due to chemical sensitivities that are increasing at an alarming rate. Many people who have several chemical sympathies find that wearing organic clothing helps them reduce their exposure to the wide range of toxic chemicals that we are unknowingly exposed to every day [3].

2.1 Facts About Conventional Cotton Cultivation

Following are some of the facts regarding the cultivation of conventional cotton:

- In California, five of the nine main pesticides used in cotton production are carcinogenic.
- Children whose parents used pesticides in their homes or gardens had a nearly seven-fold greater risk of leukaemia, according to the National Cancer Institute Study (1987).

- Each year, more than one million Americans will be found to have cancer, and 10,400 will die from pesticide-related cancers in the USA.
- More than half of cotton workers in Egypt in the 1990s showed evidence of chronic pesticide exposure, including neurological and visual problems.
- 91% of cotton workers in India were exposed to pesticides for at least 8 h each day [3].

3 Organic Clothing—Need of Future

Cotton fibre is considered sustainable because it is biodegradable and environmentally responsible in its life cycle. The primary sustainability parameters associated with cotton fibre production are land, soil, water, air quality, and energy. Cotton fibres quality may be affected due to the quality of the soil. Also, the quality of soil gets affected due to soil erosion, and synthetic chemicals, and fertilisers used for cultivation of crops. In the United States, to produce 1 kg of cotton fibre, approximately 5 g of insecticides are needed. It has been observed that nowadays there is tremendous decrease in the usage of fertilisers in cotton cultivation. Now a day for the cultivation of cotton fibre requires only 8.5% of total pesticides. In few places, because of good rainfall, naturally available water is sufficient for growing cotton. Land use for cotton production has declined significantly because of the use of modern agricultural technologies. Environmental and air pollution is a significant factor in cotton fibre production and the ginning process. The use of modern technologies, such as reduced tillage, has been successful in reducing dust emissions from the field [6, 7].

With a recent bombardment of media attention, sustainable apparel and green eco-fashion have become popular awareness. What exactly is sustainable clothing, and how does it vary from organic clothes? [8]. Although the terms sustainable clothes and organic clothing seem similar, their origins and histories are distinct. The environmental movement has resulted in sustainable clothing, as organic clothing arose as a result of the organic agriculture drive. They are both working in the same region, yet one feels like a farm and the other like a laboratory. One of the important significant differences between organic and sustainable approaches is the emphasis on the reuse and recycling of manufactured products in the sustainable method.

As far as sustainability is concerned, Green Strategy's Anna Brismar has identified, major requirements in production, and consumption. Textile and apparel products should possess eternal design features. During production, the wastage of raw material should be least, theoretically, it should be "zero." They should be used for many years through repair, redesign, and recycling. If bored with continual usage, the textile and garments should be donated to needy persons, charity organisations, and swap shops. The active life of a textile product should be prolonged by sharing/transferring it to relatives, friends, etc. When a garment is completely worn out, it should be recycled so that it can be used as a raw material in the production of new goods. Instead of buying a new garment, one should think of borrowing or swapping clothes or buying second-hand products.

Durable fabrics should be made from natural or recycled materials to reduce damage caused by the production process, fibre properties or overall environmental impact. They can help reduce waste, conserve water, reduce emissions, and regenerate the soil. As sustainability means different to different people, each stage in the textile chain concentrates on distinct facets of sustainability, such as the use of water and energy, the usage of renewable raw materials, waste reduction, use of hazard-free chemicals and dyes, welfare programs for the employees and the communities, animal welfare as a part of social responsibility, etc. [9]. In short, the entire supply chain of textiles and apparel should satisfy three foundations of long-term sustainability viz. environmental, economic, and social sustainability [10].

4 Organic Cotton

Organic farming is in harmonisation with nature. Farmers grow their crops by making use of natural systems and cycles, and it all begins with the soil! Farmers can cultivate crops that are both healthy and resilient by caring for the world beneath their feet. Crop rotation, green manures, and composting are all employed when crop rotation, green manures, and composting are used, soil with full of nutrients gives healthy crop. Furthermore, hazardous pesticides and artificial fertilisers are not permitted, biodiversity increases, and pest control using natural things is done. Non-organic cotton cultivation, on the other hand, relies on synthetic chemical inputs like nitrogen fertilisers and insecticides. These contribute to climate change and can have negative consequences for local ecosystems as well as human health [11].

Organic cotton is cultivated with environmentally friendly processes and ingredients [12]. Organic farming practices reload and preserve soil fertility, restrict the application of potentially hazardous substances and insistent herbicides and nourishments, and promote biological diversity. Certification by third-party bodies ensures that organic farmers utilise approaches and resources approved for organic production. Organic farming is founded on the notion of working with nature rather than against it. Organic farmers raise crops using biologically-based methods rather than chemically-based methods [13]. Despite the fact that organic cotton is commonly utilised in other parts of the world, in India it has great potential for farmers as well as the processors of cotton in the industry. Since all the organic cotton sold in India goes to the cotton converter industry before it is exported to the outside world.

The prime necessity of organic farm production is the certification of the farms and then the processing units have also to be certified by authentic certification agencies registered and known to the remaining world since the ultimate organic cotton product is to be used by the people resending outside India. The organic cotton produced and certified has to be followed by very stringent formalities to claim it as organic. But when it enters the processing, the complete organic structure of cotton is lost. The reason behind this is when it is brought to the colouration process, dyes used are not organic at all, as the organic dyes are not available in bulk and the cost is exorbitant. Now we hear about a lot of trials for the production of organic dyes that are going on. A lot is to be done in this field to claim that the final product used to be organic [3].

4.1 Why Wear Organic?

Remember the days of leisure suits, disco, and the peak of polyester? Those days have passed, but synthetic fabrics will live on in our nation's landfills for generations to come. As people's understanding of aesthetics and fashion has risen, cotton and natural fibres have supplanted polyester and other synthetic fibres as the preferred cloth. Concerns about the harmful effects of pesticides and insecticides on our food supply have fuelled demand for organic production and vegetables, and concern about the widespread chemicals and insecticides are used in the production of cotton and other natural fibres has fuelled demand for organic cotton and garments free of these and other toxins and chemicals [3, 8, 14].

5 Cultivation of Organic Cotton

In order to grow organic cotton, systematic approach is required. The incorporation of various processes is required for good organic cotton fibre. Organic cotton needs organic seeds, which are derived from an organic plant. Genetically engineered seeds are not used.

5.1 Seed Preparation

Seeds are generally treated with fungicides and pesticides in traditional seed preparation, and genetically modified (GM) seeds are employed in many circumstances. Organic methods look for non-chemical alternatives to pesticides and employ non-GM seeds that have not been treated.

5.2 Soil Fertility

In conventional cotton farming, which is primarily a monocrop farming culture, synthetic fertilisers are sprayed to the soil. Organic farming encourages naturally productive crops and isn't reliant on synthetic fertilisers to boost yields [15]. To maintain the soil fertility, Nitrogen level in soil, natural practices such as natural fertilisers, ash, compost, cow dunk are used for cultivation of organic cotton [16]. Soil organisms are appropriately managed to release the nutritional minerals needed to grow crops. The natural ingredients used to improve soil fertility get consumed by the soil organisms and release the nutrient products which are helpful for crop growth.

5.3   Crop Rotation

Crop rotation is useful for minimising the spreading of bugs, flies, and other wild plants. Rotation of crops also helps to maintain soil fertility. Crop rotation is also economically beneficial for farmers. It generates more income due to the diversity of crops [16]. A study on cotton conducted in Auburn, Alabama, found that the rotation of cotton crops with pulses produced yields equal to those obtained through the usage of nitrogenous fertilisers. The same efficiency can be accomplished by using natural fertilisers instead of synthetic. As a result, crop rotation is useful economically. It is also helpful biologically.

5.4   Cover Cropping

Crop cover is a supportable system for soil protection. Soil erosion abruptly reduces crop growth and leads to poor crop productivity. When ploughed in the soil, the cover crop supplies nitrogen and improves fertility by adding organic matter to subsequent cotton crops. Cover crops are frequently referred to as green manure. These green fertilisers manage several macronutrients and micronutrients of the soil, these nutrients are very favourable to the management of nitrogen which restricts the most nutrients in agricultural production.

5.5   Pest and Weed Management

Traditional farming heavily relies on hazardous chemicals and insecticides to manage pests. Aerial spraying techniques that can blow onto neighbouring agriculture and wildlife are frequently used. To aid in the maintenance of a healthy soil ecosystem, organic pest management technologies recognise beneficial insects and maintain a balance between pests and their natural exterminators [17, 18]. On organically farmed land, physical weed removal is used, and weed control is maintained through farming and hand dig out. Huge number of chemical weedkillers are used in conventional cotton farmland, which are not required in organic farming [15].

In case of organic cotton, there are several unwanted plants are growing [19]. In the conventional cotton cropping system, these weeds are treated with herbicides, but in organic agriculture, weeds are mostly controlled by rotation. In the farming of organic cotton, not all weeds are treated to be cleaned because they are favourable to cultivation. They kill the natural foes of the pests and sometimes divert attention from the cottonseed pests. Pest management processes are not difficult without pesticides being used and there are also fewer pets in organic cotton than in traditional cotton farming [16].

5.6 Harvesting

Organic methods either wait for natural seasonal freezing to cause defoliation or water management can be used to promote defoliation. Toxic chemicals are used in conventional farming to defoliate the land. Organic farming increases soil fertility, which is good for wildlife, and uses fewer sprays, which reduces pollution. As synthetic fertilisers are eliminated and organic cotton is grown using sustainable farming practices results in good ground water, surface water, air, and soil quality [15, 19].

5.7 Post-Harvesting Operations

Organic fibre products should be isolated and protected against comingling with conventional fibre products, as well as contact with forbidden materials or other contaminants, at all stages of manufacturing, transit, and storage. Seed cotton is a raw agronomic product when it is harvested. The produce is subsequently transferred to a ginning plant or kept in the field in vehicles.

When cotton is stored in modules in the field, the modules must be adequately covered to protect the cotton from rain and maintain its quality. Ginning is the harvesting process of extracting and separating lint fibres from seed and plant debris. Cotton fibre has very less saleable value unless and until, it is separated from seed and other impurities which are present along with fibre. Almost all such impurities are removed at ginning stage [20].

The yarn manufacturing process (e.g., spinning and allied activities), fabric manufacturing process (weaving, knitting, or non-woven), and textile wet processing are the three key post-harvest processes in the conversion of raw cotton fibre into a completed fabric (sequence of operations such as desizing, scouring, souring, bleaching, mercerisation, dyeing printing, and finishing). To manufacture organic textiles, certified organic cotton should be processed organically. It should be spun, woven/knitted, and processed in environmentally friendly and non-toxic ways. All processing agents must meet toxicity and degradability/liminality standards. All wet processing facilities should practise water conservation and resource management, and wastewater disposal criteria should be satisfied [21].

The Organic Trade Association (OTA) established voluntary organic standards for fibre processing (post-harvest handling, processing, recordkeeping, and labelling; "Organic Trade Association's Fibre Processing Standards") in 2004. They go over everything from organic fibre storage in the gin yard or hayloft to spinning, weaving, textile wet processing, quality control, and labelling, as well as the chemicals that can be employed. The Organic Exchange (OE) created voluntary standards, the Organic Exchange 100 standard (OE 100), and the OE Blended Standard, to help enterprises making organic products understand what they need to do to manage and document the acquisition, handling, and use of organic cotton in their products.

6 Advantages of Organic Cotton Cultivation

6.1 Environmentally Friendly Technology

The conventional method pollutes all elements of the agro-ecosystem due to the excessive usage of fertilisers and insecticides. Organic cotton production uses non-chemical inputs and reduces pollution risks. The pesticide residues in fibre may cause carcinogenesis in users. In organic farming, pest management is done through the use of biorational products and biocontrol agents which has no such consequences. The textile industry's and dyeing units' main cause of human health problems is untreated effluent, which is discharged as it is without any treatment. Even the effluent which is discarded also harms the animals, especially aquatic and marine life. This polluted water, which is being used for irrigation, affects the crop and cotton fibre yield in that area.

6.2 Cultivation Cost Reduction

Cotton production's cost–benefit ratio has been reduced thanks to advances in manufacturing technology. Farmers in Maharashtra, Punjab, and Andhra Pradesh have reportedly committed suicide as outcome of rising production costs and losses in cotton cultivation. Whereas in the case of organic farming, it creates rural jobs and uses farm resources to reduce costs.

6.3 Insecticide Resistance Management

Due to the indiscriminating application of dangerous insecticides to control cotton pests, insect resistance to the insecticide has increased, necessitating the use of additional sprays, creating a vicious cycle that increases the cost of farming. Organic cotton cultivation will contribute to reversing this trend. Evidence of a poorer choice of pest multiplication rate in organically grown cotton is the motivating factor for the progress of this protocol. Organic farming is guided by the principle of working with nature as opposed to against it. Organic cotton cultivators raise crops using growing systems that are biological rather than chemical in nature [3].

7 Advantages of Organic Cotton Fibre

The conventional way of growing cotton has serious side effects. The method followed is very dangerous to the living bodies as well as it will hamper the environment too. Around the globe, nearly 35% of total usage of pesticides are used in cotton farming only [22]. The usage of organic cotton is excellent solution to replace conventional cotton. It has got many advantages over conventional cotton. It is useful to develop the local varieties as well as make use of resources available nearby which will help to have good environmental conditions. The farming of organic cotton also helps the farmer to earn more money. Farming of organic cotton also helps to minimise the impact of hazardous chemicals on living bodies. During the cultivation there will not use of excess quantity of pesticides and fertilisers for the growth of cotton. Not only during cultivation but the conversion of fibre to finished product various harmful banned chemicals are used in conventional cotton processing. But in case of organic cotton usage of such banned chemicals will not be there to have a completely organic finished product [23].

8 Roadmap for Organic Cotton

Cotton processing must be sustainable because it is a source of income for many people. Diverse geographic regions necessitate a variety of long-term adaptive solutions to fulfil local environmental needs. Climate, natural resources, water and chemical inputs and outputs, the efficacy of agricultural assets and operations, and local pests all have different aspects and necessitate different responses.

Selecting the cotton variety to be seeded is one of the inputs that is included under technology, agriculture, and production. Only by including humans in the supply chain can sustainability be realised. Contributions from many walks of life are needed, both individual and collective. They comprehend the roles of the farmer, the manufacturer, the retailer, the customer, and the government in both domestic and international processes. Plant biotechnology is continuously evolving, and all stakeholders should be kept informed about new developments. To encourage effective land use and mechanisation of agricultural operations, biotechnology must be used to optimise plant architecture.

Climate change, unpredictable rainfall, protracted natural conditions, saline soil conditions, and other pressures all affect cotton production. Genes are responsible for drought, heat, and cold resistance, as well as improved adaptation to a wide range of environmental circumstances [24]. The cotton-growing environment necessitates the most stringent water management, soil improvement, and crop protection methods, all without jeopardising the environment's health. Pre-harvest methods, for example, involve explicit management strategies that help cotton grow and develop.

Quiet RNA interference technology is used to create seedlings that are free of gossypol. Cotton thus becomes a source of oil and protein that does not require detoxification prior to use [25]. Minor ingredients are recovered using selective separation technologies, and cotton oil is converted into oil beneficial to responders using specific enzymes. Without social consideration, sustainability cannot be achieved, and the International Labour Organization's principles must be applied to community development and the improvement of the lives of agricultural and plantation workers. Post-harvest technologies include the modernisation of the ginning industry to make it even more energy-efficient, costing based on quality, the use of environmentally friendly chemical processing, clean technologies, natural dyeing of fibres and coloured cotton to achieve zero discharge in effluent treatment, and so on.

The strategy also emphasises the importance of stakeholders at all levels of the supply chain, as product demand will only rise if sustainability issues are addressed. Within the context of sustainable cotton processing and development, everyone works individually and together to achieve sustainability. Consumers' reluctance to pay more for sustainable items is causing problems. When a concerted effort is made and only environmentally friendly products are made available to clients, these aspects will become cost effective. Governments all around the world are focusing on the qualities of sustainable development in all products. Responses to these concerns must be provided on a global scale. Share study findings from the lab with commercial markets, thereby shifting people's perspectives and attitudes towards the environment.

9 Economic Viability of Organic Cotton

Many conventional farmers have experienced a decrease in cotton yields in recent decades, despite the increased use of chemical fertilisers and pesticides. Organic farming could be a solution to this problem if it improves the ecological and socio-economic sustainability of cotton production. WWF Switzerland and Swiss Agency for Development and Co-operation has started a Research Institute of Organic Agriculture to conduct the detailed study in India, regarding the investigation of economic viability of organic cotton cultivation and its effect on farmers. Some of these points are as follows.

9.1 Control of Pests

The research group discovered that organic cotton had variable production costs that were 13–20% lower. This is primarily due to 40% lower input costs. Thus, the need for loans is much lower in organic farms. In case of organic farming more time is required for weeding rather on pest control. Even though advances in organic cultivation methods have enabled cotton yields to be comparable to conventional methods.

9.2 The Most Difficult Challenge

The most significant barrier to switching from conventional to organic cotton farming is drop in yield of cotton fibre at initial starting years, resulting in drop of profit for initial 1–3 years. To improve the routine of organic cotton agricultural systems even further, efforts must be made to develop production methods and improve marketing options, this is especially true for crops grown in alternation with cotton [3].

10 GOTS—Global Organic Textile Standards

GOTS was developed by leading standard organisations existing standards in the field of eco-textile processing and defining globally recognised standards that ensure textiles' organic status, from raw material harvesting to environmentally responsible manufacturing and labelling, to provide a reliable guarantee to the end user.

GOTS is a processing standard for organic fibres that applies to the full supply chain, from raw material (fibre) harvesting through the finished (final) product sold to the consumer. Many producers follow it around the world, and as a result, GOTS-certified products are generally accepted on the international market. Both social and environmental compliance are required for GOTS certification. Environmental and toxicological criteria must be met by dyestuffs, chemicals, and reagents used in dyeing and finishing. Wet processing units require an effluent (wastewater) treatment plant.

Despite best efforts, a cotton fibre that is free of hazardous instances cannot be labelled as organic. At most, it can be referred to as eco-friendly fibre, and the textile products made from it can be referred to as eco-friendly textiles. As a result, having a protocol in place to label a textile as organic textiles is must. Furthermore, the portfolio developed should address the product's life cycle analysis LCA as well as international acceptance. As a result of the manufacturing of organic textiles or green textiles with an eco-design reference, the "GOTS—Global Organic Textile Standard" arose.

The main goal of GOTS is to provide textile consumers with credible assurance, the standards seek to define requirements for ensuring the organic status of textiles, from raw material harvesting to socially and environmentally responsible manufacturing and labelling. The production, processing, packing, labelling, exporting, and administration of all-natural fibres are all covered by the GOTS system. The aspects covered under GOTS are technical and system aspects, environmental aspects, and social aspects.

10.1 Certification Process of GOTS

To assure the GOTS label's integrity, the certification process involves an on-site inspection and monitoring system. A GOTS-approved certifying organisation can apply for GOTS certification on behalf of any textile manufacturing, processing, or trade company. certifying bodies There are eighteen GOTS-accredited certifying bodies around the world. The certifying bodies are independent, special accredited bodies that can certify in four different scopes:

- Various mechanical textile manufacturing operations and their end products.
- Textile Wet processing and finishing operations, also the end products of the same
- Trading progressions and trading-related products
- The distribution of positive lists of chemical inputs to the chemical industry (dyes and auxiliary agents) [26].

11 Organic Production Constraints

Organic cotton cultivation has not spread to other nations for a variety of reasons. Around 19 countries began cultivating organic cotton in the 1990s. Many of these countries, however, have already discontinued producing organic cotton due to a lack of interest amongst farmers and customers in organic cotton products. Insecticides must be completely eliminated from the organic cotton cultivation process since they are dangerous to use, have a wide range of long-term impacts on pests, and are destructive to the environment.

Furthermore, due to the exorbitant expense of insecticides, many countries have abandoned the cotton producing technique. The Organic Trade Association made a determined effort in 2002/03 to uncover problems with organic cotton cultivation in the United States. The Organic Fibre Council of the Organic Trade Association undertook a study of all organic cotton growers in the United States. The ICAC polled participants on two topics:

1. Cost of producing of conventional cotton and organic cotton and
2. Organic cotton carries a price premium.

According to a 2002 survey conducted by the Organic Fibre Council, organic cotton growers' main concerns are weed control in the absence of herbicides, as well as defoliation and insect control. Some farmers were also concerned about seed treatment, which is forbidden under organic certification. In handpicked cotton, defoliation is a severe problem that does not exist in the United States, where even organic cotton is picked by machines.

Acceptable measures to promote appropriate farming methods must be put in place if organic cotton output is to increase. Some of the comments will be specific to cotton, but the great majority will apply to organic farming in general.

12   World Organic Cotton Production

If synthetic fertilisers and insecticides are not employed, varieties that perform well under ideal conditions may be unable to maintain yield levels. Organic cotton breeding stock must be organically tested. Single plants, progeny rows, or bulks should be grown under organic circumstances on a continuous basis to select for organic output.

Over the last three decades, developing varieties that are lower in stature, mature sooner, and respond well to heavy fertiliser doses has been a priority. It has been attempted to transfer effective fruiting sites closer to the main stem and on lower branches. High fertiliser responses and fruiting position shifts are favourable features when high inputs are used, but they may not be desirable when synthetic fertilisers are not used. Organic circumstances have been used to cultivate varieties that demand a lot of fertilisers. As a result, such cultivars are likely to have had more yield losses than expected, discouraging farmers from continuing to grow organically. Variety creation for organic production settings is necessary; they may not be as high yielding as conventional varieties, but they must meet certain requirements [27].

13   Organic Fibre Properties

We observe that the chunk of literature is available on the structure and properties of conventional cotton fibres [28–37], however, literature on similar studies on the organic cotton is scant. Nevertheless, Babu et al. [38], Chae et al. [39], Günaydina et al. [40] and Inoue et al. [41] have done remarkable study in the field of characterisation of organic cotton. Babu et al. [38] conducted ground-breaking research into the characterisation of conventional and organic cotton fibres in order to better comprehend the differences and reported that organic and conventional cotton fibres are comparable in appearance and have similar morphological characteristics, crystallinity, and surface chemical composition. They also found that except for Ca, organic cotton fibres have a higher concentration of metals than conventional cotton fibres. When compared to ordinary cotton fibres, organic cotton fibres have a very high percentage of Fe, Al, and Mg, and higher proportion of representative groups and slightly lower fraction of Iβ cellulose [38].

The majority of research done on the structure and characteristics of white and naturally coloured organic cotton (NaCOC) fibres reported that brown cotton has a morphology that is quite similar to white cotton, however green cotton is distinct because it contains suberin [42–49]. Chae et al. [39] found significant difference in low stress mechanical properties, when studied on KES-FB system, between organic cotton fabrics made from coyote brown and green NaCOC fibres. According to Günaydina et al. [40], the mechanical and air permeability properties of white and naturally coloured organic Turkish cotton textile fabrics employed in this study were satisfactory, indicating that the samples could be used for clothing garments.

A study on the mechanical characteristics of cotton yarns produced from organic and non-organic cotton fibres was conducted by Inoue et al. [41] who discovered that organic cotton fibres had higher breaking point strains and stresses than non-organic cotton fibres. These findings indicate that organic cotton's essential characteristics contribute to the quality of finished cotton goods. They also studied at cotton fibres grown in various regions and discovered that the moduli of cotton fibres grown in the northern and southern hemispheres were remarkably comparable, as were the yarn characteristics of yarns manufactured from cotton fibres grown in both areas. They also discovered that yarn made up of a combination of cottons exhibited higher strain, torsional stiffness, and hysteresis in the northern and southern hemispheres.

14 Ginning

Organic cotton must be ginned separately from conventional cotton, which means the gin must be completely cleaned before ginning the organic cotton, and organic cotton must be stored in a separate section of the ginning area unless a gin is specifically designated for organic production. By-products must be transferred and segregated to reduce the transmission of soil or plant-borne diseases from conventional and organically cultivated cotton. Only the first bale of cotton ginned is designated "conventional cotton" if any conventional lint remains in the ginning system after cleaning the gin [50].

15 Spinning of Organic Cotton Yarn

Organic cotton spinning is equivalent to conventional cotton spinning in terms of all spinning activities from start to finish, especially when no processing oils are used. Processing oils are frequently unnecessary due to the natural waxes on the outermost part of cotton. Ring spinning occasionally uses biodegradable oils, however these oils are removed before dyeing by scouring and bleaching the fabric. Only due precaution is taken whilst handling the material. After receiving the fibres from the farmer, the fibres are cleaned in ginning. After the ginning, the fibre bale is packaged in environmentally friendly packaging material. Because this fibre is regarded as "green fibre," green-coloured packaging material is used to facilitate bale identification. The material carriers were used for handling organic fibres and the intermediate step product must be green in colour, after reaching the bale in spinning. So that it can be identified very easily and due care can be taken to avoid the contamination of fibres and intermittent step product with a conventional cotton fibre product [51].

16 Fabric Manufacturing

As long as natural starch sizes are utilised, conventional and organic cotton weaving, knitting, and non-woven fabric manufacturing should be equivalent. Weaving preparation is a crucial step in the manufacture of woven fabric, especially for high-speed weaving, and it will continue to be so in the future, dependent on woven fabric technology. For increased strength and abrasion resistance, the yarn must be coated with starch or another "size" before weaving. Sizing is the only activity in weaving that could cause environmental issues. When starch is used for size, enzymes are typically utilised to remove it during the desizing process. The starch size can't be recycled, so it has to be handled in a way that doesn't contaminate the environment or pollute the water. Starch is broken down and rinsed away during the enzymatic process. PVA size, which is often used in traditional sizing, can be removed and recovered through scouring. As a result, it is regarded to be less harmful to the environment than "natural" starch.

Knitted textiles do not need to be sized, however knitting oils are popular. They're scoured out as part of the standard scouring process before being dyed and finished. Natural or synthetic sizes are not required in the production of non-woven fabrics, although numerous additives that must be deemed natural can be employed [52].

17 Wet Processing of Organic Cotton Fabric

From sunrise to sunset, hundreds of chemical products are used in our daily lives. These chemical products may include hygiene products, office or school goods, or our lifestyle products.

Animals and humans have been living in harmony since the beginning of time until an inventive human discovered crude oil. By carrying out the distillation of this crude oil, human beings can carry out the synthesis of chemicals that are not present in nature. More than 150 years ago, the first synthetic dye was developed by W. H. Perkin in the UK. This invention made a considerable change in the advancement of organic chemistry-based products.

The use of organic chemicals made the human lifestyle more luxurious by the use of new regenerated as well as synthetic fibres, pharmaceutical chemicals, plastics, pesticides for plants, etc. But due to these developments, human life is made luxurious, but it adversely affects human health and also the environment. This created many problems related to the environment, such as global warming, greenhouse gases, ozone depletion, toxic effluent generation, etc. Also, studies related to human health show that it is creating problems such as allergies, cancer, immunity, and genetic disorders, and reproduction. For centuries, the human population was practically constant in the world. It has exploded 10 times in 300 years. Also, chemical production has grown 400 times in 75 years.

Europe and European countries have been at the forefront of the development and production of chemicals since the early days. It's also worth mentioning that five of the world's chemical giants are situated either on the western European river, the Rhine, or one of its tributaries. On its way from the Swiss Alps to the North Sea, it travels through Switzerland, Germany, and Holland. In the beginning, large European chemical companies discharged chemicals into the river. It was no wonder that western Europe was the first to realise the negative consequences of the chemicals it was producing.

Persistent organic pollutants have been discovered in human blood samples and ordinary house dust. Residues of chemicals such as flame retardants, alkylphenol ethoxylates, short-chain para polychlorinated biphenyls, organotin, and phenolates have been found. It's also worth mentioning that over 3000 chemicals were listed in Europe in 1981 (of which were in production). About 4000 more chemicals have been registered after 1981. Out of these 34,000 high production volumes, only 3% are estimated to have full data about their harmful effects. Some have partial data, and an estimated 15% of chemicals have no data at all. This shows that the human being is today using chemicals daily where their ill-efforts are not known. It is quite obvious that if we are to continue this practice, there will be problems. Recognising this, since the 1970s, different conventions and conferences have tried to address the problem.

The textile wet processing industry is considered amongst the most popular polluting industries. Hence, it has stood out enough to be noticed by all. Different practices were done by governments and individual companies to limit contamination. European nations have passed laws limiting the utilisation of specific synthetics. Eco Labels came up and promoted their models for "green" items. Governments likewise have come out with their ecological names, like EU Flower in the EU and Eco-Mark in India. Individual organisations considered this to be a chance for a better consumer interface and have thusly thought of their lists. These lists are called restricted substances lists, or RSLs.

RSLs are a list of chemicals that are restricted for use either by legislation or voluntarily by a retailer. These apply to all consumer products that the retailer markets and they are intended to safeguard the consumer, the production centres and importantly, the environment. Furthermore, legislation has ensured that the usage of some hazardous chemicals is restricted. Retailers have found these chemicals to have a good consumer connection to the tool. Also, the Green Movement is quite strong in developed countries. An environmentally friendly product is rated higher than its non-environmental counterpart. Goods that do not match requirements are immediately exposed to the common public, which cause a bad impact on the product, resulting in a loss in terms of money.

Therefore, retailers are very careful and their RSLs are usually much stricter than legal requirements. Their policies towards environmental parameters reflect their commitment to this cause. Processors, manufacturers, and suppliers need to comprehend the requirements, comply with all legal demands, and always be aware of any changes in the same. All requirements must be communicated through the supply chain and set up a quality system that will fulfil all the requirements for quality. Testing at an accredited laboratory helps in such cases. If a problem arises, consulting with the suppliers and laboratory will also be beneficial. Often, they may have information that will be useful to all [53].

Pretreatments like as singeing, desizing, scouring, bleaching, and mercerisation are commonly used to prepare grey fabric for dyeing, printing, and finishing. These methods remove non-cellulosic contaminants from the cellulose whilst enhancing its affinity for dyes and finishes [54]. Natural waxes and contaminants are commonly removed from cloth by scouring with alkali. Bleaching is done using hydrogen peroxide rather than optical whiteners, and the temperature must be over 60 °C, which consumes more energy. Bleaching is the process of removing contaminants from a fabric's colour and turning it white. To strengthen the fabric's dimensional stability, mercerisation is utilised, which includes treating it with a strong aqueous sodium hydroxide solution. This technique also improves the lustre, strength, dye absorption, and chemical reactivity of cotton fabrics. Fabrics are mercerised to improve their appearance and strength [55]. Chemicals that are allowed in the preparation, dyeing, printing, and finishing of organic cotton textiles that comply with various voluntary organic criteria for wet finishing are listed below.

Synthetic chemicals are allowed during the processing of organic cotton:

Sodium silicate, Sodium sulphate, Aluminium silicate, Aluminium sulphate, Biodegradable soaps, Surfactants, Hydrogen peroxide, Fatty acids and their esters, Oxalic acid, Ozone, Polyethylene, Potassium hydroxide, and Sodium hydroxide.

Non-synthetic chemicals are allowed during the processing of organic cotton:

Natural dyes, Enzymes, Sodium carbonate, Sodium chloride, Acetic acid, Tannic acid, Tartaric acid, Chelating agents, Pigment-natural only, Citric acid, Clay-based scours, Iron Copper, Tin, Potassium acid tartrate, and Mined minerals.

The fabric is dyed after it has been prepared. All dyes used in organic processing should adhere to the norms of ETAD. Organic cotton dyeing should be free of heavy metals like chromium, formaldehyde, and other hazardous chemicals. Cotton textiles are dyed with a variety of dyes such as direct, reactive, vat, sulphur, azoic, and pigments [56]. Natural dyes are preferred over synthetic dyes when it comes to organic cotton. However, the availability of natural dyes is very limited, and mordants based on heavy metals are generally required for fixation of natural dyes, increasing pollution levels. Furthermore, natural dyes have poor fastness properties to a variety of agencies, including rubbing, washing, and lightfastness. As a result, the next solution is to use low-impact dyes. Some reactive azo dyes can be used to colour organic cotton. Because reductive cleavage during degradation can liberate one or more of 22 aromatic amines, azo dyes are forbidden for use in organic processing. Most EU countries, and the United States and the majority of other countries, have banned the usage of these dyes. In terms of pigments, the usage of natural pigments is recommended for organic cotton. Water or natural oil-based printing methods are permitted for organic cotton printing. Heavy metals are not permitted in the discharge printing style.

After dyeing or printing, organic fabrics can be treated with chemicals such as antimicrobials, flame retardants, anticrease/durable press agents, water repellent agents, and so on to make them more functional. Chemicals that can emit or contain formaldehyde are prohibited. Stone washing and other environmentally hazardous textile finishing procedures that require auxiliary agents and a lot of water are not allowed in organic processing. Ecolabel-certified products are meant to be market-driven tools that help consumers make informed decisions. However, because life cycle analysis is difficult, ecolabels may fail to identify products that are better for the environment than non-labelled products. Dyeing should be avoided in order to avoid pollution. However, if organic cotton textiles are to have the same aesthetics as conventional textiles, most dyeing and finishing procedures must be used; however, these can be replaced with more ecologically friendly operations and chemicals. Because some operations and treatments are unavoidable, it's vital to employ energy-efficient processing that uses fewer harmful chemicals and uses less water. Ecolabel-certified products are meant to be market-driven tools that help consumers make informed decisions. However, because to the complexities of life cycle analysis, ecolabels may fail to identify products that are better for the environment than those that are not labelled [57].

18   Application of Enzymes in Organic Cotton Processing

Cotton is a fantastically multipurpose and worldwide important fibre due to its inherent properties such as softness, breathability, absorbency, performance, comfort, and durability. Organic cotton is gaining popularity in the manufacture of the finest fabrics due to its zero toxicity and 100% "natural origin." Organic cotton is cotton that has been grown and harvested without making use of synthetic chemical inputs and then processed using "Green Technology" (eco-friendly chemicals and processes).

Cotton is one of the world's most heavily sprayed field crops, accounting for more than 10% of insecticide use and around 23% of herbicide use. As a result, organic clothing manufacturers must ensure that toxic substances, such as petroleum scouring agents, formaldehyde in resins which are used for the anti-crease or wrinkle-free finish, chlorine-containing bleaching agents, etc., are avoided at each processing step. Biodegradable oil for spinning, potato starch for sizing, H_2O_2 for bleaching, natural dyes, low salt dyes for dyeing and printing, natural thickeners for printing organic cotton fabric are all-natural alternatives. Finally, the use of enzymatic treatments as a partial substitute for chemical wet processing is urgently needed. These eco-friendly alternatives are used to minimise and remove toxic effects.

As more fibre preparation, pretreatments, and value-added finishing processes shift to biotreatment, enzymes are likely to have an even bigger impact on effluent quality. Biotreatment could also be utilised to reduce BOD and remove and/or decolourise colours from textile waste streams. Enzymes are highly effective catalysts even under mild conditions; thus, they don't require the huge energy input that chemical processes do. Biotechnology allows for the development of cleaner and more energy-efficient processes, the manufacture of higher-quality products, and the cleansing of effluents in terms of the environment.

18.1 Enzyme Source

The majority of enzymes used in cellulose pretreatments are derived from microorganisms via solid-state or submerged fermentation. Microbial cultures are obtained through random selection, screening, and assay. Natural samples, such as soil or compost material discovered near high concentrations of potential substrates, are employed as a source of cultures. Further research and standardisation of the production protocol in the chosen culture are planned. This strain is known as a Wild Type Strain. Further strain improvement is accomplished through the use of advanced genetic engineering techniques to obtain higher yields or to modify the Enzyme protein to the required selectivity of its mode of action.

18.2 Genetically Modified Organism

A "genetically modified organism" (plant, animal, or microbial) that has been transformed by genetic modification to overproduce the desired end result is referred to as a GMO (enzyme).

18.3 Use of GMOs to Produce Enzymes

When used to increase enzyme production efficiency, resulting in higher yields; increase enzyme purity by reducing or eliminating side activities; and improve the function of specific enzyme proteins, such as by increasing the temperature range over which an enzyme is active, modern biotechnology has proven to be a safe and valuable tool. As a result, better goods are created that are more efficient, often at a lower cost, and have a smaller environmental effect.

18.4 GOTS—Prohibited/Restricted Inputs in All Production Stages During Organic Cotton Processing

GOTS prohibits GMOs and their derivatives which include enzymes those are sourced from GMOs. When considering the benefits of using enzymes in the processing of organic cotton, one might conclude that biotreatment is the best option. However, the majority of enzymes on the market are derived from GMOs.

It is well understood that producing enzymes from wild strains is economically unfeasible. The primary reasons are as follows:

1. The entire cellulase is produced by wild strains. This means that, despite its high activity, the enzyme protein produced may not produce the desired results on cotton. Endo glucanases (EG), Exo glucanases, cellobiohydrolases, beta glucosidases, and other enzymes make up the cellulase enzyme. We have endoglucanases ranging from EGI to EGVIII. Because of the wide range of proteins produced by wild strains, the enzymes produced are ineffective.
2. The possible yields from wild strains are also low, making the costs prohibitive.
3. Because enzyme protein production is not consistent, even if activity remains largely constant, application results vary.
4. Because wild strains are not hardy, they are easily influenced by changes in temperature, time, and pH.
5. These strains are also easily contaminated, and they do not produce thermally or pH-stable enzyme proteins.
6. Making wild strains robust without genetic modification is a difficult task, and contamination/strain mutation becomes a regular problem for manufacturers.

18.5 Alternatives to GMO Enzymes

The followings are the alternatives to GMO Enzymes.

1. Wild-type strains
2. Random Mutagenesis—Chemical induced or UV techniques
3. Site-Directed Mutagenesis
4. Transformation—removal of the cell wall, making the protoplast and exchanging genetic material without any barrier where the protoplasts fuse.
5. Transduction—The virus is used as a vector to carry genes from the donor to the recipient. Like in bio-insulin production where restriction enzymes are used.

In addition to production from wild-type strains, GOTS must consider and approve other methods of strain modification. All of the above methods of production will be significantly more expensive, necessitating the industry's acceptance of/consideration of these costs when calculating the cost of producing organic cotton. A general estimate of the cost escalation is 10–20 times the current cost of enzymes derived from a GMO. Finally, the enzyme manufacturing industry must begin a parallel activity to produce Non-GMO enzymes with assurances from organic cotton users, particularly the Big Brands that have begun to use organic cotton, that the cost of Non-GMO enzymes in processing is understood and taken into account when designing the final product [27].

19   Garment Manufacturing

The apparel business has recently been focused on generating organic raw materials that are both environmentally and socially beneficial. Rapid fashion's negative implications are becoming increasingly apparent to consumers. Organic cotton apparel is advantageous to both customers and producers. Children's and babywear are a significant sector of the fashion industry. Organic cotton clothing has long been seen to be the finest option for new-borns. It has become a popular choice because of its hypoallergenic and pleasant properties. In addition to clothing, organic cotton is used in infant diapers and blankets. Organic cotton items such as night suits, jeans, t-shirts, beddings, towels, mattresses, pillows, socks, sheets, bathrobes, bags, underwear, and so on are popular amongst consumers of all ages.

19.1   Garment Manufacturing Process Flow Chart

From order receipt to shipment, a whole garment must go through many stages like pattern making, grading, marking, fabric spreading, cutting, bundling, stitching, pressing, folding, finishing, and packing. Both organic and conventional cotton fabrics undergo the same process.

To quickly execute an order during garment production, a process flow chart is required. A process flow chart also aids in comprehending the garment production technique, which explains how raw materials are transformed into wearable textiles. The following is a typical flow chart for garment manufacturing:

(a)   *Design*

   The buyer is in charge of the design. After placing an order, the buyer sends the merchandiser the technical sheet and illustrations for the order. This procedure can be carried out manually or with the aid of a computer.

(b)   *Sample Pattern Making*

   Each clothing style's pattern should be produced by following the technical sheet and illustrations. It's done both manually and with the help of a computer.

(c)   *Fit Sample Creation*

   The main goal of creating a fit sample is to follow the specific instructions for that garment's style. It is then submitted to the buyer to be corrected. This process is usually done manually.

(d)   *Production Pattern Making*

   Allowance is supplied here with net dimension for bulk production. Production pattern making can be done manually or with the aid of a computer.

(e)   *Grading*

During the purchase confirmation process, the buyer specifies the order's size ratio. As a result, the buyer's instructions should be followed whilst grading the order. Grading can be done by hand or by computer.

(f)   *Marker Making*

A marker is a very flat piece of paper that includes all of the components of a certain garment. It must be required to make the cutting procedure easier. The procedure of producing markers can be done manually or with the help of a computer.

(g)   *Spreading*

To effectively cut the fabric, it is spread inlay form.

(h)   *Cutting*

Fabrics must be cut according to the garment markers provided. The fabric cutting procedure can be done using different tools depending on the size of the lay and according to the pattern. Cutting process generally produces a substantial amount of fabric wastage. The cost of fabric waste for organic cotton fabric is normally higher than that for the conventional cotton fabric. This can be attributed to the higher cost of the raw material of the organic cotton fabric.

(i)   *Cut Parts Sorting or Bundling*

Cutting parts must be sorted or bundled so that they may be readily sent to the next step.

(j)   *Sewing*

This is where all the pieces of a garment are sewn together to form a full garment. Sewing is done on different sewing machines according to the stitch type required.

(k)   *Garment Inspection*

Once the stitching is finished, the garments should be inspected to ensure that they are free of flaws. The manual approach is used to inspect the garments.

(l)   *Ironing and Finishing*

Garments are steam-treated here, and any needed finishing is done here as well. The mechanical approach is used to complete this operation.

(m)   *Final Inspection*

At this point, the entire garment is evaluated to ensure that it meets the buyer's specifications. The manual approach is used for the final inspection.

(n)   *Packing*

Complete clothes are shipped using the packing details specified by the customer.

(o)   *Cartooning*

To reduce fabric damage, all garments must be cartooned according to the purchasers' instructions.

(p)   *Shipment*

Once all of the essential steps have been completed, cartoons are shipped.

If organic cotton fabric and the conventional cotton fabrics are stored or processed in the same area, there are chances that the free formaldehyde used for the conventional cotton fabric getting transferred to the organic cotton fabric products. To avoid such contamination, organic garments should be packed separately, especially in the paper bags.

19.2   Waste Generation in the Apparel Industry: An Overview

When a tangible object is deemed unwanted or unneeded, it is referred to as waste. Textile waste can be divided into two types. Pre-consumer textile waste is created during the production of fibre, yarn, fabric, and garments. Consumers generate post-consumer textile waste, which includes clothing and home fabrics. Every garment manufacturing company creates similar categories of waste, although the amounts of waste generated may vary. Fabric waste, paper garbage, polyethylene waste, chemical waste, and other accessory waste are amongst the most prevalent forms of waste. Fabric waste is generated at a higher rate than other forms of trash in every garment industry.

Sample yardage waste, swatch sample waste, sampling garment waste, left over on the fabric roll, cutting waste, cut-and-sew waste, and unsold finished garment waste are all examples of organic cotton fabric waste generated in the textile manufacturing business. Factors influencing fabric waste are marker planning, style and design requirements of garment, marker efficiency, marking loss, spreading loss, etc. When fabric waste is a factor in the fashion production process—at the designing, pattern making, and cutting phases—the planned garment is already inviolable; it cannot be modified. This implies that, in order to reduce fabric waste, it must be considered earlier in the process: in fashion design and pattern cutting. What has been developed and pattern cut governs manufacturing [58].

Prior to cutting the fabric, the pattern pieces are put out on the fabric width as precisely as possible to save fabric use and expense. This may be done manually on paper, but computer software such as OptiTex, Gerber, or Lectra is increasingly being utilised to identify the most effective cutting plan. Even with the most up-to-date software or the most seasoned and talented hand marker makers, fabric waste in adult outerwear ranges from 10 to 20%. It looks to make sense to waste 10–20% of the total cloth utilised from a strictly economic standpoint. The existing industrial models make it appear as though there is no economic need to be worried about this degree of waste [59].

19.3 Zero Waste Concept

Zero waste is a collection of waste avoidance concepts that supports resource life cycle redesign such that all goods are reused. The ultimate objective is to avoid sending waste to landfills, incinerators, or the ocean. Only 9% of plastic is indeed recycled at the moment. Material will be reused until the optimal level of consumption is reached in a zero waste system. Zero Waste refers to trash avoidance rather than waste handling at the end of the pipe. It's a whole-systems strategy that strives for a major shift in the way resources move through society, with no waste as a result. The term "zero waste" refers to more than just reducing, reusing, and recycling garbage.

The components of a design's pattern are put together in Zero Waste design in the garment industry so that no fabric is wasted during the cutting process. In this design, there are no gaps between the pattern parts. This method is based on designing pattern pieces in the style of interlocking puzzle pieces, or ingeniously using scraps of fabric to make embellishments, bias tape, and other items. Zero waste processes frequently result in the creation of one-of-a-kind garments that are tailored to the kind and size of the initial fabric. Traditional clothing such as the Indian saree, Japanese kimono, and African kangas and kitenges all use zero waste design.

The following are the factors affecting zero waste designs.

(a) **Fabric in Zero Waste Fashion Design**

Fabric width, being one of the spatial dimensions from which the garment is cut, must be a design requirement from the start. In the industry, it is usual to create a popular item in a variety of fabrics and widths. This may not be achievable in zero waste fashion design since the pattern elements will be configured differently on two distinct fabric widths. However, this might be a chance to add depth to a collection through design variations: for example, a shirt designed for 140 cm-wide fabric could be converted to 110 cm-wide fabric. A new design is developed, but the previous design's common features are kept, which is a valuable technique for adding depth to collection growth. Many fabrics are prone to width variations and other flaws during weaving or knitting, so this needs to be taken into consideration. Allowances for minor variations may be required in individual clothes of the same type. The final fabric edge, for example, may be shorter or longer than the fabric's main body; either difference might have consequences for zero waste fashion design.

(b) **Pattern Design**

Square-cut and tailored are the two major kinds that emerge. Square-cut clothes are made largely of rectangular or triangular pattern forms, with straight pattern piece edges dominating. Patterns with curved pattern edges are said to as tailored. Rectangular pattern forms appear to be more prevalent in the zero waste clothes examined, especially amongst "zero waste fashion design without fashion designers." This might imply that zero waste design results in simple box and t-shaped clothing. Offsetting two identical rectangular pattern pieces against each other, on the other hand, produce forms other than the fundamental box and t-shapes. Slashes cut into a pattern piece, along with the usage of gussets, can occasionally be used to create three-dimensional forms instead of shaped seams. By merging pattern pieces into bigger ones, straight seams may be avoided. Small fabric pieces can be used as structural elements on the inside or outside of a building. Selvedge strips, for example, can be used to support armholes or necklines, and high-end clothes are sometimes offered with a tiny portion of the garment fabric (external). Fabric rectangles are appliquéd internally to strengthen the armhole and shoulder region of an eighteenth-century shift.

(c)   *Garment Design*

The term "garment design" best expresses concerns pertaining to a garment's aesthetic and technical design. Creating many styles of clothing at the same time and cutting them all from the same piece of fabric can (but is not always) decrease fabric waste. Tailors employ this method when cutting two- or three-piece suits; the jacket and trouser pieces (and occasionally the vest) are combined on a length of cloth. An extremely complex design process, as well as the fact that an equal number of each garment would need to be cut in manufacturing, regardless of whether an equal number of orders for each garment had been received, are potential drawbacks. Simultaneous assessment of the garment's technical and aesthetic components, as well as simultaneous consideration of the garment in two and three dimensions, is required for zero waste fashion design. In traditional practice, the garment design is generally resolved first by a two-dimensional drawing, then a two-dimensional pattern is cut, and finally the three-dimensional garment is made. Furthermore, a pattern cutter may traditionally examine only a portion of a garment at a time and make the pattern parts in stages. To avoid fabric waste, the complete garment must be taken into account throughout the pattern cutting process. Zero waste fashion design may provide a chance to think more carefully about which clothing designs to sample; this would necessitate tight collaboration with the sales department [5].

19.4   *Zero Waste Fashion Techniques for Organic Cotton Clothing*

There are many strategies that may be utilised to reduce fabric waste in the clothing production process by employing certain design principles. Apparels made using zero waste fashion design processes use only the quantity of fabric required for a specific piece of clothing, resulting in practically minimal fabric waste. In general, the zero waste fashion idea refers to a system in which waste is virtually eliminated at every step of the garment's design, manufacture, and use. There are several approaches to zero waste fashion [58, 59]. Some of them are mentioned below.

(a)   *Planned Chaos*

In this approach, conventional garment pattern blocks are merged to produce the fundamental shape of the garment and all of the components are united. This approach necessitates a thorough grasp of traditional pattern cutting principles, but these rules are modified in such a way that waste is minimised. Pattern cutting can be done in two ways: Subtraction cutting, Jigsaw cutting.

Jigsaw Cutting

This method saves waste by cutting little components (pockets, collars, trimmings, and so on) from a single piece of fabric and piecing all of the pieces together like a jigsaw to produce a lovely garment. This is an environmentally friendly strategy that has mostly gone unrecognised in the industry.

Subtraction Cutting

The pattern is cut to represent the garment's negative spaces rather than the exterior forms in this manner. The garment is made up of large sheets of material with oddly shaped holes that the body passes through.

(b)   *Zero-Waste Cutting Technique "Geo Cut"*

This method employs pattern pieces cut in geometrical shapes like as squares, triangles, and circles. This method has previously been used in the construction of kimono designs, for example.

(c)   *Cut and Drape*

Cutting, draping, folding, steam moulding, and machine and hand sewing are only some of the techniques used to create the garment. In this technique, a fashion designer might experiment with the way material drapes to develop new patterns.

(d)   *Reusing Scraps of Cloth and Yarns*

After the product production process is completed, this approach deals with off cuts or leftover material, and there are a variety of ways to reuse scraps.

(e)   *Pleating*

Pleated and fashioned into a garment are rectangular bits of textile waste leftovers.

(f)   *Draping Scraps of Fabric*

The designer may drape scraps of cloth on the dress form to develop new designs. Gathers, pleats, darts, and bias can be used to create the garment without cutting off and wasting fabric.

(g)   *Replace*

The use of novel No-Waste or low-waste technologies, such as large-scale 3D printing of garments, to replace traditional cut-and-sew technology can lead to zero waste production. 3D printing, or additive manufacturing, is a technology for generating three-dimensional solid objects from a computer file. A sequential layer of material is built up until the item is created in an additive process.

19.5 Trims and Accessories

Trims and accessories are crucial fashion elements which are used to improve aesthetic and functional values of the garments. The following trims and accessories are recommended to be used for organic clothing.

Thread—Instead of conventional cotton thread or synthetic cotton thread, organic cotton thread should be used for both needle and bobbin.

Buttons—Coconut or wooden buttons are recommended instead of plastic, ceramic or heavy metal buttons.

Ribbons—Vegetable-dyed ribbons should be used in the organic cotton clothing.

20 Retailing

There are several fashion brands making a remarkable difference in the field of sustainable fashion. They are using non-toxin, GMO-free, organically farmed cotton. Many of these firms utilise natural colours, source materials from ethical manufacturers, decrease waste in their supply chain, and so on. Some of these brands are People Tree, PACT, Organic Basics, Thought, Mata Traders, Beaumont Organic, Threads for thought, Amour Vert, Groceries Apparel, Patagonia, Indigenous designs, Synergy, Bibico, Lanius, etc.

The clothing on the market now can be roughly split into two types. There are two types of cotton: conventional cotton and organic cotton. Organic cotton is unquestionably better for the environment than regular cotton. Organic cotton is gentler on the soil in which it grows, as well as everything that comes into contact with it throughout its life cycle. Furthermore, two categories of stores sell organic cotton products. Two sorts of merchants sell organic cotton, each with their unique pricing approach. High-street brands that sell organic apparel or things made with organic cotton as a secondary line are classified as Category C. (cheaper). This category includes retailers like as Gap, Marks and Spencer, and H&M. Clothing made of organic cotton is a key sector for ethical product suppliers in Category E. (expensive). This category includes retailers such as People Tree, Natural Store, and Green Apple. These two types of retailers' value chains are frequently arranged differently. The merchants in Category C buy organic fibre and fabric and have it processed according to their specifications. The procedures used to create these clothing are not always organic or environmentally friendly. The finished products are labelled "made from organic cotton." These t-shirts cost around $30 on average, which was $14 more than a regular cotton t-shirt. These merchants make a lot of money. Because they are not paying the premium prices that totally organic raw materials earn for their suppliers, they may demand comparatively high prices. The fabric they buy is only approved at the agricultural level, rather than at all stages of the supply chain. Authentication is done by Category E merchants throughout the manufacturing process. Some folks are given more than others. The average price of their t-shirts is $44, which is $28 more than regular cotton and $14 more than products sold at Category C stores. These companies may charge a premium because of their great ethical reputation, although wholesale pricing and the costs of publicising their ethics are often low. It's difficult to establish how much of the price difference is due to the organic origin of the product rather than, instance, the product's design or the company's brand image. However, demand is often limited due to their high prices, and despite huge unit profit margins, overall profitability is typically low [60].

21   Conclusion

The prime necessity of organic farm production is that organic cotton needs organic seeds, which are derived from organic plants. To maintain the soil fertility and nitrogen level in the soil, natural practises such as natural fertilisers, ash, compost, and cow dung are used for the cultivation of organic cotton. Pest management processes are not difficult without pesticides being used, and there are also fewer pets in organic cotton than in traditional cotton farming. As synthetic fertilisers, pesticides, and insecticides are eliminated and organic cotton is grown using sustainable farming practices, which results in good ground water, surface water, air, and soil quality. To manufacture organic textiles, certified organic cotton should be processed organically. The Organic Trade Association established voluntary organic standards for fibre processing. They go over everything from organic fibre storage in the gin yard or hayloft to spinning, weaving, textile wet processing, quality control, and labelling, as well as the chemicals that can be employed. But when it enters the processing, the complete organic structure of cotton is lost. The reason behind this is that when it comes to the colouration process, the dyes used are not organic at all, as the organic dyes are not available in bulk and the cost is exorbitant. The processing units have to be certified by authentic certification agencies registered and known to the rest of the world since the ultimate organic cotton product is to be used by people all over the globe.

In the case of the garment manufacturing process, from order receipt to shipment, a whole garment must go through many stages like pattern making, grading, marking, fabric spreading, cutting, bundling, stitching, pressing, folding, finishing, and packing. Both organic and conventional cotton fabrics undergo the same process. The cost of fabric waste for organic cotton fabric is normally higher than that for conventional cotton fabric. If organic cotton fabric and conventional cotton fabrics are stored or processed in the same area, there are chances that the free formaldehyde used for the conventional cotton fabric gets transferred to the organic cotton fabric products. Fabric waste, paper garbage, polyethylene waste, chemical waste, and other accessory waste are amongst the most prevalent forms of waste. Sample yardage waste, swatch sample waste, sampling garment waste, left over on the fabric roll, cutting waste, cut-and-sew waste, and unsold finished garment waste are all examples of organic cotton fabric waste generated in the textile manufacturing business. Factors influencing fabric waste are marker planning, style and design requirements of the garment, marker efficiency, marking loss, spreading loss, etc. When fabric waste is a factor in the fashion production process—in the designing, pattern making, and cutting phases—the planned garment is already inviolable; it cannot be modified. This implies that, in order to reduce fabric waste, it must be considered earlier in the process: in fashion design and pattern cutting. What has been developed and pattern cut governs manufacturing. The components of a design's pattern are put together in a Zero Waste design in the garment industry so that no fabric is wasted during the cutting process. This method is based on designing pattern pieces in the style of interlocking puzzle pieces, or ingeniously using scraps of fabric to make embellishments, bias tape, and other items. Zero waste processes frequently result in the creation of one-of-a-kind garments that are tailored to the type and size of the initial fabric.

References

1. Mor O (2008) Organic cotton farming—vision & opportunities. In: Organic textiles farming to finishing. The Textile Association (India)—Mumbai Unit, pp 5–7
2. Parchure M (2008) Perception of organic cotton farming. In: Organic textiles farming to finishing. The Textile Association (India)—Mumbai Unit, pp 1–2
3. Shah RR, Mota JV, Menkudale V (2008) Future of sustainable textile. In: Organic textiles farming to finishing. The Textile Association (India)—Mumbai Unit, pp 81–89
4. Textile Market Size|Industry Analysis Report, 2021–2028. https://www.grandviewresearch.com/industry-analysis/textile-market. Accessed 8 Oct 2021
5. Textile Exchanges (2020) Preferred fiber & materials market report 2020 Welcome to the 2020 Preferred Fiber & Materials Market Report, p 103
6. Home—CottonToday. https://cottontoday.cottoninc.com/. Accessed 13 Oct 2021
7. Cotton and Sustainability Frequently Asked Questions. https://www.cottoncampus.org/cotton-environmentally-Friendly-Sustainability/. Accessed 13 Oct 2021
8. Why Choose Organic? Hae Now. http://www.haenow.com/cart/whyorganic.php. Accessed 13 Oct 2021
9. Sustainable solutions in the textile industry. https://www.lead-innovation.com/english-blog/sustainable-solutions-in-the-textile-industry. Accessed 13 Oct 2021
10. Muthu SS (2017) Sustainability in the textile industry. Springer Singapore
11. Organic Cotton|Fashion & Textiles|Soil Association. https://www.soilassociation.org/take-action/organic-living/fashion-textiles/organic-cotton/. Accessed 13 Oct 2021
12. Kuepper G, Gegner L (2004) Organic crop production overview: fundamentals of sustainable agriculture. Spring 28
13. Pick S, Givens H. (2004) Organic cotton survey—2003 US organic cotton production & the impact of the national organic program on organic cotton farming. In: Org Trade Assoc http://www.ota.com/2004_cotton_survey.html. Accessed 13 Oct 2021
14. Yafa SH (2005) Big cotton : how a humble fiber created fortunes, wrecked civilizations, and put America on the map. Viking Penguin
15. Organic cotton—an overview—Fibre2Fashion. https://www.fibre2fashion.com/industry-article/1584/organic-cotton-an-overview. Accessed 13 Oct 2021
16. Myers D, Stolton S (1999) Organic cotton : from field to final product. Intermediate Technology
17. McWhorter CG, Abernathy JR (1992) Weeds of cotton : characterization and control. The Cotton Foundation, Memphis

18. King E, Phillips J, Coleman R (1996) Cotton insects and mites: characterization and management
19. Swezey SL, Goldman PH (1999) Organic cotton in California: technical aspects of production. In: Myers D, Stolton S (eds) Organic cotton—from field to final product. Intermediate Technology Publications, London, pp 125–132
20. Anthony WS, Mayfield WD (1994) Cotton Ginners handbook. US Dept. of Agriculture, Washington, DC
21. Tripathi KS (2005) Textile processing of the future: the state of the art. In: 4th international conference on organic textiles—INTERCOT. Organic Trade Association, PO Box 547, Greenfield, MA 01302
22. USDA—National Agricultural Statistics Service Homepage. https://www.nass.usda.gov/. Accessed 13 Oct 2021
23. Sanfilippo D, Pesticide Action Network (Group). United Kingdom. (2007) My sustainable t-shirt : a guide to organic, fair trade, and other eco standards and labels for cotton textiles
24. Saha S (2010) Role of biotechnology in sustainable development. Int J Res Ayurveda Pharm 1:43–46
25. Sunilkumar G, Campbell LM, Puckhaber L et al (2006) Engineering cottonseed for use in human nutrition by tissue-specific reduction of toxic gossypol. Proc Natl Acad Sci 103:18054–18059. https://doi.org/10.1073/PNAS.0605389103
26. Nagaraja TS (2008) Control IMO Institute for Marketcology. In: Organic textiles farming to finishing. The Textile Association (India)—Mumbai Unit, pp 59–80
27. Menezes EW (2008) Enzymes in organic cotton. In: Organic textiles farming to finishing. The Textile Association (India)—Mumbai Unit, pp 93–97
28. Ilharco LM, Garcia AN, Lopes da Silva L, Ferreira LFV (1997) Infrared approach to the study of adsorption on cellulose: influence of cellulose crystallinity on the adsorption of benzophenone. Langmuir 13:4126–4132. https://doi.org/10.1021/LA962138U
29. Ioelovich M, Leykin A (2008) Structural investigations of various cotton fibers and cotton celluloses. BioResources 3:170–177. https://doi.org/10.15376/biores.3.1.170-177
30. Kljun A, Benians TAS, Goubet F et al (2011) Comparative analysis of crystallinity changes in cellulose I polymers using ATR-FTIR, X-ray diffraction, and carbohydrate-binding module probes. Biomacromol 12:4121–4126. https://doi.org/10.1021/BM201176M
31. Niranjana AR, Divakara S, Somashekar R (2011) Characterization of field grown cotton fibres using whole powder pattern fitting method. Indian J Fibre Text Res 36:9–17
32. Harzallah O, Dre J-Y (2011) Macro and micro characterization of biopolymers: case of cotton fibre. Biotechnol Biopolym. https://doi.org/10.5772/16831
33. Parikh DV, Thibodeaux DP, Condon B (2007) X-ray crystallinity of bleached and crosslinked cottons. Text Res J 77:612–616. https://doi.org/10.1177/0040517507081982
34. Abhishek S, Samir OM, Annadurai V et al (2005) Role of micro-crystalline parameters in the physical properties of cotton fibers. Eur Polym J 41:2916–2922. https://doi.org/10.1016/J.EURPOLYMJ.2005.06.005
35. Divakara S, Niranjana AR, Somashekar R (2009) Computation of stacking and twin faults in varieties of cotton fibers using whole powder pattern fitting technique. Cellulose 16:1187–1200. https://doi.org/10.1007/s10570-009-9354-5
36. Samir OM, Somashekar R (2007) Crystal structure and elastic constants of Dharwar cotton fibre using WAXS data. Bull Mater Sci 30:503–510. https://doi.org/10.1007/s12034-007-0079-5
37. Samir OM, Somashekar R (2007) Intrinsic strain effect on crystal and molecular structure of (dch32) cotton fiber. Powder Diffr 22:20–26. https://doi.org/10.1154/1.2434790
38. Murugesh Babu K, Selvadass M, Somashekar R (2013) Characterization of the conventional and organic cotton fibres. J Text Inst 104:1101–1112. https://doi.org/10.1080/00405000.2013.774948
39. Chae Y, Lee M, Cho G (2011) Mechanical properties and tactile sensation of naturally colored organic cotton fabrics. Fibers Polym 128(12):1042–1047. https://doi.org/10.1007/S12221-011-1042-Z

40. Karakan Günaydina G, Palamutcu S, Soydan AS et al (2020) Evaluation of fiber, yarn, and woven fabric properties of naturally colored and white Turkish organic cotton. J Text Inst 111:1436–1453. https://doi.org/10.1080/00405000.2019.1702611
41. Inoue M, Yamamoto S, Yamada Y, Niwa M (2006) Effects of cultivating methods and area on the mechanical properties of cotton fiber and yarn. Text Res J 76:534–539. https://doi.org/10.1177/0040517506065592
42. Basavaradder A, Maralappanavar MS (2014) Evaluation of eco-friendly naturally coloured Gossypium hirsutum L. cotton genotypes. Int J Plant Sci 9:414–419
43. Dickerson DK, Lane EF, Rodriguez DF (1999) Naturally colored cotton: resistance to changes in color and durability when refurbished with selected laundry aids
44. Stankovič Elesini U, Pavko Čuden A, Richards AF (2002) Study of the green cotton fibres. Acta Chim Slov 49:815–833
45. Günaydin GK, Soydan AS, Palamutçu S (2018) Evaluation of cotton fibre properties in compact yarn spinning processes and investigation of fibre and yarn properties. Fibres Text East Eur 26:23–34. https://doi.org/10.5604/01.3001.0011.7299
46. Günaydin GK, Yavas A, Avinc O, et al (2019) Organic cotton and cotton fiber production in Turkey, recent developments. Springer, pp 101–125
47. Palamutcu S, Soydan AS, Avinc O et al (2019) Physical properties of different Turkish organic cotton fiber types depending on the cultivation area. Springer, Singapore, pp 25–39
48. Parmar MS, Chakraborty M (2016) Thermal and burning behavior of naturally colored cotton 71:1099–1102. https://doi.org/10.1177/004051750107101211
49. Parmar MS, Sharma RP (2002) Development of various colours and shades in naturally coloured cotton fabrics. Indian J Fibre Text Res 27:397–407
50. Wakelyn PJ, Thompson DW, Norman BM et al (2005) Why cotton ginning is considered agriculture. Cott Gin Oil Mill Press 106
51. Wakelyn PJ (1997) Cotton yarn manufacturing. ILO Encycl Occup Heal Saf
52. Diehl R (2004) Weaving preparation yesterday, today, tomorrow. International Textile Bulletin
53. Nimkar UM, Bhajekar R (2008) Global concern—ecological requirements for the textile industry. In: Organic textiles farming to finishing. The Textile Association (India)—Mumbai Unit, pp 9–12
54. Wakelyn PJ, Bertoniere NR, French AD et al (2006) Cotton fibers. In: Menachem L (ed) Handbook of fiber chemistry, 3rd ed
55. Wakelyn PJ (1994) Cotton: environmental concerns and product safety. In: Harig H, Heap SA (eds) 22nd international cotton conference. Faserinstitut Bremen, Bremen, pp 287–305
56. Cotton Incorporated (1996) Cotton Dyeing and finishing : a technical guide. Cotton Incorporated, Cary, NC
57. Gordon S (Stuart), Hsieh Y, Textile Institute (Manchester E (2007) cotton : science and technology, p 548
58. Rathinamoorthy R (2018) Sustainable apparel production from recycled fabric waste. Springer
59. Gupta L, Kaur Saini H (2020) Achieving sustainability through zero waste fashion—a review. Curr World Environ 15:154–162. https://doi.org/10.12944/cwe.15.2.02
60. Rieple A, Singh R (2010) A value chain analysis of the organic cotton industry: the case of UK retailers and Indian suppliers. Ecol Econ 69:2292–2302. https://doi.org/10.1016/j.ecolecon.2010.06.025

Printed in the United States
by Baker & Taylor Publisher Services